Second and Third Generation Bioplastics

This book provides a comprehensive overview of and state-of-art information on the production and application of second- and third-generation bioplastics, such as polylactic acid (PLA) and polyhydroxyalkanoates (PHAs). The uses of alternative raw materials are presented, and innovations applied in bioplastics production processes to reduce costs and decrease environmental impacts in a circular bioeconomy are discussed.

Second and Third Generation Bioplastics

Production, Application, and Innovation

Edited by
Luciana Porto de Souza Vandenberghe
Ashok Pandey
Ranjna Sirohi
Carlos Ricardo Soccol

CRC Press
Taylor & Francis Group
Boca Raton London New York

CRC Press is an imprint of the
Taylor & Francis Group, an **informa** business

Cover Image Credit: Shutterstock.com

First edition published 2024
by CRC Press
2385 Executive Center Drive, Suite 320, Boca Raton, FL 33431

and by CRC Press
4 Park Square, Milton Park, Abingdon, Oxon, OX14 4RN

CRC Press is an imprint of Taylor & Francis Group, LLC

ISBN: 978-1-032-36598-5 (hbk)
ISBN: 978-1-032-38221-0 (pbk)
ISBN: 978-1-003-34401-8 (ebk)

DOI: 10.1201/9781003344018

Typeset in Times
by Apex CoVantage, LLC

Contents

Preface

In recent decades a significant increase of bioplastics production was observed with a great expansion of the global bio-based market. This is a reaction to the environmental impact of traditional plastics produced from fossil-based materials. Bioplastics have emerged from the sustainability concept, where the focus was to reduce the petrochemical pollutants through the replacement of biodegradable products. Recent research has sought new possibilities to contribute to bioplastics' large-scale production and commercialization cost-effectively through second- and third-generation (2G and 3G) technologies. However, there is a lack of compiled information describing the main aspects, recent developments, patents and innovation, and perspectives about 2G and 3G bioplastics.

This book, *Second and Third Generation Bioplastics: Production, Application, and Innovation*, brings new, broad information about the subject. It is divided into 14 chapters, which describe the main commercialized bioplastics, such as polylactic acid (PLA), polyhydroxyalkanoates (PHAs), and others, obtained through the use of 2G and 3G technologies. There is no doubt that 2G and 3G bioplastics are a promising solution to enable large-scale and viable production of these bio-based materials. Second-generation bioplastics are produced from lignocellulosic biomass and non-food edible oils, while third-generation bioplastics are obtained from sugars or oils produced by microorganisms (e.g., microalgae, bacteria, mushrooms, yeasts) or from municipal waste material. The use of alternative raw materials in the production of these biomaterials will be presented, along the actual market, recent advances and innovations that are applied in 2G and 3G bioplastics' production processes to reduce the costs and decrease environmental impacts in a biorefinery and circular bioeconomy approach.

Many researchers and experts have shared their experiences on 2G and 3G bioplastic technologies, which will be useful for researchers, students, academicians, and industry experts of the area. We are grateful to reviewers for their important contribution in the critical review of the chapters. We sincerely acknowledge the Biotech Research Society, India, for giving us the opportunity and support to prepare this book under the agreement with Taylor & Francis. We thank the team of Taylor & Francis Group, including Dr. Gagandeep Singh, Senior Publisher; Dr. Madhurima Kahali, Editor (Life Science); Ms. Neha Bhatt, Editorial Assistant, CRC Press; and the entire team of the CRC Press/Taylor & Francis Group for their consistent support during the publication process.

Editors
Luciana Porto de Souza Vandenberghe
Ashok Pandey
Ranjna Sirohi
Carlos Ricardo Soccol

Editor Biographies

Luciana Porto de Souza Vandenberghe is a full professor at the Department of Bioprocess Engineering and Biotechnology, Federal University of Paraná, Brazil. She obtained her PhD in *Génie de Procédés Industriels* in 2000 at Université de Technologie de Compiègne, France. Her areas of interest include industrial enzymes, organic acids, and biopolymers.

Ashok Pandey is currently Distinguished Scientist at the Centre for Innovation and Translational Research, CSIR–Indian Institute of Toxicology Research, Lucknow, India, and Distinguished Professor in the Sustainability Cluster, School Engineering, University of Petroleum and Energy Studies, Dehradun, India.

Ranjna Sirohi is currently Assistant Professor in the School of Health Sciences and Technology, University of Petroleum and Energy Studies, Dehradun, India. Her major research interests are in bioprocess technology, food and food waste valorization, waste to wealth, –biofuels.

Carlos Ricardo Soccol is the research group leader of the Department of Bioprocesses Engineering and Biotechnology (DEBB) at Federal University of Paraná – UFPR. He has experience in bioprocesses/biological engineering, industrial biotechnology, green fuels technology, applied microbiology and fermentation technology.

Contributors

Sundaram Arvindnarayan
Anna University
Chennai, India

Mukesh Kumar Awasthi
College of Natural Resources and Environment
Northwest A&F University
Shaanxi, China

L. Bartolucci
Department of Industrial Engineering
University of Rome "Tor Vergata"
Rome, Italy

Parameswaran Binod
Microbial Processes and Technology Division
CSIR–National Institute for Interdisciplinary Science and Technology
Trivandrum, Kerala, India
Academy of Scientific and Innovative Research
Ghaziabad, India

Gustavo Amaro Bittencourt
Department of Bioprocess Engineering and Biotechnology
Federal University of Paraná, Centro Politécnico
Curitiba, Paraná, Brazil

Júlio Cesar de Carvalho
Department of Bioprocess Engineering and Biotechnology
Federal University of Paraná, Centro Politécnico
Curitiba, Paraná, Brazil

Dão Pedro de Carvalho Neto
Federal Institute of Education, Science and Technology of Paraná
Curitiba, Paraná, Brazil

Lucas Lourenço Castiglioni Guidoni
Postgraduate Program of Biotechnology
Federal University of Pelotas
Pelotas, RS, Brazil

Érico Kunde Corrêa
Federal University of Pelotas, NEPERS, at Center of Engineering
Pelotas, RS, Brazil

E. De Maina
Department of Industrial Engineering
University of Rome "Tor Vergata"
Rome, Italy

Jeyaprakash Dharmaraja
Division of Chemistry, Faculty of Science and Humanities
AAA College of Engineering and Technology
Amathur, Tamil Nadu, India

Niklas Mathias Döhler
WZB Berlin Social Science Center
Berlin, Germany

Shreya Dwivedi
Institute for Industrial Research & Toxicology
Ghaziabad, Lucknow, India

Gabriel Rossignol Frassetto
Department of Bioprocess Engineering and Biotechnology
Federal University of Paraná, Centro Politécnico
Curitiba, Paraná, Brazil

Vivek Kumar Gaur
School of Energy and Chemical Engineering, UNIST
Ulsan, Republic of Korea

Krishna Gautam
Centre for Energy and Environmental Sustainability
Lucknow, India

Luis Daniel Goyzueta-Mamani
Vicerrectorado de Investigación
Universidad Católica de Santa María
Arequipa, Peru

Pallavi Gupta
Bioscience and Biotechnology Department
Banasthali University
Rajasthan, India

Leonardo Wedderhoff Herrmann
Department of Bioprocess Engineering and Biotechnology
Federal University of Paraná
Curitiba, Paraná, Brazil

Susan Grace Karp
Department of Bioprocess Engineering and Biotechnology
Federal University of Paraná
Curitiba, Paraná, Brazil

Gopalakrishnan Kumar
Institute of Chemistry, Bioscience and Environmental Engineering, Faculty of Science and Technology
University of Stavanger
Stavanger, Norway

Yoong Kit Leong
Department of Chemical and Materials Engineering
Research Center for Smart Sustainable Circular Economy
Tunghai University
Taichung, Taiwan

Luiz Alberto Junior Letti
Department of Bioprocess Engineering and Biotechnology
Federal University of Paraná
Curitiba, Paraná, Brazil

Clara Matte Borges Machado
Department of Bioprocess Engineering and Biotechnology
Federal University of Paraná
Curitiba, Paraná, Brazil

Aravind Madhavan
School of Biotechnology
Amrita Vishwa Vidyapeetham
Amritapuri, Kerala, India

Walter José Martínez-Burgos
Department of Bioprocess Engineering and Biotechnology
Federal University of Paraná
Curitiba, Paraná, Brazil

Patricia Beatriz Gruening de Mattos
Department of Bioprocess Engineering and Biotechnology
Federal University of Paraná
Curitiba, Paraná, Brazil

P. Mele
Department of Industrial Engineering
University of Rome "Tor Vergata"
Rome, Italy

Ariane Fátima Murawski de Mello
Department of Bioprocess Engineering and Biotechnology
Federal University of Paraná
Curitiba, Paraná, Brazil

V. Mulone
Department of Industrial Engineering
University of Rome "Tor Vergata"
Rome, Italy

Ardra Nandakumar
Nara Institute of Science and Technology (NAIST)
Nara, Japan

Natália Rodrigues Nitsch
Department of Bioprocess Engineering and Biotechnology
Federal University of Paraná
Curitiba, Paraná, Brazil

Priscilla Zwiercheczewski de Oliveira
Department of Bioprocess Engineering and Biotechnology
Federal University of Paraná
Curitiba, Paraná, Brazil

Eapen Philip
Post Graduate and Research Department of Chemistry
Bishop Moore College
Mavelikara, Kerala, India

Retnam Krishna Priya
Research Department of Physics
Holy Cross College (Autonomous)
Nagercoil, Tamil Nadu, India

Reshmy R
Department of Science and Humanities
Providence College of Engineering
Chengannur, Kerala, India

Manuela Mendonça Cardozo Ribeiro Ravazoli
Department of Bioprocess Engineering and Biotechnology
Federal University of Paraná
Curitiba, Paraná, Brazil

Sutha Shobana
Green Technology and Sustainable Development in Construction Research Group, School of Engineering and Technology
Van Lang University
Ho Chi Minh City, Vietnam

Pau Loke Show
Zhejiang Provincial Key Laboratory for Subtropical Water Environment and Marine Biological Resources Protection
Wenzhou University
Wenzhou, China
and
Department of Sustainable Engineering
Saveetha School of Engineering, SIMATS
Chennai, India
and
Department of Chemical and Environmental Engineering
University of Nottingham, Malaysia
Semenyih, Selangor, Malaysia

Raveendran Sindhu
Department of Food Technology
TKM Institute of Technology
Kerala, India

Deepa Thomas
Department of Chemistry
Bishop Moore College
Mavelikara, Kerala, India

Rekha Unni
Department of Chemistry
Christian College
Chengannur, Kerala, India

Vaisakh P H
Post Graduate and Research Department of Chemistry
Bishop Moore College
Mavelikara, Kerala, India

Kim Kley Valladares-Diestra
Department of Bioprocess Engineering and Biotechnology
Federal University of Paraná
Curitiba, Paraná, Brazil

Sunita Varjani
School of Energy and Environment
City University of Hong Kong
Kowloon, Hong Kong
and
Sustainability Cluster
School of Engineering
University of Petroleum and Energy Studies
Dehradun, Uttarakhand, India

Jéssica Aparecida Viesser
Department of Bioprocess Engineering and Biotechnology
Federal University of Paraná
Curitiba, Paraná, Brazil

André Wolf
Centre for European Policy
Berlin, Germany

1 Introduction About Second- and Third-Generation Bioplastics

Priscilla Zwiercheczewski de Oliveira,
Luciana Porto de Souza Vandenberghe, and
Carlos Ricardo Soccol

CONTENTS

1.1 INTRODUCTION

Petroleum-based plastic production started at the beginning of the 20th century. Since then, plastic pollution is a serious global concern, as there is a low percentage of recycling, some of which is rejected in landfills and oceans, or other areas of the natural environment. They are a threat to sustainable development due to their incredible stability under environmental conditions, where the half-life can be over a hundred years. Under ideal conditions of exposure to ultraviolet light and heat, the half-life of decomposition of common plastic items—plastic bottles made from HDPE (high-density polyethylene) or PET (polyethylene terephthalate)—can be several decades (Geyer et al. 2017).

One of the main goals established by the United Nations is to reach a clean and sustainable environment. One solution is to reduce petroleum-based plastic consumption (Awasthi et al. 2021). However, the current challenge is to find alternatives that can work similarly to petroleum-based plastics. Bioplastics appear as the solution with similar characteristics, but with less impact on the environment. However, the functionality of these materials must approach that of existing plastics, and the production technology must be inexpensive, which is the main problem and limitation of

DOI: 10.1201/9781003344018-1

large-scale production of these materials. That is why the called second (2G)– and third (3G)–generation substrates (or biomasses) are being studied and applied as sources for 2G and 3G bioplastics development (Vandenberghe et al. 2021).

Commercially, some bioplastics are well recognized. Considering a patent search, data shows that polylactic acid–related innovation activity is still predominant. The main focus is on improving the final mechanical performance at industrial scale, suggesting that production methods and properties are under constant improvement. Depending on the end product use, adaptations are necessary to achieve specific requirements. The range of end-use areas covered by the patent search files is linked to the world's actual challenges, such as food packaging, textiles, adhesives, different devices, and tools containing petroleum-based plastics (Ibrahim et al. 2021).

Bioplastics play a crucial role within a circular economy where biorefinery approaches are employed. Some examples of bioplastics production employ agricultural wastes and sub-products, wastewaters, forestry, and others, which bring competitiveness and sustainability of new generation materials. In order to fulfill its growth potential, it is important that the bioplastics industry guarantees access to agricultural biomasses at competitive prices in sufficient quantities and quality. This requires global and integrated policies to ensure greater value creation and stronger environmental benefits (Otoni et al. 2021).

1.2 BIOPLASTICS

Bioplastics have a significant role in reducing plastic waste, and they have been intensively investigated over the last two decades to complement or replace conventional plastics in packaging (Ibrahim et al. 2021). Nowadays, bioplastics have become an alternative with similar or improved properties, compared to conventional plastics, and also great functionality (Otoni et al. 2021).

Discovering potential substrates as sources for bioplastics production is the key to increase its sustainability. So its final process costs will depend on the employed 2G or 3G substrate. In addition, it is essential to evaluate how efficient the substrate conversion to sugars is for further biopolymer, or its precursor, production. In this sense, the use of cellulosic feedstocks as 2G bioplastics represents a novel technology, while 1G, such as corn or sugarcane, are effective at converting CO_2 to carbohydrates. Considering innovations in technology, bacteria and cyanobacteria are being tested to convert methane or CO_2 into lactic acid. This prospect may be the most efficient 3G bioplastic capable of reducing the greenhouse effect (Jawarraj et al. 2020). Through these examples, innovative substrates are important, but the final bioplastics characteristics are fundamental to prosper in competitive applications (Narancic et al. 2020). Figure 1.1 shows the 1G, 2G and 3G substrates currently employed in bioplastics production.

1.2.1 First-Generation Bioplastics

First-generation substrates (1G) are feedstocks and plants directly obtained from the land crop. They are rich in carbohydrates, which are normally consumed by humans and animals. First-generation substrates have been extensively used to produce polylactic acid–bioplastic with high efficiency, which is mainly linked to the required area for cultivation and the final bioplastic quality. In this case, the major employed feedstocks are sugarcane, corn, wheat, potato, rice and vegetable oil. Critical opinions were made as they are essential crops for human and animal feed, where the destination of these feedstocks for bioplastics production creates a sort of competition. To solve this critical point, new substrate sources were found in order to replace feedstocks, and they were classified as 2G and 3G (Wellenreuther and Wolf 2020).

Starch is available in a wide range of feedstocks; the main sources are potatoes, wheat and tapioca, where starch from potato-derivatives presents the best mechanical properties (Gonzalez-Gutierrez et al. 2010). Still, starch-based bioplastics are susceptible to moisture absorption; an alternative solution is its reinforcement with natural fibers, from sugarcane bagasse, rice straw and coconut husks,

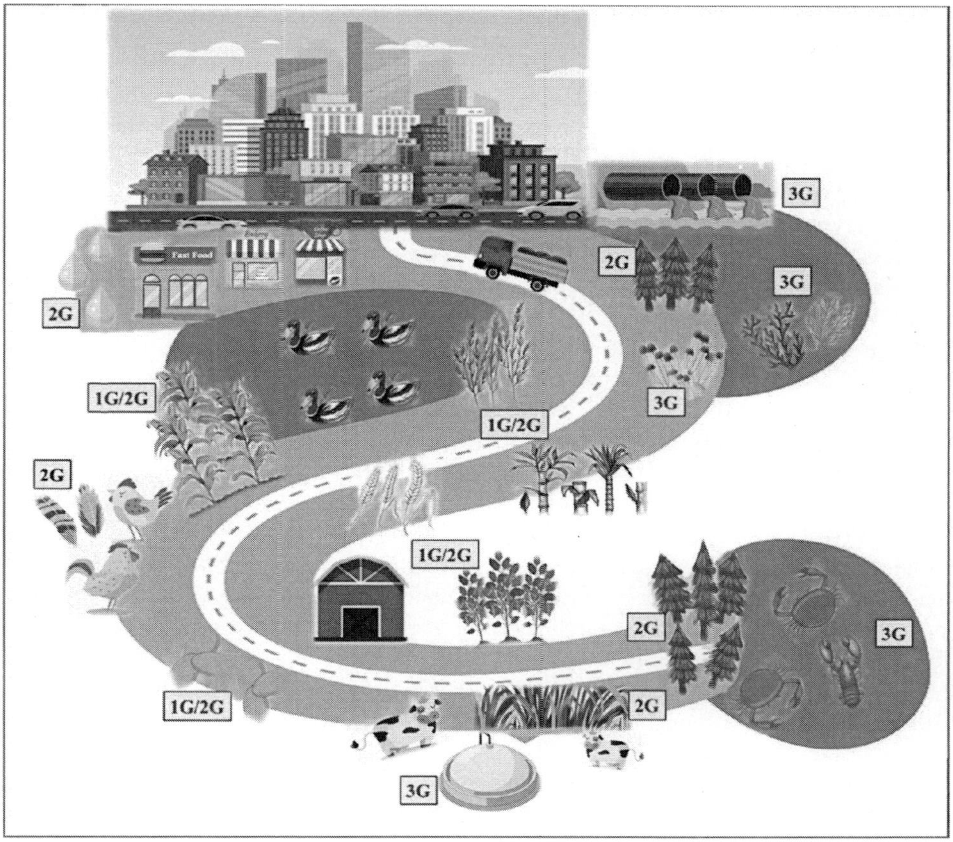

FIGURE 1.1 Representative origin of 1G, 2G, and 3G substrates for bioplastics production.

Abbreviations: 1G = first-generation substrate; 2G = second-generation substrate; 3G = third-generation substrate.

Source: Adapted from SPI (2016).

for example. Then, enhanced properties such as moisture tolerance and tensile strength are observed (Ghasemlou et al. 2022; Babalola and Olorunnisola 2019).

1.2.2 Second-Generation Bioplastics

Second-generation substrates (2G) are composed of lignocellulosic materials and vegetable oil residues, originated and/or extracted from crops and plants that are inadequate for human nutrition and animal feed. Some examples are feathers, wood, rotation crops, wheat straw, sugarcane bagasse, corncobs, palm empty fruit bunches and barter grass (Vandenberghe et al. 2021). These materials must pass through pretreatment and/or enzymatic hydrolysis to obtain sugar-rich hydrolysates that are employed in bioplastics' production. Different and advanced pretreatment is now available, involving green solvents, for efficient preparation of lignocellulosic materials for further enzymatic treatment. Also, highly performant enzymes are being developed for optimal release of sugars that are employed in further fermentative processes (Govil et al. 2020; Nduko and Taguchi 2021). These approaches define what is called the 2G biorefineries under bioeconomic and sustainable processes.

1.2.3 Third-Generation Bioplastics

Third-generation substrates involve microbial biomasses, chitin and municipal wastes (organic wastes and wastewaters). Algae and microalgae are in the spotlights in this group, due to their versatility and their capacity to consume these substrates and, concomitantly, produce medium- to high-value biomolecules, including bioplastics. An example is the production of polyhydroxyalkanoates. This possibility caught the scientific community's attention and economic interest because large volumes of these organic wastes and wastewaters are generated. So sustainable solutions for bioplastic production have appeared. There is no need for fertilizers or land for cultivation, and it is estimated that this bioplastic derived would have better biodegradability (Yashavanth et al. 2021). The actual focus is to process the seaweed biomass in order to obtain polysaccharide-rich fractions. Through optimized processes, single-use bioplastics could be applied in food industries and the biomedical and cosmetic areas. The most studied microalgae species so far are *Chlorella* sp. and *Spirulina* sp. (Khyalia et al. 2022).

Wastewaters are rich in organic matter and salts. Activated sludge generated during the wastewater treatment is very abundant and could be employed in the production of PHBs from pure or mixed microbial cultures via anaerobic fermentation, where volatile fatty acids are generated and could be converted to PHA (Koller and Obruca 2022).

Chitin is usually found in chitosan form, a tough polysaccharide found in crustaceans shells, exoskeletons and fungal mycelium. It is considered a renewable polysaccharide and bioplastics development from this material is still in the beginning. It is already known that is possible to obtain three different chitosans: those that are derived from marine animals; the one obtained via enzymatic from animal chitin; and fungal chitosan, which in turn is considered of high purity (Esposti et al. 2021). Due to its biocompatibility, biodegradability and nontoxicity, biomedical applications of these bioplastics are expected.

Methane gas is a newer and greener alternative substrate that is intensively studied. It can be obtained from different fermentation residues, and it can be reused in composting, fertilization, drying and incineration (Benavides et al. 2018). The actual challenge is to improve the converting waste possibilities into methane, and then its conversion into bioplastics, using the consolidated structure of biogas production (Jawarraj et al. 2020).

1.3 THE MAIN PRODUCED 2G AND 3G BIOPLASTICS

1.3.1 Polylactic Acid

Polylactic acid (PLA) is the most studied biopolymer and the most commercially known biopolymer. Lactic acid monomer is mainly produced via fermentation using sugar-rich 1G/2G substrates, and then polymerized to PLA. PLA has two active optical isomers, poly-D and poly-L-lactic acid, where the polymerization process is highly dependent on the isomer purity (Oliveira et al. 2022). It has several attractive mechanical properties compared to other bioplastics. It is a clear, colorless, shiny bioplastic whose properties are similar to common petroleum-based plastics such as polyethylene, polypropylene or polystyrene.

PLA is in high demand at industrial scale due to its similarity to petroleum-based plastics, not only for its properties, but also because it can also be processed on existing machinery without adaptations (Farah et al. 2016; Wellenreuther et al. 2022). PLA products with different molecular weights due to the optical isomers are available, and they are suitable in a variety of polymer processing applications such as injection molding, fiber spinning, extrusion thermoforming, and as a nucleating agent. It can be used in packaging and food service utensils and has durable applications in automobiles, electronics and textiles. Depending on specific requirements, it can be designed to biodegrade quickly or remain stable for several years (Stefaniak and Masek 2021).

1.3.2 Polyhydroxyalkanoates

Polyhydroxyalkanoates (PHAs) are linear polyesters, which are produced via bacterial fermentation using sugar or lipids. PHAs are naturally found as metabolic products or as intracellular granules produced under stressful growth conditions. For example, under nutrients' limitation and carbon excess, some bacteria such as *Alcaligenes eutrophus, Azotobacter vinelandii, Bacillus* sp. and *Streptomyces* sp. accumulate PHA as stored feed material (Meereboer et al. 2020).

Considering the biotechnology advancements, different types of PHAs can be produced with improved functional properties. PHAs are excellent candidates for an extensive variety of applications in thermoplastics, elastomers, lubricants, glues and adhesives. They also find many non-traditional applications among polymers such as in animal feed, cell regeneration in humans and animals, denitrification and cosmetics. PHA is mainly produced from 2G/3G substrates that present high sugar and high fat content, including gaseous content. PHA yield and productivity are sensible to fermentation parameters, substrate type and microbial strains (Alsafadi et al. 2020; Akhlaq et al. 2022).

PHA is widely used in the medical industry due to its singular ductility. Its combination with other biopolymers allows for designing specific properties. The most suitable composition is PLA-PHA biopolymers for improvements in flexibility, heat resistance, and weather and ultraviolet light resistance. The fast nucleation of the melt and the improved barrier properties demonstrate that these hybrid PLA-PHA bioplastics can be processed faster than petroleum-based polymers (Guo et al. 2022).

PHAs are the unique bioplastics able to be compostable at home and completely degrade in the marine environment. A variety of PHA-based biopolymers are available at commercial scale, including poly-3-hydroxybutyrate (P3HB) and copolymers such as poly-3-hydroxybutyrate-co-valerate (P3HB-co-HV), poly-3-hydroxybutyrate-co-hexanoate (P3HB-co-HH), and poly-3-hydroxybutyrate-co-4-hydroxybutyrate (P3HB-co-4HB; Jian et al. 2020).

1.3.3 Polyhydroxybutyrate

Polyhydroxybutyrate (PHB) is a non-transparent polyester co-polymer of PHA, and it is very similar to polypropylene. Its production, using carbon-rich 2G/3G substrates, through specific bacteria is currently being developed at industrial scale due to their energy storage mechanism, in pellet form, that can be up to 80% of their own cell weight (Koller and Obruca 2022).

PHBs are mainly films that can be processed using established machinery, such as for extrusion and injection molding. In addition to the quick biodegradation in industrial composting (including methane conversion into PHB), as well as in domestic compost, in the soil or in seawater (Liu et al. 2020; Pérez et al. 2019). Its mechanical properties can be adapted to exhibit excellent flexibility and hardness. Main applications are generally packaging and food packaging (Ibrahim et al. 2021; Markl et al. 2018).

1.3.4 Polybutylene Succinate

Polybutylene succinate (PBS) is a biodegradable aliphatic polyester that is usually obtained from succinic acid and 1,4-butanediol via condensed polymerization. This bioplastic has been studied since it can be obtained via fermentation using 2G substrates rich in starch, glucose and xylose. The known bacteria strains that are able to produce succinic acid are *Mannheimia succiniciproducens* and *Anaerobiospirillum succiniciproducens*. The final properties are similar to polypropylene (Aliotta et al. 2022).

1.4 2G and 3G Bioplastics Advantages and Disadvantages

Bioplastics offer valuable advantages when compared to conventional plastics such as renewability, reduced carbon footprint, utilization of agro-industrial wastes and less generation of residues during

TABLE 1.1
Advantages and Disadvantages during Bioplastics Production

Substrate	Advantages	Disadvantages
Feedstock (1G)	Huge availability, stablished technology in large scale	Large usage of lands, fertilizers, water. Lignocellulosic biomass needs pre-treatment, slow production time due to saisonability
Organic waste (2G)	Abundant, low-cost, contribution for disposal and environmental issues, residues are bioconverted to valuable biomolecules, high nutritive content for micro-organisms	Pre-treatments can be complex and costly
Wastewater (3G)	High availability, low-cost, waste can be re-used	Requires new technology, production yield depends on previous treatment
Algae (3G)	Non-land dependence, fast growth with high yields, possible integration with wastewater systems, bioremediation	At large scale process, the technology can be expensive, genetic engineering may be required

Source: Adapted from Coppola et al (2021).

their production. They are non-toxic, free of phthalates and bisphenol A, and their production does not involve food application. A disadvantage is the two to three times higher cost of production, therefore it is crucial to establish a simple, rapid and large-scale technique for bioplastics manufacture (Dhaka et al. 2022; Liwarska-Bizukojc 2022). Table 1.1 displays the advantages and disadvantages according to 1G, 2G or 3G substrates.

1.5 LIFE CYCLE OF 2G AND 3G BIOPLASTICS

Specialists are working on the ability to reuse biopolymers to increase their potential recycling (Liwarska-Bizukojc 2021). The purpose at the end of bioplastics intended use is to ensure that the cycle is closed, which means cycling the bioplastic product back at the end of its life to be reused. Landfill disposal or incinerating are not recommended: bioplastics are more intended to be recycled or composted (Lamberti et al. 2020).

Bioplastics can be recycled, but recycling needs to be done in separate streams. If the degradable material enters the stream of conventional plastics and is completely degraded in the recycling process, it can change the characteristics and specifications of the conventional material with which it is mixed (Narancic et al. 2020; Singathi et al. 2022).

Composting is carried out under aerobic conditions, where the bioplastic is the substrate for microorganisms under certain temperature and humidity. Industrial composting is highly recommended, due to existing specialized companies that have control to guarantee quick biodegradation, around 6–12 weeks. Temperature is the key factor, between 50°C and 70°C, and the final products are heat, water, CO_2 and microbial biomass (European Bioplastics 2022).

1.5.1 Standards and Certification

As 2G/3G bioplastics are not yet easily distinguishable from petroleum-based plastics, it is necessary to provide a mechanism to ensure their quality (Moreno et al. 2021). This is done through a standardization and certification system. Until now, there are no international regiments strictly for bioplastics. Some local standardization bodies are making efforts to, at least, create rules considering important aspects of compostable characteristics (European Bioplastics 2022). Descriptions are presented in Table 1.2.

TABLE 1.2
Currently Standards Applied by Region

Reference	Standard	Entity
ASTM D6400	Definition for compostable plastic: "a plastic that undergoes degradation by biological processes during composting to produce CO_2, water, inorganic compounds and biomass at a rate consistent with other known compostable materials that leave no visible, distinguishable or toxic residue" (Biodegrade + Disintegration + Eco-toxicity).	ASTM American Society for Testing and Materials (USA)
ISO 17088:2012	Aspects related to compostability: (a) biodegradation; (b) disintegration during the compound; (c) negative effects on the composting process and facility; (d) negative effects on the quality of the resulting compost, including the presence of high levels of regulated metals and other harmful components.	ISO International Organization for Standardization (International)
EN 13432	Aspects related to compostability: (a) Chemical testing: Disclosure of all constituents, threshold values for heavy metals must be respected; (b) Biodegradability under controlled composting conditions (oxygen consumption and CO_2 production): Proof must be made that at least 90% of the organic material is converted to CO_2 within 6 months; (c) Disintegration: After 3 months of composting and further sieving through a 2 mm sieve, no more than 10% residue can remain, compared to the original mass; (d) Practical composting test in a semi-industrial (or industrial) composting facility: No negative influence on the composting process is allowed; e) Ecotoxicity test: Examination of the effect of the resulting compound on plant growth (agronomic test).	CEN European Committee for Standardization (European Union)
Based on EN 14995 or ISO 17088	Certification for compostable plastics	DIN Certco (European Union)
Based on EN 13432 or ASTM D6400	Certification for compostable packaging	DIN Certco (European Union)
Based on EN 13432	Certification for compostable plastics, for plastics that can be composted in home compost, for plastics that biodegrade in soil and for plastics that biodegrade in water	Vincotte (European Union)
–	Certification for compostable plastic	BPI Biodegradable Products Institute (USA)
–	Certification for compostable plastic	JBPA Japan BioPlastics Association (Japan)

Source: Adapted from European Bioplastics (2022).

1.6 COMMERCIAL APPLICABILITY

Currently, bioplastics are widely used in packaging (including flexible and rigid packaging), consumer goods, cosmetics, tissue engineering, pharmaceutics, bio-implants, medical and biomedical procedures, automotive, construction, agriculture and electronics (Lackner 2015).

Bioplastics are usually applied in blends, specially PLA-PHA or PLA-PHB blends. In agriculture, they are alternatives to replace high-density polyethylene in grow bags and mulch films. Bioplastics-based grow bags are biodegradable and eco-friendly to the water bodies. In mulch films, they can improve humidity retention, and soil structure (Jian et al. 2020; Moreno et al. 2021).

2G/3G bioplastics packaging offers no interaction with food, through their good performance regarding barrier properties, mainly being impenetrable to odors. Also, they can be processed on almost all the usual machines. Some packaging applications are already known, such as cups for coffee and coffee capsules, shopping bags, plates for frozen food, and cutlery (Ibrahim et al. 202_).

Advancements in bioplastics properties allow the blend application in the medical and biomedical areas. Development of drug delivery systems and therapeutic devices for tissue engineering, such as implants and scaffolds, are observed (Narancic et al. 2020). Concerning applications involving structure sustentation, cellulose and its derivatives are under test for implants, tissue and neural engineering. It is possible due to its organized and fibrillar structure, where the fibrils are the elementary structure (Chandel et al. 2018; Ghasemlou et al. 2022; Lei et al. 2022).

PHA biocompatibility makes it ideal for medical applications, such as post-surgical ulcer treatment, heart valves, artificial blood vessels and artificial nerve conduits (Guo et al. 2022). PLA is widely used as a surgical thread, as a carrier for implanted active substances and resorbable implants such as screws and pins that are degraded by the metabolism, avoiding removal surgeries (Ahmad et al. 2022).

In personal care, 2G/3G bioplastics are resorbable; for example, thermoplastic starch is an alternative to gelatin, used as a pill coating. PLA and its copolymers can be used in lipstick cases and shampoo bottles (DeStefano et al. 2020). Utensils can be made from PLA, such as desk utilities, adhesive tapes and food storage containers based on PLA-PHA blends, and glue sticks (Coppola et al. 2021). Baby toys are made of PLA or starch-based bioplastics, since they have to be as healthy as possible. Another available option is PHA-beach toys, considering their biodegradation in seawater (Markl et al. 2018).

1.7 CHALLENGES

2G and 3G bioplastics challenges are linked to innovation, as they must present good performance over petroleum-based plastics, reach a low-cost production, and be eco-friendly to integrate the bioeconomy (Awasthi et al. 2021; SYSTEMIQ 2022).

The main challenging aspect can be cited as the long path to reach large-scale market consolidation, because bioplastics are two to four times more expensive than petroleum-based plastics. Upstream technology is still evolving and there are uncertainties about post-use processing; for example, PLA recycling to its former oligomer, and then the possibility of this oligomer being repolymerized is under validation to elucidate the reverse context of bioplastics production (Döhler et al. 2022; Leal Filho et al. 2021).

1.8 MARKET

The bioplastics market is driven by factors including mandates and regulations, ecological awareness among consumers, and companies focused on sustainability, technological stabilization and cost reduction (FNR 2020).

Substrates from 2G and 3G still represent a very small part of the total plastic produced annually, around 2%. However, 2G/3G bioplastics production has continuously grown, which can reach 3% of the plastic market by 2026. This slight growth is driven by the global demand for sophisticated products, advances in research and the efforts for a greener cycle process (FNR 2020). The European Bioplastics organization in cooperation with the nova-Institute, released the information about the market scenario in 2021 as well as the projection by 2026 (European Bioplastics 2022), considering global capacity production, region capacity production, bioplastic type and the market consumer demand by segment.

All data were collected from the global plastic market, which means there are comparative data between petroleum-based plastics, bio-based plastics and bioplastics. This chapter focuses on bioplastic (2G and 3G) market analysis; they can be referred to as biodegradable plastics in graphs for better understanding. Global bioplastics production capacities in 2021 were around 2.42 million tonnes and are set to increase by approximately 7.59 million tonnes in 2026 (Figure 1.2).

As mentioned before, 2G/3G bioplastics production is increasing, which is promoted by their commercial market valorization. A clear growth of 2G/3G bioplastics production is observed, corresponding to more than 64% (over 1.5 million tonnes) of the global bioplastics production capacities. The production is expected to keep growing, increasing to around 5.3 million by 2026 (Figure 1.3).

Considering the specific type of bioplastic (Figure 1.4), PLA production is estimated to grow till 2026, due to commercial movements observed mainly in Asia, with new production sites (Oliveira et al. 2022). PBS is projected to have impressive growth, as far as the studies and tests advance. Poly-butylene-adipate-co-terephthalate (PBAT), has aroused industrial interest, and high production growth is also expected. However, PBAT is considered a biodegradable plastic, yet it has a fraction derived from petroleum (Jian et al. 2020). Although studies have been made, for the moment, PBAT is not considered a 2G/3G bioplastic.

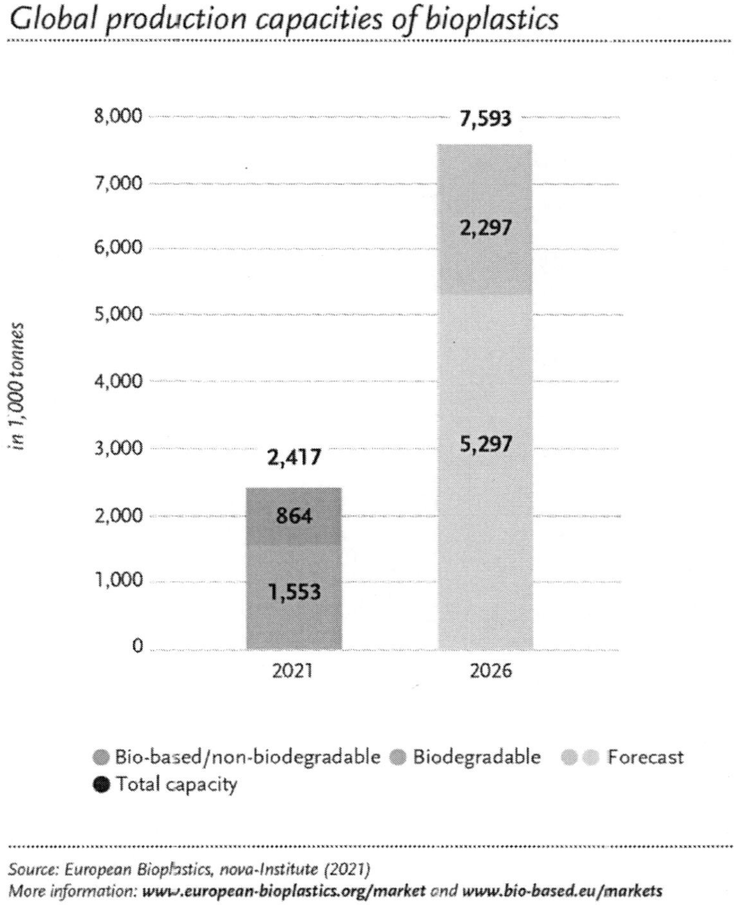

FIGURE 1.2 Global production capacity in 2021 and in 2026; 2G and 3G bioplastics are represented by "biodegradable."

Source: European Bioplastics (2022), with permission for reproducing.

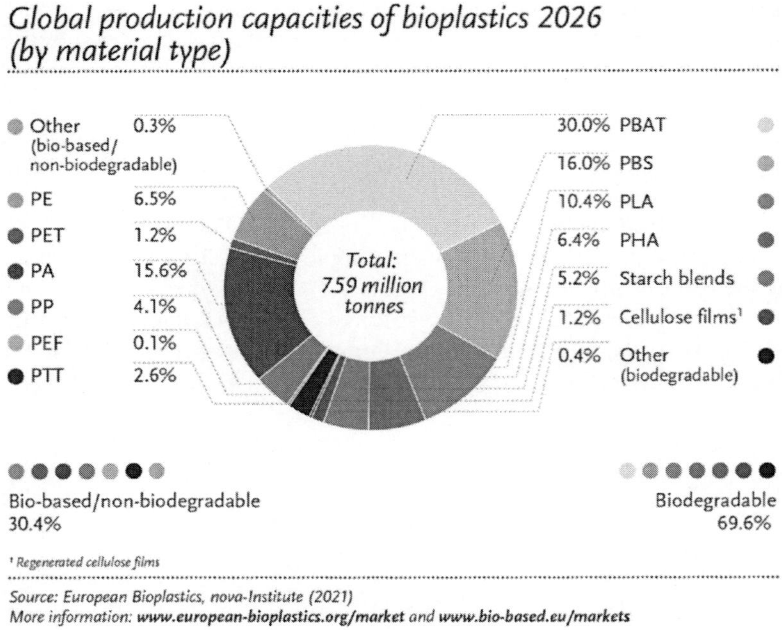

FIGURE 1.3 Global bioplastics production capacity in 2026 by material type; 2G and 3G bioplastics are represented by "biodegradable" on the right.

Source: European Bioplastics (2022), with permission for reproducing.

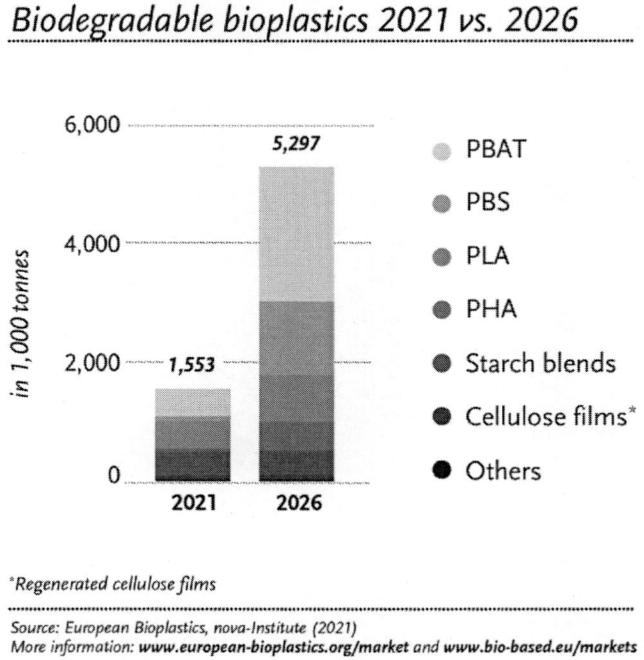

FIGURE 1.4 Second- and third-generation bioplastics production in 2021 and their estimated production in 2026; 2G and 3G bioplastics are represented by "biodegradable."

Source: European Bioplastics (2022), with permission for reproducing.

Bioplastics have a wide range of applications. Second- and third-generation bioplastics are amplifying their market application, as seen in Figure 1.5. Packaging remains the largest market segment, with 50% (including flexible and rigid) in 2021, followed by agriculture, consumer goods and others.

Regarding regional capacity production, Asia has strengthened its position as a major 2G/3G bioplastics production hub. In 2021, with a global production of 1.55 million tonnes, Asia produced 52.1% of this total. Contrarily, other regions, such as Europe and North America, are expected to reduce their production by 2026. Asia will significantly concentrate its production around 77.4%, from a world production of 5.30 million tonnes (Figure 1.6).

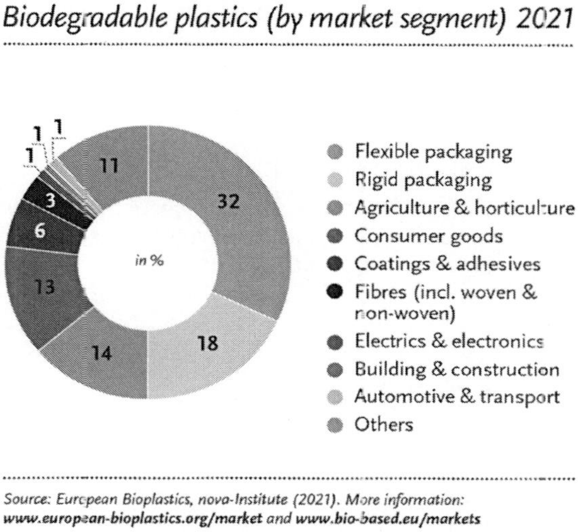

FIGURE 1.5 Second- and third-generation bioplastics application by market segment in 2021; 2G and 3G bioplastics are represented by "biodegradable."

Source: European Bioplastics (2022), with permission for reproducing.

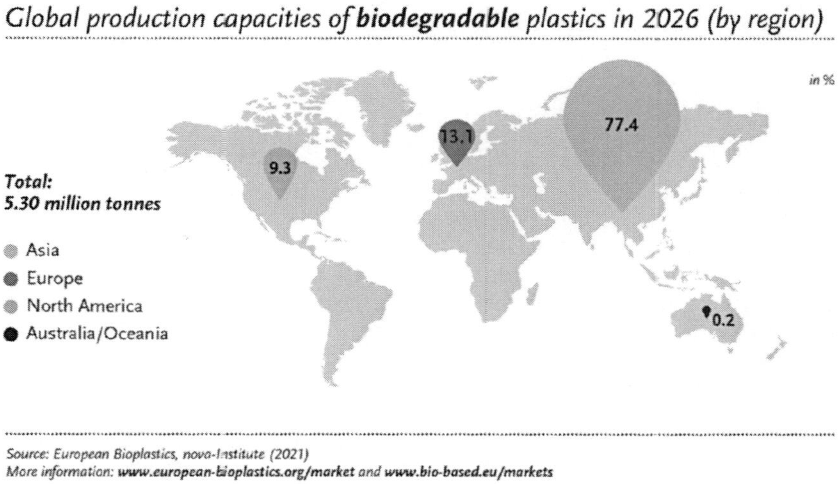

FIGURE 1.6 Second- and third-generation bioplastics global predicted production in 2026; 2G and 3G bioplastics are represented by "biodegradable."

Source: European Bioplastics (2022), with permission for reproducing.

1.9 SUMMARY AND CONCLUSIONS

The synthesis of 2G and 3G bioplastics is the actual emerging trend in the global market. Through biotechnology studies, these biopolymers, made from different renewable sources, are being tested and commercialized. Also, biotechnology advancements help to better understand biodegradation mechanics via microorganisms or enzymes. In this sense, knowledge can design new methods to produce 2G/3G bioplastics at lower costs without losing quality. They can replace synthetic nanomaterials with the advantage of being developed with desired properties (thermal and mechanical) according to final product application, where, this final product is composed of green material, contributing to the environment. Considering their contribution to the circular economy, it is possible to highlight the following aspects: reduction of carbon footprint using 2G and 3G substrates, bringing nutrients back into the soil, increase of bioplastics utility and functionality.

REFERENCES

Ahmad, A.; Banat, F.; Alsafar, H.; Hasan, S.W. (2022). An overview of biodegradable poly (lactic acid) production from fermentative lactic acid for biomedical and bioplastic applications. *Biomass Conv. Bioref*, v.13, 1–20. https://doi.org/10.1007/s13399-022-02581-3

Akhlaq, S.; Singh, D.; Mittal, N.; Srivastava, G.; Siddiqui, S.; Faridi, S.A.; Siddiqui, M.H. (2022). Polyhydroxybutyrate biosynthesis from different waste materials, degradation, and analytic methods: a short review. *Polymer Bulletin*, v.79, 1–33. https://doi.org/10.1007/s00289-022-04406-9

Aliotta, L.; Seggiani, M.; Lazzeri, A.; Gigante, V.; Cinelli, P. (2022). A brief review of Poly (Butylene Succinate) (PBS) and its main copolymers: synthesis, blends, composites, biodegradability, and applications. *Polymers (Basel)*, v.14(4), 844. https://doi.org/10.3390/polym14040844

Alsafadi, D.; Ibrahim, M.I.; Alamry, K.A.; Hussein, M.A.; Mansour, A. (2020). Utilizing the crop waste of date palm fruit to biosynthesize polyhydroxyalkanoate bioplastics with favorable properties. *Sci. Total Environ.*, v.737, 139716. https://doi.org/10.1016/j.scitotenv.2020.139716

Awasthi, K.; Akhtar, S.; Khan, M.K.A. (2021). Bioplastic: an accost towards sustainable development. *NeuroPharmac J.*, v.4, 162–168. https://doi.org/10.37881/1.614

Babalola, O.A.; Olorunnisola, A.O. (2019). Evaluation of coconut (*Cocos nucifera*) husk fibre as a potential reinforcing material for bioplastic production. *Mat. Sci.*, v.11, 195–200. https://doi.org/10.21741/9781644900178-14

Benavides, P.T.; Dunn, J.B.; Han, J.; Biddy, M.; Markham, J. (2018). Exploring comparative energy and environmental benefits of virgin, recycled, and bio-derived PET bottles. *ACS Sustain. Chem. Eng.*, v.6(8), 9725–9733. https://doi.org/10.1021/acssuschemeng.8b00750

Chandel, A.K.; Garlapati, V.K.; Singh, A.K.; Antunes, F.A.F.; da Silva, S.S. (2018). The path forward for lignocellulose biorefineries: bottlenecks, solutions, and perspective on commercialization. *Biores. Technol.*, v.264, 370–381. https://doi.org/10.1016/j.biortech.2018.06.004

Coppola, G.; Gaudio, M.T.; Lopresto, C.G.; Calabro, V.; Curcio, S.; Chakraborty, S. (2021). Bioplastic from renewable biomass: a facile solution for a greener environment. *Earth Syst. Environ.*, v.5, 231–251. https://doi.org/10.1007/s41748-021-00208-7

DeStefano, V.; Khan, S.; Tabada, A. (2020). Applications of PLA in modern medicine. *Eng. Reg.*, v.1, 76–87. https://doi.org/10.1016/j.engreg.2020.08.002

Dhaka, V.; Singh, S.; Anil, A.G.; Naik, T.S.S.K.; Garg, S.; Samuel, J.; Kumar, M.; Ramamurthy, P.C.; Singh, J.J. (2022). Occurrence, toxicity and remediation of polyethylene terephthalate plastics. A review. *Environ. Chem. Letters*, v.20, 1777–1800. https://doi.org/10.1007/s10311-021-01384-8

Döhler, N.; Wellenreuther, C.; Wolf, A. (2022). Market dynamics of biodegradable bio-based plastics: projections and linkages to European policies. *EFB Bioecon. J.*, v.2, 100028. https://doi.org/10.1016/j.bioeco.2022.100028

Esposti, M.D.; Morselli, D.; Fava, F.; Bertin, L.; Cavani, F.; Viaggi, D.; Fabbri, P. (2021). The role of biotechnology in the transition from plastics to bioplastics: an opportunity to reconnect global growth with sustainability. *FEBS Open Bio.*, v.11(4), 967–983. https://doi.org/10.1002/2211-5463.13119

European Bioplastics (2022). www.european-bioplastics.org (Access July 2022).

Farah, S.; Anderson, D.G.; Langer, R. (2016). Physical and mechanical properties of PLA, and their functions in widespread applications—a comprehensive review. *Adv. Drug Deliv. Rev.*, v.107(15), 367–392. http://dx.doi.org/10.1016/j.addr.2016.06.012

FNR—Fachagentur Nachwachsende Rohstoffe. (2020). Plants raw materials products. Agency for Renewable Resources, Federal Republic of Germany. https://fnr.de/

Geyer, R.; Jambeck, J.R.; Law, K.L. (2017). Production, use, and fate of all plastics ever made. *Sci. Adv.*, v.3(7). https://doi.org/10.1126/sciadv.1700782

Ghasemlou, M.; Daver, F.; Murdoch, B.J.; Ball, A.S.; Ivanova, E.P.; Adhikari, B. (2022). Biodegradation of novel bioplastics made of starch, polyhydroxyurethanes and cellulose nanocrystals in soil environment. *Sci. Total Environ.*, v.815, 152684. https://doi.org/10.1016/j.scitotenv.2021.152684

Gonzalez-Gutierrez, J.; Partal, P.; Garcia-Morales, M.; Gallegos, C. (2010). Development of highly-transparent protein/starch-based bioplastics. *Biores. Technol.*, v.101(6), 2007–2013. https://doi.org/10.1016/j.biortech.2009.10.025

Govil, T.; Wang, J.; Samanta, D.; David, A.; Tripathi, A.; Rauniyar, S.; Salem, D.R.; Sani, R.K. (2020). Lignocellulosic feedstock: a review of a sustainable platform for cleaner production of nature's plastics. *J. Cleaner Prod.*, v.270, 122521. https://doi.org/10.1016/j.jclepro.2020.122521

Guo, W.; Yang, K.; Qin, X.; Luo, R.; Wang, H.; Huang, R. (2022). Polyhydroxyalkanoates in tissue repair and regeneration. *Eng. Regen.*, v.3(1), 24–40. https://doi.org/10.1016/j.engreg.2022.01.003

Ibrahim, N.I.; Shahar, F.S.; Sultan, M.T.H.; Md Shah, A.U.; Safri, S.N.A.; Yazik, M.H.M. (2021). Overview of bioplastic introduction and its applications in product packaging. *Coatings*, v.11(1423). https://doi.org/10.3390/coatings11111423

Jawarraj, K.; Shrestha, N.; Chilkoor, G.; Dhiman, S.S.; Islam, J.; Gadhamshetty, V. (2020). Valorization of methane from environmental engineering applications: a critical review. *Water Res.*, v.187, 116400. https://doi.org/10.1016/j.watres.2020.116400

Jian, J.; Xiangbin, Z.; Xianbo, H. (2020). An overview on synthesis, properties and applications of poly(butylene-adipate-*co*-terephthalate)—PBAT. *Adv. Ind. Eng. Polymer Res.*, v.3(1), 19–26. https://doi.org/10.1016/j.aiepr.2020.01.001

Khyalia, P.; Gahlawat, A.; Jugiani, H.; Kaur, M.; Laura, J.S.; Nandal, M. (2022). Review on the use of microalgae biomass for bioplastics synthesis: a sustainable and green approach to control plastic pollution. *Pollution*, v.8(3), 844–859. https://doi.org/10.22059/POLL.2022.334756.1273

Koller, M.; Obruca, S. (2022). Biotechnological production of polyhydroxyalkanoates from glycerol: a review. *Biocat. Agri. Biotechnol.*, v.42, 102333. https://doi.org/10.1016/j.bcab.2022.102333

Lackner, M. (2015). Bioplastics. In: *Kirk-Othmer Encyclopedia of Chemical Technology*. John Wiley & Sons, Inc. https://doi.org/10.1002/C471238961.koe00006

Lamberti, F.M.; Ramírez, L.A.R.; Wood, J. (2020). Recycling of bioplastics: routes and benefits. *J. Pol. Environ.*, v.28, 2551–2571. https://doi.org/10.1007/s10924-020-01795-8

Leal Filho, W.; Salviac, A.L.; Bonoli, A.; Saari, U.A.; Voronova, V.; Klõga, M.; Kumbhar, S.S.; Olszewski, K.; De Quevedo, D.M.; Barbir, J. (2021). An assessment of attitudes towards plastics and bioplastics in Europe. *Sci. Total Environ.*, v.755(1), 142732. https://doi.org/10.1016/j.scitotenv.2020.142732

Lei, C.; Wei, Y.; Qian, Y.; Wang, Q.; Zhu, P.; Qiu, G.; Chen, G. (2022). Large-scale manufacture of recyclable bioplastics from renewable cellulosic biomass derived from softwood kraft pulp. *ACS Applied Polymer Mat.*, v.4(2), 1334–1343. https://doi.org/10.1021/acsapm.1c01729

Liu, L.Y.; Xie, G.J.; Xing, D.F.; Liu, B.F.; Ding, J.; Ren, N.Q. (2020). Biological conversion of methane to polyhydroxyalkanoates: current advances, challenges, and perspectives. *Environ. Sci. Ecotechnol.*, v.2, 100029. https://doi.org/10.1016/j.ese.2020.100029

Liwarska-Bizukojc, E. (2021). Effect of (bio)plastics on soil environment: a review. *Sci Total Environ.*, v.15(795), 148889. https://doi.org/10.1016/j.scitotenv.2021.148889

Liwarska-Bizukojc, E. (2022). Application of a small scale-terrestrial model ecosystem (STME) for assessment of ecotoxicity of bio-based plastics. *Sci. Total Environ.*, v.828, 154353. https://doi.org/10.1016/j.scitotenv.2022.154353

Markl, E.; Grünbichler, H.; Lackner, M. (2018). PHB—bio based and biodegradable replacement for PP: a review. *Nov. Tech. Nutri. Food Sci.*, v.2(4), 206–209. https://doi.org/10.31031/NTNF.2018.02.000546

Meereboer, K.W.; Misra, M.; Mohanty, K. (2020). Review of recent advances in the biodegradability of polyhydroxyalkanoate (PHA) bioplastics and their composites. *Green Chem.*, v.22, 5519–5558. https://doi.org/10.1039/d0gc01647k

Moreno, E.; Beltrán, F.R.; Arrieta, M.P.; Gaspar, G.; Muneta, L.M.; Carrasco-Gallego, R.; Yáñez, S.; Hidalgo-Carvajal, D.; Orden, M.U.D.L.; Urreaga, J.M. (2021). Technical evaluation of mechanical recycling of PLA 3D printing wastes. *Proceedings*, v.69, 19. https://doi.org/10.3390/CGPM2020-07187

Narancic, T.; Cerrone, F.; Beagan, N.; O'Connor, K.E. (2020). Recent advances in bioplastics: application and biodegradation. *Polymers Basel*, v.12(4), 920. https://doi.org/10.3390/polym12040920

Nduko, J.M.; Taguchi, S. (2021). Microbial production of biodegradable lactate-based polymers and oligomeric building blocks from renewable and waste resources. *Frontiers Bioeng. Biotechnol.*, v.8, 618077. https://doi.org/10.3389/fbioe.2020.618077

Oliveira, P.Z.; Vandenberghe, L.S.P.; Mello, A.F.M.; Soccol, C.R. (2022). A concise update on major poly-lactic acid bioprocessing barriers. *Biores. Technol. Reports*, v.18, 101094. https://doi.org/10.1016/j.biteb.2022.101094

Otoni, C.G.; Azeredo, H.M.C.; Mattos, B.D.; Beaumont, M.; Correa, D.S.; Rojas, O.J. (2021). The food-materials nexus: next generation bioplastics and advanced materials from agri-food residues. *Adv. Mat.*, v.33(43), 2102520. https://doi.org/10.1002/adma.202102520

Pérez, R.; Cantera, S.; Bordel, S.; García-Encina, P.A.; Muñoz, R. (2019). The effect of temperature during culture enrichment on methanotrophic polyhydroxyalkanoate production. *Inter. Biodet. Biodeg.*, v.140, 144–151. https://doi.org/10.1016/j.ibiod.2019.04.004

Singathi, R.; Raghunathan, R.; Krishnan, R.; Rajendran, S.K.; Baburaj, S.; Sibi, M.P.; Webster, D.C.; Sivaguru, J. (2022). Towards upcycling biomass-derived crosslinked polymers with light. *Angewandte Chemie*, v.61(31), e202203353. https://doi.org/10.1002/anie.202203353

SPI—The Plastics Industry Trade Association. (2016). Bioplastics simplified. Bioplastics Division, Washington, DC 20005–3686. www.plasticsindustry.org/sites/default/files/Bioplastics%20Simplified_0.pdf

Stefaniak, K.; Masek, A. (2021). Green copolymers based on poly(lactic acid)—short review. *Materials*, v.14(5254). https://doi.org/10.3390/ma14185254

SYSTEMIQ. (2022). ReShaping plastics: pathways to a circular, climate neutral plastics system in Europe. https://plasticseurope.org/knowledge-hub/systemiq-reshaping-plastics/

Vandenberghe, L.P.S.; Oliveira, P.Z.; Bittencourt, G.A.; Mello, A.F.M.; Vásquez, Z.S.; Karp, S.G.; Soccol, C.R. (2021). The 2G and 3G bioplastics: an overview. *Biotechnol. Res. Innov.*, v.5(1), e2021004. http://dx.doi.org/10.4322/biori.202104

Wellenreuther, C.; Wolf, A. (2020). Innovative feedstocks in biodegradable bio-based plastics: a literature review. HWWI Research Paper, No. 194, Hamburgisches WeltWirtschaftsInstitut (HWWI), Hamburg. http://hdl.handle.net/10419/228761

Wellenreuther, C.; Wolf, A.; Zander, N. (2022). Cost competitiveness of sustainable bioplastic feedstocks—A Monte Carlo analysis for polylactic acid. *Cleaner Eng. Technol.*, v.6, 100411. https://doi.org/10.1016/j.clet.2022.100411

Yashavanth, P.R.; Meenakshi, D.; Soumen, K.M. (2021). Recent progress and challenges in cyanobacterial autotrophic production of polyhydroxybutyrate (PHB), a bioplastic. *J. Environ. Chem. Eng.*, v.9, 105379. https://doi.org/10.1016/j.jece.2021.105379

2 Bioplastics
Classification and Properties

Ardra Nandakumar

CONTENTS

2.1 INTRODUCTION

Plastic pollution has become a problem worldwide due to the very same features that launched plastics into becoming a staple in our everyday lives; their durability and inertness. As of 2022, scientists have estimated that the ocean contains at least 45,000 individual pieces of plastic per square mile (Phillips 2022). To put it forth more visually, 40% of all our oceans are contaminated with plastics from the micro to the macro scale, with the most prominent outcome of this being the "Great Pacific Garbage Patch." The continued presence of plastics in the ocean and soil has also led to the formation and spread of microplastics, which are far more difficult to clean up owing to their small size and pervasive nature. This in turn has culminated in the contamination of groundwater reserves and thereby human and environmental health risks. Since a majority of plastics are derived from fossil fuels, the alarming rate of depletion of these raw materials is also a cause for concern. Owing to all these factors, the quest for the alleviation of plastic pollution along with finding the perfect replacement for conventional plastics has been going on for years, and one such outcome of this search was bioplastics (Damide 2021).

Bioplastics have become the need of the hour in terms of having a polymer that can emulate the properties of common plastics, like polypropylene (PP) and polyethylene (PE), and yet at the same time be easily degraded or recycled without a myriad of side effects. With the growing concern for the environment that has spread among the masses in modern times, a rise in the variety and application of widely promoted bioplastics can also be seen.

2.2 WHAT ARE BIOPLASTICS?

Bioplastics are defined as polymers that are either biodegradable or bio-based or both, according to European Bioplastics. However, IUPAC does not condone the use of this term, as they have claimed

DOI: 10.1201/9781003344018-2

that it can be quite misleading, due to its broad nature, and promotes the use of the term "bio-based polymer" instead for polymers such as bio-polyethylene terephthalate (bio-PET) that are not biodegradable (Vert et al. 2012). Another reason for this, is due to the existence of plastics such as polybutylene adipate-co-terephthalate (PBAT) and polycaprolactone (PCL), that are derived from fossil fuels and yet can be degraded completely and naturally in the environment over a short period of time i.e., they are biodegradable (Ferreira et al. 2019). While the major utility of bioplastics lies in their ability to be the greener substitute to plastics, recent techniques have also helped them find a niche in the field of biomedicine as grafts, sutures and drug delivery modules (Elieh-Ali-Komi and Hamblin 2016). Although the current market prices of bioplastics when compared to conventional plastics is a major drawback in their ubiquitous implementation, the multifaceted possibilities they provide may outweigh the negatives.

2.3 PROPERTIES OF BIOPLASTICS

Due to the broad nature of the definition of a bioplastic, there are few properties that are shared by all bioplastics. Yet the most common ones we see in recent times can be grouped based on some shared similarities in terms of features that give these polymers an advantage over conventional plastics in the first place. These characteristics along with their ability to closely mimic conventional plastics allow them to be implemented in a wide range of areas. Table 2.1 contains a quick rundown of the major properties of some representative bioplastics seen today.

2.3.1 Biodegradability

A material is said to undergo biodegradation when it is degraded by microbes into non-toxic products like biomass, carbon dioxide and water under normal environmental conditions. The mechanism of biodegradation usually involves three major steps: (1) biodeterioration, (2) biofragmentation and (3) assimilation. However, prior to biodegradation, abiotic degradation also plays a role in determining the rate of biodegradation. Factors including, but not limited to, shear/tension forces, light, temperature and oxidative reactions contribute to abiotic degradation. Biodeterioration is the superficial breakdown of materials as a result of the growth of microorganisms on or within the material, which leads to degradation via mechanical, chemical or physical means. The accumulation of microbes depends on the constitution of the material and whether it can serve as a suitable source of nutrition for the microbes present in the environment. Biofragmentation takes place when microbes secrete enzymes that act upon polymers at the structural level to break them down into smaller compounds or oligomers that can cross the cell membrane and/or cell wall. And finally, during assimilation, microbes integrate the polymeric materials into their cells. The process of assimilation is what allows microbes to produce energy and maintain activities like growth and reproduction. In the official definition of biodegradation by ASTM D5338 and ASTM D6400, no time limit has been specified, but in general, if a material undergoes 90% decomposition within 6 months or less it is called biodegradable; any longer and it is termed durable (ASTM 2021). Efforts are being made to bring in standards for biodegradable materials such as those implemented for the aerobic biodegradation of plastics (ISO17556) and for mulch films (EN17033; ISO 2019). One important distinction that has to be made from the beginning is that not all materials made from renewable resources (i.e., bio-based) are biodegradable, as exemplified by the case of bio-PET (Prieto 2016).

2.3.2 Compostability

According to the European Committee for Standardization (CEN) in 2000, a material is termed compostable if it fulfills the following criteria:

- Biodegradable under composting conditions.
- Completely disintegrates at the end of the composting process and there is a complete absence of visual pollution in the final compost obtained.

TABLE 2.1
Properties of the Most Commonly Used Bioplastics

Polymer	Raw Material	Synthesis	Biodegradable	Compostable	Generation	Reference
Bio-PET	Starch	Aqueous phase reforming and transesterification	No	No	1G	(Prieto 2016)
TPS	Starch	Using thermal and mechanical energy with a plasticizer	Yes	Yes	1G	(Przybyłek et al. 2018)
PLA	Lactate from starch	Polycondensation or ring-opening Polymerization	No	Yes (industrial composting)	1G/3G (if microbes are used to produce the monomers)	(Carlota 2019; Williams 1981)
Cellulose acetate	Crop residues	Cellulose with acetic acid	Yes	Yes (industrial composting)	2G	(Ach 1993)
Starch-lignin	Crop residues	Kraft method	Yes	Yes (industrial composting)	2G	(Yang et al. 2019)
Chitosan-based bioplastics	Crustacean shells	Demineralization or deproteinization using chemicals and/or microbes	Yes	Yes	2G/3G	(Shimahara and Takiguchi 1988; Degli Esposti et al. 2021)
PVA-alginate	Seaweed and fossil fuels	Alkali extraction and desiccation post blending	Yes	Yes (PVA portion requires industrial composting)	3G	(Eghbalifam et al. 2015)
Casein plastics	Milk and cheese	Precipitation using heat or acid	Yes	Yes	2G	(Wagh et al. 2014)
PHA	Waste sludge, crop residues, used oil, etc.	Using microbes	Yes	Yes	3G	(Tan et al. 2014)

Note: This table encompasses some representative bioplastics that are in common use today along with their major substrates, properties and routes of synthesis. From this table, it can be seen that although bioplastics like bio-PET are made from renewable sources, this does not automatically imply that they are biodegradable. Bioplastics from fossil fuels, like PBAT, can be more biodegradable than the former.

- There is no negative effect on either the composting process or the final compost (e.g., devoid of traces of heavy metals and visible residues) such that the agronomic value of the compost is not negatively impacted.

Unlike biodegradability, a compostable material should more often than not be degraded in 3 months or less into nutrient-rich biomass under composting conditions. Furthermore, standards such as the ASTM D6400 standard for compostability have to be met for a bioplastic to be considered compostable (ASTM 2021). The two major types of composting systems in place are (1) garden/home composting, where the maximum temperature can reach 40°C; and (2) industrial composting, which is a much more mechanical and controlled process. Industrial composting uses professional composting plants to deal with significant amounts of waste. The controlled optimized biotechnological processes that take place ensure good emission control, compost quality and degradation. Moreover, these systems usually contain a pre-processing phase to ensure that all materials reach ideal composting conditions. The temperature in these systems can go up to 50–70°C, which allows a shift in the microbial population over time, from mesophilic to thermophilic, hence allowing a diverse microbial community to thrive in the compost and increase the rate of successful degradation. The latest technologies such as windrow composting, aerated static piles and tunnel composting are also used to further enhance the process (European Bioplastics 2009). On the other hand, while home composting may require reduced initial investments and may be equipped in many households, the inability of the system to reach high temperatures during the process makes it inefficient when it comes to dealing with waste containing high energy like dairy, meat and some bioplastics. Seasonal fluctuations and the difficulty in monitoring the process in every household also prove to be disadvantages. To this day, legal bodies refuse to acknowledge home composting as a legal waste treatment option, but this is now changing as home compostability certifications are currently being offered by various companies and boards. The problem with the ASTM standard arises when the type of composting the bioplastic would have to undergo is not specified. However, the European standard (EN 13432) specifies that the bioplastic is degraded via industrial composting and hence proves to be more reliable (European Committee for Standardization 2000).

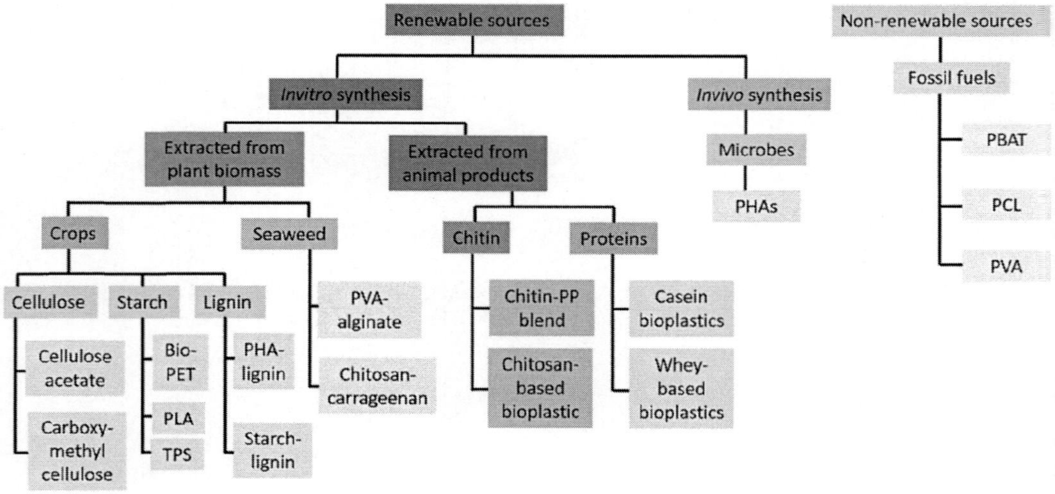

FIGURE 2.1 Types of bioplastics based on raw material and production. This figure categorizes well-known bioplastics based on their raw materials, and production based on in vivo or in vitro synthesis.

2.3.3 Raw Material

The raw material used for the synthesis of a bioplastic more often than not determines its other properties. This is not always the case, as showcased by bio-PET, which is not biodegradable although it is derived from biodegradable monomers (Prieto 2016). But the raw material does determine the inherent sustainability of the bioplastic using life cycle analyses (LCAs), which help to determine the environmental impact that the pre-processing of the raw material contributes to the overall greenness of the bioplastic. Over the years, advancements made in biotechnology and manufacturing processes have allowed a wider variety of raw materials to be used as the substrate for bioplastic production, as broadly categorized in Figure 2.1, and a few of which have been listed as follows:

1. *Starch*: Accounting for 85%–90% of all the bioplastics in the world, starch-based bioplastics may be derived directly from starch or from moieties derived from starch. Heating helps water disrupt the starch molecules, and when all the water is removed via drying, the disordered starch molecules entangle and form a film. Blends of starch with other natural or synthetic polymers also fall under this category. Bioplastics derived from starch alone are not viable as they are neither resistant to humidity nor do they show the properties required for melt process. However, starch is considered a good filler for petrochemical-based plastics as it possesses adequate thermal stability and is biodegradable (Fabunmi et al. 2007). Examples: thermoplastics starch (TPS), bio-PET (Fabunmi et al. 2007; Prieto 2016).
2. *Cellulose*: These bioplastics are manufactured from the cellulose fibers of softwood trees, which are processed to produce a pulp, where water is then removed and the remainder used for the manufacturing of cellulose esters and in turn cellulose-based bioplastics. An important point of interest is the presence of the β(1,4) linkages between the glucose molecules of cellulose that can only be broken by specific symbiotic microbes, as in the case of ruminants (Moohan et al. 2019). A major advantage of using cellulose as a raw material is its solubility in various media through a process that is reversible, hence bringing regenerated cellulose into the picture as a viable alternative. Coatings derived from cellulose for food items are edible and increase shelf life, thereby allowing it to be a contender as a replacement for conventional plastics used in single-use packaging, such as low-density polyethylene (LDPE). The fact that cellulose is biodegradable, renewable and non-toxic further pushes it to the forefront as a potential replacement (Tajeddin 2014). Examples: carboxymethyl cellulose, cellulose acetate (Tajeddin 2014).
3. *Protein*: Using a process similar to that of making cheese, protein-based bioplastics are derived from milk, gluten and other sources of protein. Total milk protein can rarely be used as the presence of lactose forms crystals during the drying process, which makes the resulting film non-homogenous. To avoid this, casein and whey proteins are used for film casting. Additionally, plasticizers such as glycerol and sorbitol may also be added to make the films more flexible by increasing their tensile strength and elastic modulus (Wagh et al. 2014). The macromolecular network in a protein-based material is stabilized by disulfide bonds, hydrophobic interactions and hydrogen bonds. A significant advantage of such polymers is their high rate of biodegradation, and their unique structure also offers a major scope for creating biodegradable materials with various functions, such as controlled release medication. Studies have also shown that perishable food covered in these edible protein films showed lower rates of oxidation as compared to LDPE due to their superior oxygen permeability barrier, although the addition of plasticizers can decrease the strength of this barrier, with the exception of clay (Wagh et al. 2014). Examples: casein bioplastics, whey-based bioplastics (Wagh et al. 2014).

4. *Lignin*: Lignin is a polyphenolic macromolecule, usually obtained as a by-product of cellulose production along with hemicellulose. Aside from alkalis, lignin is insoluble in most other solvents such as hot water and acids due to its complex structure and many hydroxyl groups that makes it a very polar molecule. These strong intermolecular interactions also prevent lignin from forming homogenous mixtures with most other polymers, hence complicating the production of lignin blends. On the other hand, this property also makes lignin recalcitrant to enzymes and other chemical factors (Yang et al. 2019). For these reasons, lignin biopolymer blends are seen as a more attractive option, considering that it is a widely available material with mechanical properties suitable for polymer applications. The miscibility of the blends can be improved using plasticizers. Furthermore, the ability of the phenolic groups of lignin to scavenge free radicals confers the blend with good antioxidant properties as well. Lignin can be used for the production of materials that require high strength and/or stiffness, such as carbon fibers, due to its high glass transition temperature. Lignin polymers may also be used as a fire retardant owing to their high charring capacity and on the flip side, for the purification of wastewater as well, due to their impressive adsorption capacity (Mandlekar et al. 2018). Examples: polyhydroxyalkanoate (PHA)-lignin polymer blends, starch-lignin polymer blends (Mandlekar et al. 2018).
5. *Chitin*: Chitin can be found in many organisms, such as in the exoskeletons of arthropods and in the cell walls of fungi; however, with the growing popularity of aquacultures, the most common source of its extraction is from the shells of crabs, prawns and other shellfish. This biopolymer is formed from N-acetyl-D-glucosamine bonded via β(1,4) linkages, thus making it insoluble in many solvents excluding strong caustic solutions. These linkages also confer significant flexibility and structural strength to the polymer. Although there is no shortage in terms of raw material, the use of hazardous organic solvents for its extraction as well as the poor quality of the final product due to loss in mass during processing makes it less than favorable (Shimahara and Takiguchi 1988). Chitin bioplastics can find applications in packaging, biomedical and optical materials owing to properties such as their low density, non-toxicity, biocompatibility and biodegradability (Elieh-Ali-Komi and Hamblin 2016). Chitosan scaffolds created via electrospinning can facilitate wound healing due to their antimicrobial nature that boosts fibroblast proliferation. Chitin is also being looked into as one of the replacements for single-use plastics, as it can break down easily even in conditions such as cold seawater in a month or two (Souza et al. 2011). The photonic properties of chitin seen in nature, such as in the iridescent wings of butterflies, as well as its transparency, also promise a key role for this polymer in the development of optical devices and sensors. Chitin can also be used in the form of blends with materials like starch to improve their water resistance (Elieh-Ali-Komi and Hamblin 2016). Examples: chitin blends with PP, chitosan-based polymers (Elieh-Ali-Komi and Hamblin 2016).
6. *Agar, alginate and carrageenan*: Agar, alginate and carrageenan are the main products of seaweed, and the properties of these polymers make seaweed a promising raw material for the production of bioplastics. Furthermore, seaweed is inexpensive and easily available in the sea, hence avoiding the food versus feed problem unlike in the case of cellulose or starch. Of the three, alginate is the most common polysaccharide and forms as much as 40% of the mass in brown seaweed (Phaeophyceae; Saji et al. 2022). It is composed of monomers of β-D-mannuronic acid and α-L-guluronic acid, which can form gels rapidly in the presence of divalent ions that aid in their association. Carrageenan is formed by a linear chain of partially sulfonated galactans and is usually found in red seaweed (Rhodophyceae). There are three types of carrageenan, each with varying chemical structures and gelation properties, and this diversity can be attributed to the differences in their anhydro bridges and sulfate groups. Agar is the most commercially used gelling agent due

to its ability to form reversible gels on heating or cooling the aqueous solution. The gelling properties of agar have also been shown to vary based on the growth conditions and chemical composition of the growth media of the seaweed (Şahin-Cebeci 2021). Polymers extracted from seaweed, phycocolloids, are being considered as potential replacements to the chemical compounds used in active packaging to improve the shelf life of the product while also improving its safety by avoiding the use of chemicals that may affect the product in case of leaching. They are also an excellent substitute for conventional plastics since they are derived from a feedstock of non-food crops that are easy to process to give rise to bioplastics with good mechanical strength and suitable gas barrier properties. An added advantage is that they are edible on top of being easily degraded. When carrageenan was added to chitosan polymers, it was seen to improve the flexibility as well as the tensile strength of the polymer while also providing resistance to microbes like *Bacillus subtilis* (Olaimat et al. 2014). PVA/alginate blends have also been manufactured to improve the tensile modulus of the polymer, although, in the process, the polymer did lose some of its flexibility, but not to an extent that affected its prior applications. Examples: chitosan-carrageenan bioplastics, PVA-alginate bioplastics (Olaimat et al. 2014).

2.3.4 Production

While all bioplastics have to undergo some form of processing before they can be used in various applications, some polymers can be extracted directly from the raw material, whereas some others would have to undergo artificial polymerization or biological polymerization within organisms that feed on the raw material.

Biopolymers, such as lignin, starch and chitin, derived from polysaccharides are usually classified as biopolymers derived directly from biomass. Starch extraction requires the least amount of processing since pure starch itself has many desirable properties, such as its lack of taste, smell, toxicity and solubility in alcohol and cold water. The steps start with the disintegration of the biomass into a fine powder, like sawdust, which is then mixed with hot water to initiate gelatinization and remove any impurities, after which the water is siphoned off. Following multiple repeats of this process, the purified white starch can then be used to prepare the films (Marichelvam et al. 2022). TPS is formed by blending starch with a plasticizer, but thermomechanical pressure is used to forcibly create the hydrogen bonds between starch and the plasticizer and in the process break the hydrogen bonds between the hydroxyl groups within starch. Bio-PET can be considered an exception in the category of starch-based polymers, as these bioplastics require external polymerization wherein only the monomers for said polymerization are derived from starch. For this reason, bio-PET has come under a lot of scrutiny because even though the monomers are derived from starch, the polymerization takes place through aqueous phase reforming and transesterification, just as in the case of their fossil fuel–based counterparts. This implies that the end products are virtually indistinguishable from each other in terms of both physical and chemical properties (Prieto 2016).

In the case of cellulose, it is derived from biomass and is often made to undergo processes such as acylation and esterification with derivatives of carboxylic acids to synthesize polymers like cellulose acetate and methyl cellulose. The cellulose is extracted from the biomass using the caustic soda method, the kraft method or the sulfite method. These processes require the treatment of biomass, such as woodchips, with alkaline chemicals like sodium sulfite to dissociate cellulose from other components like hemicellulose and lignin (Kunze and Fink 2005) and typically act as the rate-limiting step in cellulose bioplastic manufacturing in terms of both yield and cost. Furthermore, these processes are said to impact the environment in terms of releasing large amounts of odorous gases and great volumes of liquid by-products (Hoffman et al. 2015).

Although lignin extraction can also be achieved using the kraft method, the harsh chemicals used in this process cleave the lignin backbone and can inadvertently lead to the addition of new functional groups as well. On the other hand, processing via the use of organic solvents has been shown to yield polymers that are more structurally similar to native lignin. The cost of lignin extraction cannot be made up by the use of lignin directly in bioplastics and so lignin blends are considered the economically more viable option (Gregorová et al. 2005).

Chitin extraction is a two-step process that includes demineralization and deproteinization using either biological or chemical methods. The biological method uses lactic acid bacteria and proteolytic bacteria for each process or even mixed cultures of the two strains, whereas the chemical method employs a combination of hydrochloric acid for demineralization, organic solvents and sodium hydroxide for the removal of lipids and deproteinization. The biological method is usually preferred since the chemical method uses an excessive amount of hazardous solvents that lead to a significant loss of mass post-processing (Shimahara and Takiguchi 1988).

Biopolymers from seaweed can be extracted using either alkaline treatments or large volumes of hot water. Agar, for example, can be extracted by dissolving the biomaterial in hot water, which helps to concentrate the polymer thereby allowing gelation on cooling (Şahin-Cebeci 2021). Carrageenan can be extracted using alkaline solvents, after which the solution can either be refined through alcohol precipitation or semi-refined by re-dissolving it in an alkali solution, like potassium chloride, after which it can be dried to give rise to strong gels (Şahin-Cebeci 2021). Alginate can be extracted from washed seaweed by an initial pre-treatment with hydrochloric acid, which produces alginic acid, that is neutralized with either sodium hydroxide or sodium carbonate to produce sodium alginate. This sodium alginate can be further purified through another round of precipitation, with acids or salts and neutralization. Alginate extraction from contaminated seaweed has proved to be a tedious task, since cytotoxic materials may also be co-purified and form impurities in the product (Şahin-Cebeci 2021).

Casein-based bioplastics are made by first denaturing the protein using heat to allow the formation of disulfide bonds, which contribute to the film-like structure of these polymers. The thickness of these polymers can be increased with the addition of plasticizers, like glycerol or sorbitol, but they may also tend to reduce the tensile strength (Wagh et al. 2014). Recent techniques involving treatment with high-pressure carbon dioxide are also being researched in the hopes that they will decrease the solubility while increasing the hydrophobicity of the material. Although there is no concrete information on traditional polymer processing techniques for protein-based plastics, studies have shown that the use of various additives can help the mixing process to optimize steps like injection molding. It was found that the use of reducing agents showed a major improvement in both the mixing process, and the mechanical properties of the resulting polymer by reducing the molecular weight of the aggregates. Urea can also be used in the manufacturing process to further enhance denaturation and enable the formation of new interactions between segments (Felix et al. 2017).

Fermentative lactate can be used to produce lactide, which can be made to undergo ring opening polymerization to give rise to polylactic acid (PLA). In terms of expense, solution polycondensation is the preferred route, although it also makes obtaining high-molecular-weight, solvent-free PLA difficult. On the other hand, while ring opening polymerization requires a catalyst and is more complicated and expensive, it also provides more control, thereby allowing the production of specific high-molecular-weight PLA (Singhvi et al. 2019). More recently, enzymatic polymerization is being considered as an alternative as it is environmentally friendly and the use of specific enzymatic reactions can control the finer structures within the polymer, but studies on this are still scarce (Singhvi et al. 2019).

Some biopolymers can be produced in vivo through microbial fermentation, and one such key example is that of PHAs. They are naturally occurring aliphatic polyesters that were first discovered in 1926. Each monomer has a side chain R-group that can be a saturated or unsaturated alkyl group, or even branched and substituted alkyl groups (Tan et al. 2014). PHAs are produced by microbes as a carbon reserve for long-term survival under nutrient-scarce conditions. The PHA biosynthetic pathway is intrinsically linked to major metabolic pathways like the Krebs cycle and glycolysis. PHA production is hence induced by altering the media conditions so that essential nutrients like phosphorus or nitrogen are limited in the presence of excess carbon sources (Tan et al. 2014). Key enzymes required for PHA synthesis include 3-ketothiolase, which is inhibited by high levels of coenzyme A in nutrient-rich conditions, encoded by *phaA*, NADPH-dependent acetoacetyl-CoA reductase encoded by the gene *phaB*, and PHA synthase, which performs the polymerization step, encoded by *phaC*. Previously, it was thought that the polymerization by PHA synthase might constitute the rate-limiting step in the production, but this has since been disproved. The raw materials used for PHA production can range from sugars to oils and even sludge (Saratale et al. 2021). Substrates like palm oil usually give rise to medium-chain-length PHAs like poly(3-hydroxyhexanoate), whereas substrates like sugars derived from bagasse give rise to poly(3-hydroxybutyrate) (Saratale et al. 2021). Since the chain of the polymer, as well as the branching, can be controlled by the type of microorganism used or the substrate, a large variety of these polymers can be synthesized and tailor-made for specific applications.

2.4 CLASSIFICATION OF BIOPLASTICS

Although rapid industrialization did lead to the current issue of plastic pollution, it has also allowed for the invention of a variety of bioplastics, some that can even be tailor-made for processes as specific as 3D printing. The complexity and diversity of bioplastics may seem overwhelming, but nonetheless, along with the categories based on the properties mentioned above, they can also be classified into some major categories too, as follows.

2.4.1 Drop in Bioplastics

Any non-biodegradable material that is synthesized from a renewable source is technically known as a bio-based drop-in chemical. These materials are usually the bio-based counterparts of preexisting conventional plastics that already have a well-established market. This provides the advantage of helping to reduce the carbon footprint, in terms of sourcing the substrate for monomer production from a biological source, and also provides an economical solution to boot (Ibrahim et al. 2021). Additionally, with the advent of new technologies related to biomass processing, novel monomers can be more easily synthesized from biomass than from fossil fuels, and this can help in manufacturing a range of polymers with more diverse properties. Furthermore, since they are entering into a preconceived area, they currently have the strongest hold on the bioplastic market. However, since the end product is indistinguishable in terms of physical properties and chemical structure, they are still as inert as the materials they are meant to substitute, hence not quite helping to close the loop. Moreover, the processing of the raw materials requires far more initial investment and hikes the market price beyond economic viability (Barrett 2018).

Recently, a new subgroup of drop-in bioplastics known as "smart" drop-in bioplastics has entered the market, in which the production steps are more efficient by at first processing the material using novel pathways till a certain stage, after which it can be easily incorporated into a conventional pathway, hence reducing the overall number of steps. For a bioplastic to be termed a smart drop-in, it must fulfill at least two of the following criteria (Carus et al. 2017):

- The Biomass Utilization Efficiency (BUE) is significantly higher.
- The overall toxicity of the process is lower, both in terms of the chemicals used, as well as the by-products produced.
- The pathways are less complex and faster.
- The manufacturing process is more energy efficient.

Examples of smart drop-in chemicals include new methods to produce acetic acid and succinic acid, which can then help produce "smarter" bio-PET and bio-PE (Carus et al. 2017).

(a) (b)

FIGURE 2.2 Symbols for (a) biodegradable and (b) compostable bioplastics. Although the symbol in (a) is used by some companies to denote the biodegradability of a product, it does not automatically indicate that a product fulfills any standard. On the other hand, the symbol shown in (b) is used to denote compostability and this mark implies fulfillment of the EN 13432 standard by CEN.

2.4.2 Biodegradable Bioplastics

Although bioplastics can be categorized in many ways, the major terms used by consumers with respect to bioplastics in their quest to be more environmentally conscious are usually biodegradable and compostable (European Bioplastics 2021). To indicate if a product is biodegradable, some companies use the symbol shown in Figure 2.2(a) on their products.

As mentioned in Section 2.3.1, not all bioplastics derived from renewable resources are biodegradable. For example, although starch-based polymers are biodegradable, blended polymers of starch are not wholly biodegradable since only the starch component usually undergoes biodegradation (Jiang et al. 2020). Bioplastics synthesized from cellulose, like cellulose acetate, often require highly controlled thermophilic conditions for biodegradation and may even show different rates of biodegradability owing to the various acetyl groups that can cause variations in the crystallinity and degree of polymerization. Polymers derived from lignin are difficult to enzymatically depolymerize, owing to the strong carbon-carbon and ether linkages. While some organisms, like white rot fungi, can break it down, most microbes can only modify it.

PHAs can undergo surface degradation in the soil, owing to the presence of microbes that can secrete PHA depolymerases and lipases. Furthermore, this rapid surface degradation can lead to an increase in surface area, thereby providing more room for bacterial attachment, hence improving biodegradation in almost a positive feedback loop–like manner (Mukai et al. 1993).

2.4.3 Compostable Bioplastics

The standards pertaining to compostable plastics, such as EN 13432, are usually more specific to the testing conditions. PLA requires commercial composting systems that can provide the high sustained temperatures needed for its complete degradation, but polymers like chitin can be decomposed in the soil and ocean. This is in part based on how ubiquitous the enzymes are, since chitinases are found in many microbes, unlike proteinase K, an enzyme that is not readily available in the natural environment, hence making the industrial composting of PLA more feasible (Williams 1981; Elieh-Ali-Komi and Hamblin 2016). Following this thread, PHAs can also be easily degraded in both home and industrial composting systems, and are even certified by a third party, Vinçotte, on top of the usual certifications from ASTM and the European Standards (PHA Certifications–Danimer Scientific 2017). In summary, the major examples of compostable plastics include PHAs, PLA and chitin-based polymers. Figure 2.2(b) depicts the symbol used by EN 13432 to label a bioplastic as compostable. Along with the groups mentioned above, bioplastics can also be categorized based on the raw material into larger groups in terms of the source of the feedstock, into first (1G), second (2G) and third (3G) generation bioplastics, as mentioned in Chapter 1.

2.5 CHALLENGES

By putting bioplastics into well-defined categories, consumers can be made aware of the various properties of the product, hence allowing them to make environmentally sound choices. However, one of the major issues faced in the systematic nomenclature and classification of bioplastics is the phenomenon of "greenwashing." Greenwashing is a term coined as far back as 1986 by Jay Westerveld to define the act of falsely advertising a product or service to be more environmentally beneficial than it actually is. Companies usually employ this by making false claims or leaving out certain information to capitalize on the consumer's hunt for environmentally friendly items. A side effect of the skepticism brought about by greenwashing is that consumers also tend to be jaded toward real green products, which can cause a real hindrance to the economic viability of sustainable solutions (Nandakumar et al. 2021).

A major example of greenwashing is the advertisement for oxo-biodegradable plastics that may have the term "biodegradable" in their name, but in reality the polymers only degrade physically mainly due to the addition of metal compounds. Since this polymer is not chemically broken down, not only can it not be taken up by microbes, but the small size can also contribute to microplastic contamination of the soil and groundwater resources (Thomas et al. 2012).

The major drawback of bioplastics seems to be the high cost of production. Although LCAs imply that the production processes of bioplastics are more detrimental to the environment than conventional plastics, many of these analyses do not take into account processes such as the collection and sorting of waste that are eased when using compostable garbage bags or disposable cutlery (Gironi and Piemonte 2011). Furthermore, technological advancements are also allowing bioplastic production systems to undergo rapid maturation. For example, as a means of reducing the economic and environmental costs borne due to PHA production, wastewater from sewage treatment plants are being studied as a novel substrate; also methods like aqueous two-phase extraction and the use of mealworms are being considered for PHA extraction to replace the use of toxic solvents, like chloroform and methanol (Wang et al. 2016; Ong et al. 2018). Moreover, governments are also pushing for the use of sustainable products and are starting to hold companies accountable for their environmental impacts, providing benefits to greener companies, thereby incentivizing the creation of more sustainable solutions in the long run (Taebi and Safari 2017).

2.6 SUMMARY AND CONCLUSION

With the rise of environmental conscience on a global scale, people are now looking into the various options they can turn to as a means of curbing plastic pollution. In spite of the numerous types of bioplastics available today, most of them can be categorized based on the properties of the material as well as the production processes behind the scenes. In this way, ambiguity regarding properties like biodegradability of some bioplastics, such as bio-PET, can be cleared and greenwashing tactics can be abolished. Although bioplastics themselves are environmentally friendly, substrate acquisition and processing, as seen through LCA analyses, might not be so sustainable in the long run. However, novel solutions are being introduced at an exponential rate, thereby increasing the possibility that bioplastics may soon become a ubiquitous commodity. Availability and proper dissemination of this information may also allow governments and organizations to make informed choices toward a sustainable society.

REFERENCES

Ach, Alexander. 1993. "Biodegradable Plastics Based on Cellulose Acetate." *Journal of Macromolecular Science, Part A* 30 (9–10): 733–40. https://doi.org/10.1080/10601329308021259.

ASTM. 2021. "ASTM D6400–19 Standard Specification for Labeling of Plastics Designed to Be Aerobically Composted in Municipal or Industrial Facilities." *ASTM*. www.astm.org/Standards/D6400.htm.

ASTM. 2022. "Standard Test Method for Determining Aerobic Biodegradation of Plastic Materials Under Controlled Composting Conditions, Incorporating Thermophilic Temperatures." 2021. Accessed October 9, 2022. www.astm.org/d5338-15r21.html.

Barrett, Axel. 2018. "What Are Drop-In Bioplastics?" *Bioplastics News*. Accessed August 28, 2022. https:// bioplasticsnews.com/2018/08/28/drop-ins-bioplastics/.

Carlota, V. 2019. "Is PLA Filament Actually Biodegradable?—3D natives." *3D Natives*. www.3dnatives.com/ en/pla-filament-230720194/#!

Carus, Michael, Lara Dammer, Ángel Puente, A. Raschka, and O. Arendt. 2017. "Bio-Based Drop-in, Smart Drop-in and Dedicated Chemicals." *Renewable Carbon* (745623): 1–3.

Damide, Celine. 2021. "Bioplastics: Plant or Plastic?" *Brightcore*. www.brightcoreenergy.com/post/ bioplastics-plant-or-plastic.

Danimer Scientific. 2017. "PHA Certifications." *Danimer Scientific*. Accessed April 30, 2020. https://danimer-scientific.com/pha-the-future-of-biopolymers/pha-certifications/.

Degli Esposti, Micaela, Davide Morselli, Fabio Fava, Lorenzo Bertin, Fabrizio Cavani, Davide Viaggi, and Paola Fabbri. 2021. "The Role of Biotechnology in the Transition from Plastics to Bioplastics: An Opportunity to Reconnect Global Growth with Sustainability." *FEBS Open Bio* 11 (4): 967–83. https:// doi.org/10.1002/2211-5463.13119.

Eghbalifam, Naeimeh, Masoud Frounchi, and Susan Dadbin. 2015. "Antibacterial Silver Nanoparticles in Polyvinyl Alcohol/Sodium Alginate Blend Produced by Gamma Irradiation." *International Journal of Biological Macromolecules* 80: 170–76. https://doi.org/10.1016/j.ijbiomac.2015.06.042.

Elieh-Ali-Komi, Daniel, and Michael R Hamblin. 2016. "Chitin and Chitosan: Production and Application of Versatile Biomedical Nanomaterials." *International Journal of Advanced Research* 4 (3): 411–27. https://pubmed.ncbi.nlm.nih.gov/27819009.

European Bioplastics. 2009. "Industrial Composting Fact Sheet 2009." *European Bioplastics*. Accessed July 19, 2022. https://docs.european-bioplastics.org/2016/publications/fs/EUBP_fs_industrial_composting.pdf.

European Bioplastics. 2021. "Claims on Biodegradability and Compostability on Products." *European Bioplastics*. Accessed July 20, 2022. https://docs.european-bioplastics.org/publications/Claims_on_ biodegradability_and_compostability_on_products_and_packaging_210108.pdf.

European Committee for Standardization. 2000. "CEN—EN 13432—Packaging—Requirements for Packaging Recoverable Through Composting and Biodegradation—Test Scheme and Evaluation Criteria for the Final Acceptance of Packaging | Engineering360." *European Committee for Standardization*. Accessed April 30, 2020. https://standards.globalspec.com/std/323935/EN 13432.

Fabunmi, Olayide, Lope Tabil, Satyanarayan Panigrahi, and Peter Chang. 2007. "Developing Biodegradable Plastics from Starch." RRV07130, ASABE/CSBE North Central Intersectional Meeting. Accessed August 07, 2019. https://doi.org/10.13031/2013.24179.

Felix, M, V Perez-Puyana, A Romero, and A Guerrero. 2017. "Development of Protein-Based Bioplastics Modified with Different Additives." *Journal of Applied Polymer Science* 134 (42): 45430. https://doi. org/10.1002/app.45430.

Ferreira, Filipe V, Luciana S Cividanes, Rubia F Gouveia, and Liliane M F Lona. 2019. "An Overview on Properties and Applications of Poly(Butylene Adipate-Co-Terephthalate)—PBAT Based Composites." *Polymer Engineering & Science* 59 (s2): E7–15. https://doi.org/10.1002/pen.24770.

Gironi, F., and V. Piemonte. 2011. "Bioplastics and Petroleum-Based Plastics: Strengths and Weaknesses." *Energy Sources, Part A: Recovery, Utilization, and Environmental Effects* 33 (21): 1949–59. https://doi. org/10.1080/15567030903436830.

Gregorová, A, Z Cibulková, B Košíková, and P Šimon. 2005. "Stabilization Effect of Lignin in Polypropylene and Recycled Polypropylene." *Polymer Degradation and Stability* 89 (3): 553–58. https://doi. org/10.1016/j.polymdegradstab.2005.02.007.

Hoffman, Emma, Meagan Bernier, Brenden Blotnicky, Peter G Golden, Jeffrey Janes, Allison Kader, Rachel Kovacs-Da Costa, Shauna Pettipas, Sarah Vermeulen, and Tony R Walker. 2015. "Assessment of Public Perception and Environmental Compliance at a Pulp and Paper Facility: A Canadian Case Study." *Environmental Monitoring and Assessment* 187 (12): 766. https://doi.org/10.1007/s10661-015-4985-5.

Ibrahim, Nor, Farah Shahar, Mohamed Thariq Hameed Sultan, Ain Md Shah, Syafiqah Safri, and Muhamad Hasfanizam Mat Yazik. 2021. "Overview of Bioplastic Introduction and Its Applications in Product Packaging." *Coatings* 11: 1423. https://doi.org/10.3390/coatings11111423.

ISO. 2019. "ISO17556 International Standard Aerobic Biodegradability of Plastic." Accessed July 21, 2022. https://www.iso.org/standard/74993.html.

Jiang, Tianyu, Qingfei Duan, Jian Zhu, Hongsheng Liu, and Long Yu. 2020. "Starch-Based Biodegradable Materials: Challenges and Opportunities." *Advanced Industrial and Engineering Polymer Research* 3 (1): 8–18. https://doi.org/10.1016/j.aiepr.2019.11.003.

Kunze, Jürgen, and Hans-Peter Fink. 2005. "Structural Changes and Activation of Cellulose by Caustic Soda Solution with Urea." *Macromolecular Symposia* 223: 175–88. https://doi.org/10.1002/masy.200550512.

Mandlekar, Neeraj, Aurélie Cayla, François Rault, Stéphane Giraud, Fabine Salaün, Giulio Malucelli, and Jin-Ping Guan. 2018. "An Overview on the Use of Lignin and Its Derivatives in Fire Retardant Polymer Systems." In *Lignin—Trends and Applications*. InTech. https://doi.org/10.5772/intechopen.72963.

Marichelvam, M K, P Manimaran, M R Sanjay, S Siengchin, M Geetha, K Kandakodeeswaran, Pawinee Boonyasopon, and Sergey Gorbatyuk. 2022. "Extraction and Development of Starch-Based Bioplastics from Prosopis Juliflora Plant: Eco-Friendly and Sustainability Aspects." *Current Research in Green and Sustainable Chemistry* 5: 100296. https://doi.org/10.1016/j.crgsc.2022.100296.

Moohan, John, Sarah A Stewart, Eduardo Espinosa, Antonio Rosal, Alejandro Rodríguez, Eneko Larrañeta, Ryan F Donnelly, and Juan Domínguez-Robles. 2019. "Cellulose Nanofibers and Other Biopolymers for Biomedical Applications. A Review." *Applied Sciences* 10 (1): 65. https://doi.org/10.3390/app10010065.

Mukai, Katsuyuki, Yoshiharu Doi, Yukio Sema, and Kosuke Tomita. 1993. "Substrate Specificities in Hydrolysis of Polyhydroxyalkanoates by Microbial Esterases." *Biotechnology Letters* 15 (6): 601–4. https://doi.org/10.1007/BF00138548.

Nandakumar, Ardra, Jo-Ann Chuah, and Kumar Sudesh. 2021. "Bioplastics: A Boon or Bane?" *Renewable and Sustainable Energy Reviews* 147: 111237. https://doi.org/10.1016/j.rser.2021.111237.

Olaimat, Amin N., Yuan Fang, and Richard A. Holley. 2014. "Inhibition of Campylobacter Jejuni on Fresh Chicken Breasts by κ-Carrageenan/Chitosan-Based Coatings Containing Allyl Isothiocyanate or Deodorized Oriental Mustard Extract." *International Journal of Food Microbiology* 187: 77–82. https://doi.org/10.1016/j.ijfoodmicro.2014.07.003.

Ong, Su Yean, Idris Zainab L, Somarajan Pyary, and Kumar Sudesh. 2018. "A Novel Biological Recovery Approach for PHA Employing Selective Digestion of Bacterial Biomass in Animals." *Applied Microbiology and Biotechnology* 102 (5): 2117–27. https://doi.org/10.1007/s00253-018-8788-9.

Phillips, Eric. 2022. "100+ Plastic in the Ocean Statistics and Facts 2021–2022." Accessed July 11, 2022. *Dripfina*. https://dripfina.com/plastic-in-the-ocean-statistics/.

Prieto, Auxiliadora. 2016. "To Be, or Not to Be Biodegradable . . . That Is the Question for the Bio-Based Plastics." *Microbial Biotechnology* 9 (5): 652–57. https://doi.org/10.1111/1751-7915.12393.

Przybyłek, A, M Sienkiewicz, J Kucińska-Lipka, and H Janik. 2018. "Preparation and Characterization of Biodegradable and Compostable PLA/TPS/ESO Compositions." *Industrial Crops and Products* 122: 375–83. https://doi.org/10.1016/j.indcrop.2018.06.016.

Şahin-Cebeci, Oya Irmak. 2021. Chapter 4 – "Seaweed Polysaccharides: Structure, Extraction and Applications." In *Polysaccharides: Properties and Applications*. Wiley-Scrivener.

Saji, Sijin, Andrew Hebden, Parikshit Goswami, and Chenyu Du. 2022. "A Brief Review on the Development of Alginate Extraction Process and Its Sustainability." *Sustainability* 14 (9): 5181. https://doi.org/10.3390/su14095181.

Saratale, Rijuta Ganesh, Si Kyung Cho, Ganesh Dattatraya Saratale, Gajanan S Ghodake, Ram Naresh Bharagava, Dong Su Kim, Supriya Nair, and Han Seung Shin. 2021. "Efficient Bioconversion of Sugarcane Bagasse into Polyhydroxybutyrate (PHB) by *Lysinibacillus* Sp. and Its Characterization." *Bioresource Technology* 324: 124673. https://doi.org/10.1016/j.biortech.2021.124673.

Shimahara, Kenzo, and Yasuyuki B T. 1988. "Preparation of Crustacean Chitin." *Biomass Part B: Lignin, Pectin, and Chitin* 161: 417–23. https://doi.org/10.1016/0076-6879(88)61049-4.

Singhvi, M S, S S Zinjarde, and D V Gokhale. 2019. "Polylactic Acid: Synthesis and Biomedical Applications." *Journal of Applied Microbiology* 127 (6): 1612–26. https://doi.org/10.1111/jam.14290.

Souza, Claudiana, Bianca Almeida, Rita Colwell, and Irma Rivera. 2011. "The Importance of Chitin in the Marine Environment." *Marine Biotechnology (New York, N.Y.)* 13: 823–30. https://doi.org/10.1007/s10126-011-9388-1.

Taebi, Behnam, and Azar Safari. 2017. "On Effectiveness and Legitimacy of 'Shaming' as a Strategy for Combatting Climate Change." *Science and Engineering Ethics* 23 (5): 1289–306. doi:10.1007/s11948-017-9909-z.

Tajeddin, Behjat. 2014. "Cellulose-Based Polymers for Packaging Applications." *Lignocellulosic Polymer Composites: Processing, Characterization, and Properties*, 477–98. https://doi.org/10.1002/9781118773949.ch21.

Tan, Giin-Yu, Chia-Lung Chen, Ling Li, Liya Ge, Lin Wang, Indah Mutiara Ningtyas Razaad, Yanhong Li, Lei Zhao, Yu Mo, and Jing-Yuan Wang. 2014. "Start a Research on Biopolymer Polyhydroxyalkanoate (PHA): A Review." *Polymers* 6: 706–54. https://doi.org/10.3390/polym6030706.

Thomas, Noreen L, Jane Clarke, Andrew R McLauchlin, and Stuart G Patrick. 2012. "Oxodegradable Plastics: Degradation, Environmental Impact and Recycling." *Proceedings of the Institution of Civil Engineers—Waste and Resource Management* 165 (3): 133–40. https://doi.org/10.1680/warm.11.00014.

Vert, Michel, Yoshiharu Doi, Karl-Heinz Hellwich, Michael Hess, Philip Hodge, Przemyslaw Kubisa, Marguerite Rinaudo, and François Schué. 2012. "Terminology for Biorelated Polymers and Applications (IUPAC Recommendations 2012)." *Pure and Applied Chemistry* 84 (2): 377–410. https://doi.org/10.1351/PAC-REC-10-12-04.

Wagh, Y R, Heartwin A Pushpadass, F Magdaline Eljeeva Emerald, and B Surendra Nath. 2014. "Preparation and Characterization of Milk Protein Films and Their Application for Packaging of Cheddar Cheese." *Journal of Food Science and Technology* 51 (12): 3767–75. https://doi.org/10.1007/s13197-012-0916-4.

Wang, Qilin, Jing Sun, Chang Zhang, Guo-Jun Xie, Xu Zhou, Jin Qian, Guojing Yang, Guangming Zeng, Yiqi Liu, and Dongbo Wang. 2016. "Polyhydroxyalkanoates in Waste Activated Sludge Enhances Anaerobic Methane Production through Improving Biochemical Methane Potential Instead of Hydrolysis Rate." *Scientific Reports* 6 (1): 19713. https://doi.org/10.1038/srep19713.

Williams, D F. 1981. "Enzymic Hydrolysis of Polylactic Acid." *Engineering in Medicine* 10 (1): 5–7. https://doi.org/10.1243/EMED_JOUR_1981_010_004_02.

Yang, Jianlei, Yern C Ching, and Cheng H Chuah. 2019. "Applications of Lignocellulosic Fibers and Lignin in Bioplastics: A Review." *Polymers* 11 (5): 751. doi:10.3390/polym11050751.

3 Second-Generation Bioplastics from Lignocellulosic Materials

Kim Kley Valladares-Diestra, Luis Daniel Goyzueta-Mamani, Dão Pedro de Carvalho Neto, Patricia Beatriz Gruening de Mattos, and Carlos Ricardo Soccol

CONTENTS

3.1 INTRODUCTION

Nowadays, plastics have become a daily necessity in human life. Plastics are produced mainly from fossil sources, contributing greatly to the release of greenhouse gases and the dispersion of microplastics that generate negative impacts on ecosystems. Bioplastics arise as a sustainable response to plastic production worldwide, having as main characteristics the least environmental impact and produced from renewable sources. However, the production of first-generation (1G) bioplastics from sources that are rich in starch, sugars or fatty acids generates strong competition with the food industry, which promotes an increase in the prices of food products. For this reason, second-generation (2G) bioplastics, which are produced from non-edible raw materials, emerge as a proposal to supply the large bioplastics market (Vandenberghe et al. 2021).

The production of 2G bioplastics is mainly from non-edible, bio-based sources such as lignocellulosic biomass and non-edible oils, rich in carbohydrates and fatty acids, respectively. The production of 2G bioplastics can be through chemical processes or biotechnological processes; depending on the type of process applied, it is possible to obtain biodegradable or non-biodegradable 2G bioplastics (Amadu et al. 2021). These bioplastics have fewer negative impacts on the environment due to the reuse of by-products as raw materials (Rahman and Bhoi 2021). For these reason, 2G bioplastics

DOI: 10.1201/9781003344018-3

produced from lignocellulosic biomass are generating great attention in the scientific and commercial spheres, with great prospects for growth in financial flow volumes in the world plastics market.

Lignocellulosic biomass, the main source in the production of 2G biomolecules, is highly distributed worldwide. There are two main industrial producers of lignocellulosic biomass, the agricultural industry and the forestry industry, both with large volume of biomass production. Normally, these two industries produce plant residues (agro-industrial and wood residues), the main sources of lignocellulosic material. These materials represent a great renewable source of carbon due to their compositional richness of polymers such as proteins, polysaccharides and lignin (Raj et al. 2022). However, due to its recalcitrant nature, lignocellulosic biomass needs a pretreatment stage, which increases the digestibility of its structural polymers, achieving a greater release of monomers (fermentable sugars and phenolic monomers; Valladares-Diestra et al. 2021). In this sense, different pretreatments have been developed to increase the yields of obtaining monomers, which are the basis for the synthesis of bioplastics.

Prokaryotic microorganisms are the most used in the 2G bioplastics production, mainly in the production of polyhydroxyalkanoates (PHAs; Lee et al. 2021). In addition, there are yeasts capable of accumulating omega-hydroxy fatty base acids to obtain bioplastics of lipid nature. On the other hand, the production of 2G bioethanol, from hyperproducing yeasts, will allow the chemical synthesis of bioplastics such as bio-polyethylene, bio-polypropylene, bio-polyethylene-terephthalate and bio-polytrimethylene terephthalate (Rahman and Bhoi 2021). Most 2G bioplastics can be used and applied in the same way as traditional plastics and, in some cases, 2G bioplastics have superior characteristics such as mechanical resistance and better barrier properties.

The future of 2G bioplastics production is totally linked to the development of the circular bioeconomy and the implementation of biorefineries capable of producing these polymers at industrial scale. In this way, the use of lignocellulosic biomass is very important, since the development based on high-performance renewable biological raw materials is an important factor for the sustainable growth of the bioplastics industry.

3.2 SECOND-GENERATION BIOPLASTICS

2G bioplastics are a family of polymers that are produced from non-edible plant sources. Within these raw materials, two large groups can be highlighted: non-edible oils and lignocellulosic biomass (agro-industrial by-products and derivatives of forest origin). Second-generation bioplastics reduce environmental impact due to the low emission of greenhouse gases compared to plastics of fossil origin and are even less polluting than 1G bioplastics (Vandenberghe et al. 2021). In addition, the production of 2G bioplastics is considered sustainable because substrates from renewable sources are used in their process. On the other hand, although all 2G bioplastics have a bio-based origin, not all of them are biodegradable. For this reason, these bioplastics are separated into two large groups: biodegradable and non-biodegradable (Rahman and Bhoi 2021).

There are two major processes to produce 2G bioplastics, which depend mainly on the substrate used, such as lignocellulosic biomass and non-combustible oils (in some cases, waste cooking oil can be used). The production pathway from lignocellulosic biomass necessarily goes through a pretreatment stage, where the release of the monomers of the lignocellulosic structural fibers is achieved. Due to its complex composition, lignocellulosic biomass can be derived into different types of monomers after pretreatment, with monosaccharides (cellulose and hemicellulose) and phenolic monomers (lignin) being the main groups of monomers released. These monomers have different chemical compositions, the monosaccharides or fermentable sugars, which can be used in fermentative bioprocesses with the application of microorganisms for the synthesis of bioplastics (PHA, among others) or to produce bioplastics precursor molecules (lactic acid and bioethanol; Raj et al. 2022; Rahman and Bhoi 2021). As seen in Figure 3.1, most 2G bioplastics synthesized from microorganisms are biodegradable, while bioplastics derived from chemical synthesis using 2G bioethanol or phenolic monomers are normally non-biodegradable.

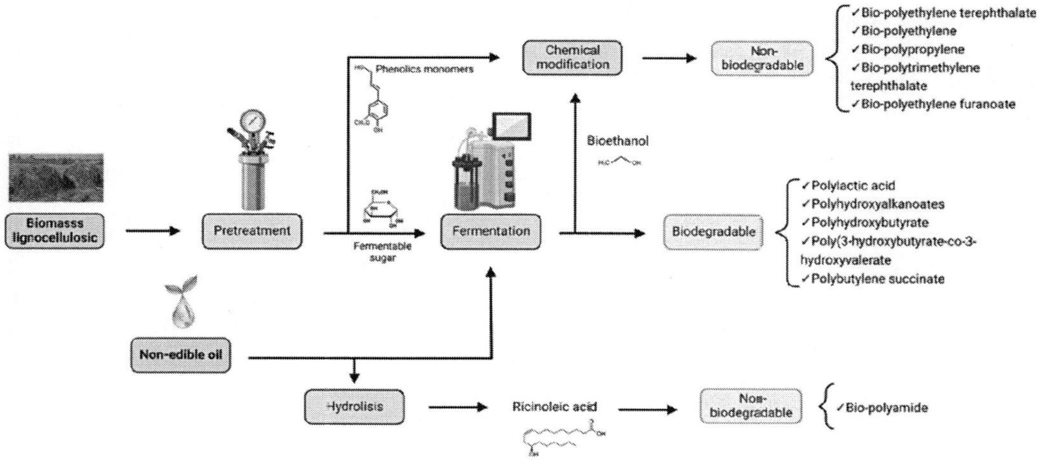

FIGURE 3.1 Second-generation bioplastic production processes.

For their part, non-edible oils and waste cooking oil, composed mainly of triacylglycerols (rich in aliphatic chain fatty acids), are a useful carbon source in the production of bioplastics. The production of 2G bioplastics from these substrates can be carried out through two different pathways. The first is the chemical pathway based on the hydrolysis of the oil. Castor oil is the most used non-edible oil to obtain ricinoleic acids (Rahman and Bhoi 2021). The ricinoleic acids go through various stages of catalysis and hydrolysis to finally generate a plastic polymer of bio-polyamide commonly known as bio-nylon. The second follows the fermentation pathway, using the rich composition of fatty acids in the production of omega-hydroxy fatty acids, monomers with high polymerization efficiency for the formation of bioplastics. The fermentative stage normally uses microorganisms such as yeasts, capable of accumulating large amounts of omega-hydroxy fatty acids. In this way, 2G bioplastics can be obtained from different sources of non-edible oils, as shown in Figure 3.1.

The three main 2G biodegradable bioplastics that are industrially produced are polylactic acid (PLA), polyhydroxyalkanoates (PHA) and polyhydroxybutyrate (PHB). The most important non-biodegradable 2G bioplastics are mainly derived from chemical treatments of 2G bioethanol such as bio-polyethylene, bio-polypropylene, bio-polyethylene-terephthalate and bio-polytrimethylene terephthalate. In addition, bio-polyamide is the main 2G bioplastic derived from non-edible oils (Rahman and Bhoi 2021; Vandenberghe et al. 2021).

Being more abundant and well distributed worldwide, lignocellulosic residues are outlined with a greater projection in the production of 2G bioplastics compared to non-edible oils. But due to the costs of pretreatment and polymerization stages of the monomers by chemical reactions or in fermentative processes, the production of 2G bioplastic is not yet economically viable (Brodin et al. 2017). However, with the new concepts of circular economy and the application of integrated biorefineries, is possible that in the future 2G bioplastic production plants will be implemented, allowing the reuse and reintegration of the previously called "lignocellulosic waste" as high potential raw material in the industry.

3.3 PRETREATMENT OF LIGNOCELLULOSIC BIOMASS FOR BIOPLASTICS PRODUCTION

Lignocellulosic biomass emerges as a feasible solution in the production of bioplastics, mainly 2G bioplastics. However, due to its composition of carbohydrates and lignin polymers, lignocellulosic biomass has a recalcitrant resistance, which makes it difficult to disintegrate into fermentable sugar monomers or simpler chemical compounds. The pretreatment fulfills the main function of increasing the digestibility of the lignocellulosic biomass polymers as cellulose, hemicellulose and/or lignin. Within this stage, the lignocellulosic biomass faces different conditions and chemical catalysts that interact with the destabilized structural polymers, thus increasing the sensitivity and accessibility of these compounds in the digestion process. Thus it is possible to obtain monomeric compounds after pretreatment, such as fermentable sugars (glucose or xylose), depending on the degree of severity of the applied method (Lorenci Woiciechowski et al. 2020).

There are different methods for the pretreatment of lignocellulosic biomass, such as physical or chemical methodologies, with physicochemical methodologies being the most widely applied. In recent years different catalysts with better yields and more environment friendly solvents have been developed. Physicochemical methodologies still present a large sub-production of toxic components for the environment and health. For this reason, biological methods are being studied and developed to avoid the formation of toxic by-products, with low energy consumption and milder operating conditions (Kumar et al. 2020). However, the longer pretreatment time and the low obtained yields make biological pretreatments rarely applied. Due to this, there is still great interest in the development and evaluation of different pretreatment methodologies for their use and application in each type of lignocellulosic biomass to maximize the use of all its chemical components.

In this way, different pretreatments can be applied, as shown in Figure 3.2. Methods of a chemical nature (alkaline, acid and organic solvent reagents) or methods of a physical nature that employ severe pressure and temperature conditions (hydrothermal, steam explosion and microwave pretreatment) have been applied in different lignocellulosic biomass. Each method has advantages and disadvantages related mainly to the specific need of the process and the employed raw material. So some strategies may present high costs depending on the equipment or chemicals used as catalysts.

3.3.1 Chemical Pretreatment

Alkaline pretreatment: Alkaline methods hydrolyze the ester-type bonds that bind the lignin and hemicellulose fraction. The released sugars are mostly pentoses such as xylose. In addition, oligomers such as xylooligosaccharide can be detached from hemicellulose fraction (Valladares-Diestra et al. 2022). The most used chemical reagents in this type of pretreatment are sodium hydroxide, calcium hydroxide, potassium hydroxide and ammonium hydroxide. Low-severity pretreatment is normally used in this methodology, with high catalyst concentrations, low reaction temperatures and low pressure. Alkaline pretreatment does not require specific reactors with corrosion resistance. However, the disadvantages of this method are the longer reaction times and the need for neutralization, causing the formation of salts, which can be impregnated in the biomass. Saratale and Oh (2015) evaluated different alkaline catalysts (NaOH, KOH, NaOCl) in the pretreatment of

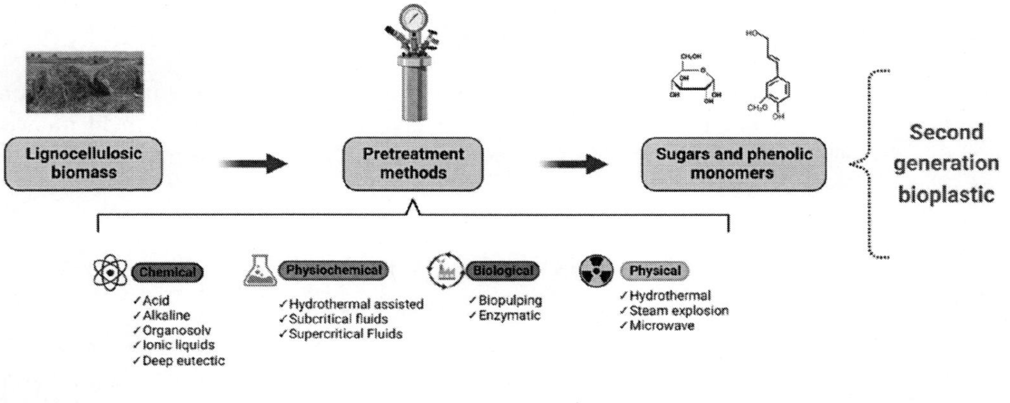

FIGURE 3.2 Different pretreatment used in biomass lignocellulosic for second-generation bioplastic production.

different biomass (paddy straw, wood straw, soybean husk, sunflower husk), obtaining high hydrolysis yield (84%) in the pretreatment of paddy straw with 2% of NaOH, 703 mg of reducing sugars were released per gram of substrate. These results allowed the production of 11.4 g/L of PHB with a total of 75.5% accumulation in the dry biomass at 48 h of fermentation of the hydrolysate with the *Ralstonia eutropha* ATCC 17699 strain.

Acid pretreatment: Acid methods are the most studied in the hydrolysis of lignocellulosic biomass. This pretreatment promotes the hydrolysis of polymers, mainly carbohydrates such as cellulose and hemicellulose. Sulfuric acid, hydrochloric acid and trifluoroacetic acid are among the most used inorganic acids (Gírio et al. 2012). However, the use of inorganic acids may lead to corrosion of the used equipment and has low hydrolysis selectivity, promoting the formation of highly toxic compounds derivates from the pretreatment of lignocellulosic biomass (Kumar et al. 2020; Gírio et al. 2012). On the other hand, organic acids, such as acetic, citric, formic and oxalic acid, are less aggressive, avoiding the generation of toxic by-products. Sharma and Bajaj (2015) used sulfuric acid in the pretreatment of rice straw, applying sequential pretreatment with a first concentration of 1% acid, followed by the use of 5% acid in the second stage. The hydrolysate rich in fermentable sugars allowed the production of 10.6 g/L of PHB with the use of the *Bacillus cereus* PS 10 strain.

New solvents pretreatment: Organosolv pretreatment uses short-chain aliphatic alcohol-type solvents (ethanol, methanol), polyhydric alcohols (ethylene glycol, glycerol, triethylene glycol) and alkylene carbonates. In certain cases, it is possible to use catalysts of an acidic or alkaline nature to improve the performance of the pretreatment. This type of pretreatment promotes the solubilization of bonds that bind lignin and hemicellulose. The main disadvantages in organosolv pretreatment are the volatility of the solvent and its flammability (Kumar et al. 2020). Ionic liquids are solvents that are composed entirely of ions, are non-volatile and readily dissolve in lignocellulosic biomass. Due to their both ionic and organic properties, these solvents can effectively reduce biomass recalcitrance by disrupting the lignin/carbohydrate complex.

Deep eutectic solvents (DES) have great potential as a method to be applied in the extraction of lignocellulosic fractions, since they have ideal physical and chemical properties for the dissolution of polysaccharides, such as cellulose and hemicellulose. DES is considered an alternative to ionic liquids (Vigier et al. 2015). DES generally have one component that acts as a hydrogen bond acceptor and another component that acts as a hydrogen bond donor. In addition, DES are easily synthesized, renewable in nature, biocompatible, biodegradable and cost-effective. Ionic liquids and DES are included within the green solvents.

3.3.2 Physical Pretreatment

Hydrothermal pretreatment: This method uses water as the catalyst for hydrolysis of lignocellulosic biomass. Under conditions of high temperature (150–220°C) and pressure (145–290 psi), which is caused by water vapor, a decrease in pKw is observed, leading to the formation of hydronium ion derivates from the water. These ions catalyze the hydrolysis of acetyl groups that are present in hemicellulose (Valladares-Diestra et al. 2022). Also, the pretreatment increases the acetic acid concentration and acid cleavage of ether bonds, releasing pentose sugars (xylose). This pretreatment, also called autohydrolysis, uses high pressures in closed systems, avoiding the evaporation of water at high temperatures. Yin et al. (2019) pretreated poplar at 200°C for 30 min, obtaining 530 g of fermentable sugars per g of substrate after enzymatic hydrolysis. Through the fermentation using the hydrolysates, it was possible to obtain 637.5 mg/L of poly(3-hydroxybutyrate-co-3-hydroxyvalerate) (PHBV).

3.3.3 Physicochemical Pretreatment

Subcritical or supercritical fluids: The pretreatment of lignocellulosic biomass by means of subcritical or supercritical fluids can be carried out with or without the use of catalysts. In general, this method destabilizes the hemicellulose and lignin fractions, noting their solubility in liquid media, thus most of the products obtained are derived from these polymers of lignocellulosic biomass (Liu et al. 2019).

Microwave-assisted extraction: This is an alternative method for the extraction and separation of lignocellulosic fractions, being able to generate a selective isolation of lignocellulosic fractions, presenting a high extraction yield and low cost without significant changes in molecular structures (Fu et al. 2019). Due to its selective heating capacity, this pretreatment leads to a high solubility of polymers, creating hot spots, which in alkaline solutions encourage the interaction of alkali and fibers. This generates an explosive effect that separates the fibers from the lignocellulosic biomass, increasing their sensitivity to enzymatic digestion.

3.3.4 Biological Pretreatment

3.3.4.1 Biopulping Pretreatment

Biopulping methods are generally applied to lignocellulosic biomass to release cellulose fractions and, in some cases, hemicellulose. The process consists mainly of the degradation of lignin and hemicellulose fractions by the action of wood-degrading fungi and in some cases bacteria or filamentous fungi. These organisms can selectively degrade the cell wall, promoting the production of pulp rich in cellulose. Once the lignocellulosic substrate is colonized, the production and release of different enzymes for the hydrolysis of the substrate begins (Singh 2018).

3.3.4.2 Enzymatic Pretreatment

The direct enzyme treatment of lignocellulosic materials can be more efficient and take less time than biopulping. Enzymatic treatments reduce the use of aggressive chemicals and exhibit high extraction efficiencies under optimal conditions. The high specificity of enzymes allows the degradation of specific lignocellulosic fiber.

Pretreatment is one of the most important stages for the efficient conversion of lignocellulosic biomass into bioplastics. The choice of the pretreatment method within a bioprocess can greatly influence the final cost of the 2G bioplastic (Prasad et al. 2016). For this reason, the combination of physicochemical and enzymatic treatments is the most applied, since they show great potential in the recovery process of lignocellulosic fractions, achieving a progressive reduction in the use of chemical products, reaching optimal performance, productivity and lower environmental impact.

3.4 BIOTECHNOLOGY PROCESS FOR SECOND-GENERATION BIOPLASTICS PRODUCTION

Bioplastics, a broad, unregulated term covering materials with diverse properties and applications, can be classified according to their nature as (1) bio-based polymers, synthetized macromolecules from hydrolyzed biomasses or constitutive monomers produced via microbial fermentation; or (2) biopolymers, macromolecules (e.g., starch, cellulose, proteins) produced by living organisms and then further purified (García-Depraect et al. 2021).

The PHA family is a group of biomolecules of economic relevance to the plastic industry. PHA is a heterologous group of polyesters accumulated in cytoplasmatic inclusions as carbon reserve during unbalanced growth conditions in over 90 genera of bacteria, archaea and lower eukaryotes presenting diverse modes of nutrition. Nitrogen, phosphorus, magnesium and sulfur limiting conditions, in the presence of excess carbon source, result in a superior acetyl-CoA flux due to reduced amino acid synthesis. Accumulated acetyl-CoA molecules are condensed into acetoacetyl-CoA, which is then submitted to a stereospecific reduction to (R)-3-hydroxybutyryl-CoA, finally allowing the incorporation of the enantiomers to the elongating polymeric chain through the action of PHB polymerases. Although the short-, medium-, and long-chain-length PHA biosynthesis from carbohydrates is the pathway commonly used in industry, these molecules can also be generated from alkane oxidation and fatty-acid β-oxidation or de novo synthesis (Luengo et al. 2003).

PHA-producing prokaryotes are vast and include both gram-positive and gram-negative species, such as *Bacillus megaterium, Ralstonia eutropha, Pseudomonas oleovorans, Pseudomonas putida* and *Lysinibacillus* sp. Alongside the PHA production accumulation feature, the capability to achieve high cell density growth within a short period of time; grow under microaerophilic conditions; produce under open, non-sterile conditions; and tolerate inhibitory compounds (e.g., acetic acid, furfural, and 5-hydroxymethylfurfural) derived from acidic and alkaline pre-treatments of lignocellulosic biomasses are also prospected (Lee et al. 2021). Although effective, the screening and isolation of PHA-producing strains are time-consuming and limited to culturable microorganisms. In this sense, culture-independent (e.g., metagenomic, metabolomic, and transcriptomic) strategies are also being applied. Vuong et al. (2021) carried out a large-scale mining of over 17,000 genome sequences of bacterial and archaea in the IMG/M database for PHA synthetase (PhaC) gene. The study revealed promising PHA-producing candidates, several of them taxa not yet described as intracellular biopolymer production, establishing a targeted strategy for starter-culture screening.

Industrial biotransformation of PHA is performed in bioreactors, where abiotic factors are controlled and aseptic conditions are ensured to reduce contamination and achieve optimal production conditions. PHA production using pure biocatalyst cultures are characterized by two distinct phases: growth associated and non-growth associated. During the first phase, an unrestricted growth is guaranteed without nutritional restraints in order to obtain biomass. Although the presence of all nutrients directs metabolism toward cell growth, a minor accumulation of PHA is achieved due to deviation of tricarboxylic acid intermediates in some bacterial and archaeal species (Mohapatra et al. 2017). This PHA accumulation provides an energy surplus that can be redirected once more to the generation of active biomass or act as an electron acceptor under anaerobic conditions (Koller and Muhr 2014). The non-growth associated phase is characterized by a pronounced PHA accumulation due to starvation induction due to restriction on growth-essential nutrients and high availability of carbon, as mentioned previously.

PHA production can be conducted in discontinuous (batch or fed-batch) or single-, two- or multistage continuous processes. During discontinuous processes, the fermentation takes place in a single bioreactor, where the starter culture is added to a substrate and, after total consumption, the PHA-accumulated cells are recovered. Although it is an easy-to-operate method, batch fermentations are subject to inhibitory effects from precursors, even in low concentrations (<1 g/L), and low conversion yield (Koller 2018), which makes it economically unfeasible.

On the other hand, continuous systems are characterized by a chemostat condition, where chemical environment conditions (e.g., pH, redox, volume, C/N ratio) remain at a "steady status" due to constant addition of fresh medium and harvest of PHA-accumulated cells. Although this single-stage, continuous fermentation presents advantages as to the volumetric productivity of PHA in relation to the batch process, microorganisms bearing a non–growth-associated metabolism show limited application because the conservation of substrate condition may compromise either cellular growth or PHA accumulation, once it occurs in two distinct phases (Koller and Muhr 2014). In order to solve this issue, two- and multistage fermentations were proposed where the biomass can be continuously generated under optimal conditions in one fermentation vessel, while the non–growth-associated PHA accumulation can take place in other fermentation vessel(s) with nutrient-limiting conditions and below the maximum growth rate.

In general lines, large-scale PHA production is achieved through microbial fermentation of pure carbon sources or precursors (e.g., lauric acid, 1,4-butanediol). The biomass inclusions are then separated from the fermenting broth through press filtration and centrifugation and submitted to cell lysis to expose the intracellular PHA inclusions. Further steps of ultrafiltration, precipitation and centrifugation are then necessary to recover a high-purity polymer. Amongst the commercially available it is possible to highlight PHB, PHBV, and poly[(R)-3-hydroxybutyrate-co-4-hydroxybutyrate] (P3HB4HB). Although this process is well established in the industry, recent studies and technologies have been conducted using biorefinery approaches in order to reduce operational costs associated to pure substrates, which can represent up to 45% of the total (Kourmentza et al. 2015).

Saratale et al. (2021) demonstrated the effectiveness of the PHB conversion from sugarcane bagasse hydrolysates without detoxification by *Lysinibacillus* sp. The optimized conditions showed a PHB accumulation of 61.5% (weight/weight of dry biomass), showing a great potential of industrial-scale manufacturing due to the great availability of sugarcane bagasse and reduction of costs given the lack of additional detoxification steps in pre-treatment processes.

Despite the numerous and expensive downstream processes allied to the low-economic viability in comparison to petroleum-based plastics, several industries continue to exploit the microbial-derived PHA due to sustainable appeal and advances provided by applied research and patents developed in the past decade.

3.5 APPLICATION OF SECOND-GENERATION BIOPLASTICS

The use of 2G bioplastics offers many benefits over petroleum originated plastics and over the 1G bioplastics. Lignocellulosic biomass can be used to obtain biopolymers, such as PLA, polyvinyl alcohol (PVA), PHA and PHB. These polymers can replace petroleum polymers in various packaging applications because of their similar characteristics (Bhatia et al. 2018; Kumar and Kim 2019). The application of bioplastics varies according to their production and composition, as shown in Table 3.1.

A high-performance bioplastic was synthesized by in situ lignin strategy from wood powder. This bioplastic presents a high mechanical strength, ultraviolet-light resistance, and high water and thermal stability, similar characteristics to petroleum plastic. Moreover, this bioplastic can be easily biodegraded or recycled, reducing the environmental waste footprint. Consequently, it is a strong candidate for replacing widely used petrochemical plastics such as acrylonitrile butadiene styrene (ABS), with application in agricultural mulching films, construction and automotive parts (Xia

TABLE 3.1
Second-Generation Bioplastics Production and Applications

Substrate	Treatment	Bioplastic	Application	Reference
Wood powder	In situ lignin regeneration strategy	Lignocellulosic bioplastic	Replacement for ABS in agricultural mulching films, construction and automotive parts	(Xia et al. 2021)
Lignocellulosic biomass	Integrated process: coproduction of FDCA and 1,5-pentanediol	FDCA as an eco-friendly, bio-based plastic monomer	Water bottle production and food packing	(Kim et al. 2020)
Cellulose nano whiskers from ramie fibers and lactic acid	Interfacial ring-opening polymerization	PLA	Food and biomedical packing, bottle labels and drug delivery	(Goffin et al. 2011)
Waste cooking oil	*Paracoccus sp.* strain LL1 as a single-cell factory	PHA and carotenoids		(Kumar and Kim 2019)
Coffee waste oil	Fermentation by engineered *Ralstonia eutropha*	PHA	Biomedical field	(Bhatia et al. 2018)

et al. 2021). Another eco-friendly bioplastic was produced from lignocellulosic biomass, designated as 2,5-furandicarboxylic acid (FDCA), which presents superior performance when compared with conventional plastics, as much in mechanical strength as in barrier properties.

Another strategy for bioplastics is the production of biofilms. Some modified bioplastic films with different applications were produced from agricultural residues. A xylan composite film was fabricated using a solution casting method and selectively modified for xylanase treatment. The modified xylan composite film was more abundant in available surface area and reactive hydroxyl groups, which showed high potential for the subsequent chemical grafting/enzyme immobilization to extend their applications such as heavy metal detection and biological treatment of wastewater. Several biopolymers should be mentioned when it comes to bioplastics production and application, especially due to their potential.

The features of PLA such as biodegradability, biocompatibility, crystallinity and thermal and mechanical properties make them ideal materials for packaging in the food and biomedical industries (Jabeen et al. 2015). It is also highlighted that PLA can be used for different applications such as tissue engineering, nanocomposites and drug delivery. The most recent studies on PLA production aim to improve its characteristics through reinforcement in order to increase its applicability. The PLA film had its microstructure improved by the addition of cellulose nanocrystals (CNC) and surfactant-modified cellulose nanocrystals (s-CNC). These CNCs are prepared by the solvent casting method and dispersed in the PLA matrix. The resulting biofilm had lower water permeability and better barrier properties against oxygen, which make the biofilms more resistant and better to be applied in food packaging. PLA production was associated with solid biofuel production generating a more efficient and sustainable process which respects the circular economy.

2G bioplastics have great potential to be applied to replace common plastics, especially in packing and food materials. According to their production and origin, bioplastics have different characteristics, which allow for different applications and adaptations. However, studies are necessary to make the production technique economically viable on a large scale.

3.6 MARKET AND COMMERCIAL PRODUCTS

Plastics based on and produced from biomass residues are gaining attention due to their sustainable matter and eco-friendliness. Thus, nowadays, a variety of bioplastic materials are produced for different applications and purposes (Figure 3.3).

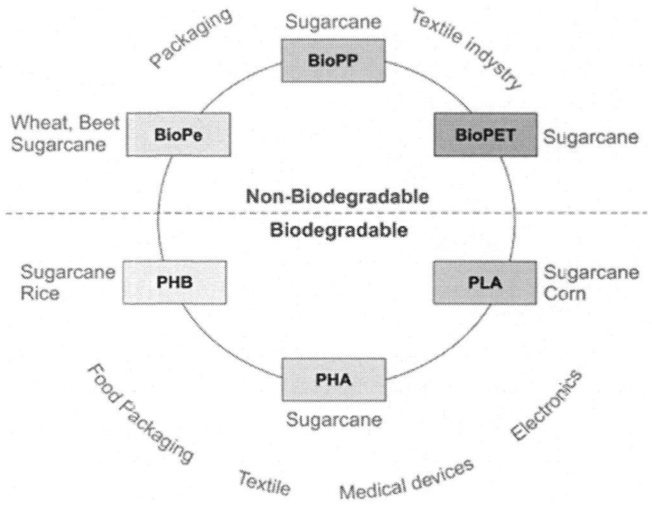

FIGURE 3.3 Most-used bioplastics from lignocellulosic sources.

3.6.1 Food Packaging

The constant need for more low ecological impact alternatives and easy customization can be obtained from degradable bioplastics guaranteeing the quality of the product (Jabeen et al. 2015). The predicted annual growth in this area is about 18%.

Examples of these products are currently used by competitive international brands, such as KLM (cups with PLA coating; Jager 2010), McDonald's (bowls made with PLA), PepsiCo's Frito-Lay (bags made with PLA; Weston 2012), Walmart (trays coated with cellulose) and Boulder Canyon (metalized cellulose film; European Bioplastics 2021).

Due to the demand, companies that produced bioplastic-based packing announced scaling up production of 60–75 kt in the near future. Hence, ambitious projects such as the Alliance to End Plastic Waste are currently spending USD 1.5 billion on sustainable bio-based plastics (Rosenboom et al. 2022); Nestle is also investing USD 2 billion in research and development of food-grade and sustainable plastics, and Toyota has a compromise to buy 25% of the bio-PE from a Brazilian plant. Principal players in the production market are Danimer Scientific (US) and RWDC Industries (Singapore).

Still, different types of food require specific features to ensure protection, such as oxide/oxygen ratio, light protection, odor barriers and permeability.

3.6.2 Agricultural Purposes

The replacement of seedling bags made of polyethene for PHA's options is increasing because of the non-toxic effect, biodegradation, moisture retention and root friendliness.

The research on PHA has increased notoriously in the past two decades. The production in 2014 was estimated to be 54 kt, and a CAGR of 6.3% is expected to reach approximately USD 119.15 million by 2025 (Research and Markets 2017), although its production is still considered more expensive: USD 4.3–5.26/kg compared to USD 1.6–1.9/kg of conventional polyethene terephthalate and polystyrene (Neelamegam et al. 2018).

3.6.3 Medical Applications

This is a promising field with several applications for bioplastics. The use of biodegradable plastics leads to the development of drug delivery and tissue engineering.

Bacterial cellulose is the main example for medical purposes, being used as a composition for the fabrications of orthopedic/dental implants (Nemati and Gholami 2021). Patented products available in the market using cellulose and nanocellulose from bacterial sources are Bioprocess, XCell, Biofill, TianAn Biological Materials (China), PHBOTTLE (Spain), Metabolix, Inc. (US), BioMatera (Canada), Biomer (Germany) and Kaneka (Japan). Tepha (US) is a producer company of PHA for medical applications (Czaja et al. 2006).

The bioplastic market's annual growth rates are estimated to hover around 30% until 2025 (Coppola et al. 2021). The trend of production in 2018 was 7.5 million tonnes, and by 2023 a production of 9.1 million tonnes is expected, with a market capitalization of USD 1.7 billion (Chinthapalli et al. 2019). Fortunately, the European growth rate is about 10% since the market regulations demand eco-friendly and sustainable products. However, the panorama could change positively to a global growth of 10%–20% if a political intervention occurs.

Nevertheless, these materials are not as popular as petroleum-based plastics due to a price difference of USD 0.6–0.9/pound compared to USD 2.25–2.75/pound (PHA-based bioplastics). Completely bio-based plastics are produced at a scale of 2 million tonnes/year, considered part of the future circular economy in order to avoid fossil resources (Yarnold et al. 2019).

In the United States, the generation of approximately 1 billion dry tons of lignocellulosic biomass was registered until 2014, and an increment to 1.6 billion tons by 2030 was predicted, quantities of biomass that can be used for the generation of bioplastics.

TABLE 3.2
Pathways Targeted and Involved for the Bioplastic Production Enhancement

Pathway	Characteristics	Reference
Threonine bypass by flux balance analysis	Engineering a strain TB17 for maximum PHB production (35.92 g L^{-1})	(Lin et al. 2015)
Aldehyde dehydrogenase	Promotes the conversion of aldehyde to carboxylic acid converting 3 HPA to 3 HP	(Li et al. 2016)
NAD^+/NADH	Necessary for 3 HP production, NAD^+ is dependent on aldehyde dehydrogenases	(Li et al. 2013)
Fumarate reductase (frdABCD), alcohol/aldehyde dehydrogenase (adhE), and pyruvate formate lyase (pflB)	Inactivation of these fumarate reductase encoding genes to lactic acid selective production with >99% purity	(Zhou et al. 2003)

3.7 INNOVATION AND RECENT ADVANCEMENTS

Due to its massive production and costs, substituting and total replacing plastics with lignocellulosic-based plastics will be challenging. Additionally, lignocellulosic hydrolysates can contain residues and contaminants that could inhibit cellular growth during fermentation, compromising yields. Thus constant screening and biomolecular research of microorganisms are being performed in order to achieve optimal conversion yields in fermentations (Raj et al. 2022). For example, during the downstream purification process, some operations units or their enhancement can be applied, such as adding adsorption and electrodialysis, to avoid using expensive and hazardous organic acids.

The configuration improvement of bioplastics has headed to the use and application of 3D printers to optimize materials based on bio-PE, PLA and PHA fibers. The evaluation of mechanical properties and water absorption are the main goals of this printing technology. The use of nanocellulose as an addition to bioplastics has demonstrated the improvement of mechanical strength as well as thermal and barrier properties. Also, some other strategies are being studied and developed concerning improving the mechanical properties, such as adding fillers in the bioplastic matrix. The use of graphene oxide has shown optimal results in suitable solubilization in bioplastic matrix, biocompatibility and surface area. Among the enhanced properties of bioplastic using graphene are mechanical and thermal stability, barrier properties (Bher et al. 2019), surface hydrophobicity, biodegradability, antioxidant and antimicrobial (Carvalho and Conte Junior 2020).

Currently, metabolic engineering and genome-scale metabolic studies are being carried out to improve the efficiency of microbial cell factories, aiming to enhance the production of monomeric units for bioplastic production (Table 3.2). These studies targeted genes (i.e., transformations of acetoacetyl-CoA reductase and PHB synthase enzymes), increasing the production of lactate, succinate and PHB bioplastic monomeric units.

3.8 CONCLUSIONS AND PERSPECTIVES

The production of 2G bioplastics is a sustainable and less polluting alternative compared to traditional plastics and 1G bioplastics. Lignocellulosic biomass is the most convenient substrate to produce 2G bioplastics and for their scaling up to an industrial scale. However, the development of new and better technologies in the pretreatment of lignocellulosic biomass will guarantee better performance and economic competitiveness of 2G bioplastics in the plastics market. Second-generation bioplastics have characteristics similar to traditional plastics, and in some cases they have much better compatibility and physicochemical characteristics. This makes 2G bioplastic even more interesting and potentially in demand in the pharmaceutical industry. The development of the circular bioeconomy could generate a boost in the production of bioplastics, with the implementation of lignocellulosic biorefineries allowing industrial-scale production in the coming years.

REFERENCES

Amadu, Ayesha Algade, Shuang Qiu, Shijian Ge, Gloria Naa Dzama Addico, Gabriel Komla Ameka, Ziwei Yu, Wenhao Xia, et al. 2021. "A Review of Biopolymer (Poly-β-Hydroxybutyrate) Synthesis in Microbes Cultivated on Wastewater." *Science of the Total Environment* 756: 143729. doi:10.1016/j.scitotenv.2020.143729.

Bhatia, Shashi Kant, Jung-Ho Kim, Min-Sun Kim, Junyoung Kim, Ju Won Hong, Yoon Gi Hong, Hyun-Joong Kim, et al. 2018. "Production of (3-Hydroxybutyrate-Co-3-Hydroxyhexanoate) Copolymer from Coffee Waste Oil Using Engineered Ralstonia Eutropha." *Bioprocess and Biosystems Engineering* 41 (2): 229–235. doi:10.1007/s00449-017-1861-4.

Bher, Anibal, Ilke Uysal Unalan, Rafael Auras, Maria Rubino, and Carlos E Schvezov. 2019. "Graphene Modifies the Biodegradation of Poly(Lactic Acid)-Thermoplastic Cassava Starch Reactive Blend Films." *Polymer Degradation and Stability* 164: 187–197. doi:10.1016/j.polymdegradstab.2019.04.014.

Brodin, Malin, María Vallejos, Mihaela Tanase Opedal, María Cristina Area, and Gary Chinga-Carrasco. 2017. "Lignocellulosics as Sustainable Resources for Production of Bioplastics—A Review." *Journal of Cleaner Production* 162: 646–664. doi:10.1016/j.jclepro.2017.05.209.

Carvalho, Anna Paula Azevedo de, and Carlos Adam Conte Junior. 2020. "Green Strategies for Active Food Packagings: A Systematic Review on Active Properties of Graphene-Based Nanomaterials and Biodegradable Polymers." *Trends in Food Science & Technology* 103: 130–143. doi:10.1016/j.tifs.2020.07.012.

Chinthapalli, Raj, Pia Skoczinski, Michael Carus, Wolfgang Baltus, Doris de Guzman, Harald Käb, Achim Raschka, and Jan Ravenstijn. 2019. "Biobased Building Blocks and Polymers—Global Capacities, Production and Trends, 2018–2023." *Industrial Biotechnology* 15 (4): 237–241. doi:10.1089/ind.2019.29179.rch.

Coppola, Gerardo, Maria Teresa Gaudio, Catia Giovanna Lopresto, Vincenza Calabro, Stefano Curcio, and Sudip Chakraborty. 2021. "Correction to: Bioplastic from Renewable Biomass: A Facile Solution for a Greener Environment." *Earth Systems and Environment* 5 (2): 231–251. doi:10.1007/s41748-021-00280-z.

Czaja, W, A Krystynowicz, S Bielecki, and R Brownjr. 2006. "Microbial Cellulose—the Natural Power to Heal Wounds." *Biomaterials* 27 (2): 145–151. doi:10.1016/j.biomaterials.2005.07.035.

European Bioplastics. 2021. "Bioplastics Market Data 2021." https://www.European-Bioplastics.Org/Market/#.

Fu, Gen-Que, Ya-Jie Hu, Jing Bian, Ming-Fei Li, Feng Peng, and Run-Cang Sun. 2019. "Isolation, Purification, and Potential Applications of Xylan." In *Production of Materials from Sustainable Biomass Resources*, edited by Zhen Fang, Richard L. Smith Jr, and Xiao-Fei Tian, 3–35. Springer Nature. doi:10.1007/978-981-13-3768-0_1.

García-Depraect, Octavio, Sergio Bordel, Raquel Lebrero, Fernando Santos-Beneit, Rosa Aragão Börner, Tim Börner, and Raúl Muñoz. 2021. "Inspired by Nature: Microbial Production, Degradation and Valorization of Biodegradable Bioplastics for Life-Cycle-Engineered Products." *Biotechnology Advances* 53: 107772. doi:10.1016/j.biotechadv.2021.107772.

Gírio, Francisco M, Florbela Carvalheiro, Luís C Duarte, and Rafał Bogel-Łukasik. 2012. "Chapter 1 Deconstruction of the Hemicellulose Fraction from Lignocellulosic Materials into Simple Sugars." *D-Xylitol Fermentative Production, Application and Commercialization*, 3–37. doi:10.1007/978-3-642-31887-0.

Goffin, Anne-Lise, Jean-Marie Raquez, Emmanuel Duquesne, Gilberto Siqueira, Youssef Habibi, Alain Dufresne, and Philippe Dubois. 2011. "From Interfacial Ring-Opening Polymerization to Melt Processing of Cellulose Nanowhisker-Filled Polylactide-Based Nanocomposites." *Biomacromolecules* 12 (7): 2456–2465. doi:10.1021/bm200581h.

Jabeen, Nafisa, Ishrat Majid, and Gulzar Ahmad Nayik. 2015. "Bioplastics and Food Packaging: A Review." *Cogent Food & Agriculture* 1 (1): 1117749. doi:10.1080/23311932.2015.1117749.

Jager, A. 2010. "IngeoTM Polylactide Een Natuurlijke Keus, Presentation given at VMT Bioplastics Market by Type, Application, and Geography – Forecast and Analysis 2023–2027." https://www.technavio.com/report/bioplastics-market-industry-analysis (accessed February 23, 2023).

Kim, Hyunwoo, Shinje Lee, Yuchan Ahn, Jinwon Lee, and Wangyun Won. 2020. "Sustainable Production of Bioplastics from Lignocellulosic Biomass: Technoeconomic Analysis and Life-Cycle Assessment." *ACS Sustainable Chemistry & Engineering* 8 (33): 12419–12429. doi:10.1021/acssuschemeng.0c02872.

Koller, Martin. 2018. "A Review on Established and Emerging Fermentation Schemes for Microbial Production of Polyhydroxyalkanoate (PHA) Biopolyesters." *Fermentation* 4 (2): 30. doi:10.3390/fermentation4020030.

Koller, Martin, and A. Muhr. 2014. "Continuous Production Mode as a Viable Process-Engineering Tool for Efficient Poly(Hydroxyalkanoate) (PHA) Bio-Production." *Chemical and Biochemical Engineering Quarterly* 28 (1): 65–77.

Kourmentza, C, I Ntaikou, G Lyberatos, and M Kornaros. 2015. "Polyhydroxyalkanoates from Pseudomonas Sp. Using Synthetic and Olive Mill Wastewater under Limiting Conditions." *International Journal of Biological Macromolecules* 74: 202–210. doi:10.1016/j.ijbiomac.2014.12.032.

Kumar, Bikash, Nisha Bhardwaj, Komal Agrawal, Venkatesh Chaturvedi, and Pradeep Verma. 2020. "Current Perspective on Pretreatment Technologies Using Lignocellulosic Biomass: An Emerging Biorefinery Concept." *Fuel Processing Technology* 199.

Kumar, Prasun, and Beom Soo Kim. 2019. "*Paracoccus* Sp. Strain LL1 as a Single Cell Factory for the Conversion of Waste Cooking Oil to Polyhydroxyalkanoates and Carotenoids." *Applied Food Biotechnology* 6 (1): 53–60. doi:10.22037/afb.v6i1.21628.

Lee, Hye Soo, Hong-Ju Lee, Sang Hyun Kim, Jang Yeon Cho, Min Ju Suh, Sion Ham, Shashi Kant Bhatia, et al. 2021. "Novel Phasins from the Arctic *Pseudomonas* Sp. B14–6 Enhance the Production of Polyhydroxybutyrate and Increase Inhibitor Tolerance." *International Journal of Biological Macromolecules* 190: 722–729. doi:10.1016/j.ijbiomac.2021.08.236.

Li, Ying, Mingyue Su, Xizhen Ge, and Pingfang Tian. 2013. "Enhanced Aldehyde Dehydrogenase Activity by Regenerating NAD+ in *Klebsiella pneumoniae* and Implications for the Glycerol Dissimilation Pathways." *Biotechnology Letters* 35 (10): 1609–1615. doi:10.1007/s10529-013-1243-1.

Li, Ying, Xi Wang, Xizhen Ge, and Pingfang Tian. 2016. "High Production of 3-Hydroxypropionic Acid in *Klebsiella pneumoniae* by Systematic Optimization of Glycerol Metabolism." *Scientific Reports* 6 (1): 26932. doi:10.1038/srep26932.

Lin, Zhenquan, Yan Zhang, Qianqian Yuan, Qiaojie Liu, Yifan Li, Zhiwen Wang, Hongwu Ma, Tao Chen, and Xueming Zhao. 2015. "Metabolic Engineering of *Escherichia coli* for Poly(3-Hydroxybutyrate) Production via Threonine Bypass." *Microbial Cell Factories* 14 (1): 185. doi:10.1186/s12934-015-0369-3.

Liu, Xiao, Weiqi Wei, and Shubin Wu. 2019. "Subcritical CO2-Assisted Autohydrolysis for the Co-Production of Oligosaccharides and Fermentable Sugar from Corn Straw." *Cellulose* 26 (13–14): 7889–7903. doi:10.1007/s10570-019-02626-3.

Lorenci Woiciechowski, Adenise, Carlos José Dalmas Neto, Luciana Porto de Souza Vandenberghe, Dão Pedro de Carvalho Neto, Alessandra Cristine Novak Sydney, Luiz Alberto Junior Letti, Susan Grace Karp, Luis Alberto Zevallos Torres, and Carlos Ricardo Soccol. 2020. "Lignocellulosic Biomass: Acid and Alkaline Pretreatments and Their Effects on Biomass Recalcitrance—Conventional Processing and Recent Advances." *Bioresource Technology* 304: 122848. doi:10.1016/j.biortech.2020.122848.

Luengo, José M., Belén García, Angel Sandoval, Germán Naharro, and Elías R. Olivera. 2003. "Bioplastics from Microorganisms." *Current Opinion in Microbiology* 6 (3): 251–260. doi:10.1016/S1369-5274(03)00040-7.

Mohapatra, Swati, Sudipta Maity, Hirak Ranjan Dash, Surajit Das, Swati Pattnaik, Chandi Charan Rath, and Deviprasad Samantaray. 2017. "Bacillus and Biopolymer: Prospects and Challenges." *Biochemistry and Biophysics Reports* 12: 206–213. doi:10.1016/j.bbrep.2017.10.001.

Neelamegam, Annamalai, Huda Al-Battashi, Saif Al-Bahry, and Sivakumar Nallusamy. 2018. "Biorefinery Production of Poly-3-Hydroxybutyrate Using Waste Office Paper Hydrolysate as Feedstock for Microbial Fermentation." *Journal of Biotechnology* 265: 25–30. doi:10.1016/j.jbiotec.2017.11.002.

Nemati, Elham, and Ahmad Gholami. 2021. "Nano Bacterial Cellulose for Biomedical Applications: A Mini Review Focus on Tissue Engineering." *Journal of Advances in Applied NanoBio-Technologies* 2 (4): 93–101. https://dormaj.org/index.php/AANBT/article/view/454.

Prasad, Aashish, Maria Sotenko, Thomas Blenkinsopp, and Stuart R. Coles. 2016. "Life Cycle Assessment of Lignocellulosic Biomass Pretreatment Methods in Biofuel Production." *The International Journal of Life Cycle Assessment* 21 (1): 44–50. doi:10.1007/s11367-015-0985-5.

Rahman, Md Hafizur, and Prakashbhai R Bhoi. 2021. "An Overview of Non-Biodegradable Bioplastics." *Journal of Cleaner Production* 294: 126218. doi:10.1016/j.jclepro.2021.126218.

Raj, Tirath, K Chandrasekhar, A Naresh Kumar, and Sang Hyoun Kim. 2022. "Lignocellulosic Biomass as Renewable Feedstock for Biodegradable and Recyclable Plastics Production: A Sustainable Approach." *Renewable and Sustainable Energy Reviews* 158: 112130. doi:10.1016/j.rser.2022.112130.

Research and Markets. 2017. "Global Polyhydroxyalkanoate (PHA) Market Analysis & Trends—Industry Forecast to 2025." https://www.researchandmarkets.com/reports/5241294/global-polyhydroxyalkanoate-pha-market-by-type#src-pos-1.

Rosenboom, Jan-Georg, Robert Langer, and Giovanni Traverso. 2022. "Bioplastics for a Circular Economy." *Nature Reviews Materials* 7 (2): 117–137. doi:10.1038/s41578-021-00407-8.

Saratale, Ganesh, and Min-Kyu Oh. 2015. "Characterization of Poly-3-Hydroxybutyrate (PHB) Produced from *Ralstonia eutropha* Using an Alkali-Pretreated Biomass Feedstock." *International Journal of Biological Macromolecules* 80: 627–635. doi:10.1016/j.ijbiomac.2015.07.034.

Saratale, Rijuta Ganesh, Si Kyung Cho, Ganesh Dattatraya Saratale, Gajanan S. Ghodake, Ram Naresh Bharagava, Dong Su Kim, Supriya Nair, and Han Seung Shin. 2021. "Efficient Bioconversion of Sugarcane Bagasse into Polyhydroxybutyrate (PHB) by *Lysinibacillus* Sp. and Its Characterization." *Bioresource Technology* 324: 124673. doi:10.1016/j.biortech.2021.124673.

Sharma, Priyanka, and Bijender Kumar Bajaj. 2015. "Production of Poly-β-Hydroxybutyrate by *Bacillus cereus* PS 10 Using Biphasic-Acid-Pretreated Rice Straw." *International Journal of Biological Macromolecules* 79: 704–710. doi:10.1016/j.ijbiomac.2015.05.049.

Singh, Shalini. 2018. "White-Rot Fungal Xylanases for Applications in Pulp and Paper Industry." In *Fungal Biorefineries*, edited by S Kumar, P Dheeran, M Taherzadeh, and S Khanal, 47–63. Springer. doi:10.1007/978-3-319-90379-8_3.

Valladares-Diestra, Kim Kley, Luciana Porto de Souza Vandenberghe, Luis Alberto Zevallos Torres, Arion Zandoná Filho, Adenise Lorenci Woiciechowski, and Carlos Ricardo Soccol. 2022. "Citric Acid Assisted Hydrothermal Pretreatment for the Extraction of Pectin and Xylooligosaccharides Production from Cocoa Pod Husks." *Bioresource Technology* 343: 126074. doi:10.1016/j.biortech.2021.126074.

Valladares-Diestra, Kim Kley, Luciana Porto de Souza Vandenberghe, Luis Alberto Zevallos Torres, Verônica Sayuri Nishida, Arion Zandoná Filho, Adenise Lorenci Woiciechowski, and Carlos Ricardo Soccol. 2021. "Imidazole Green Solvent Pre-Treatment as a Strategy for Second-Generation Bioethanol Production from Sugarcane Bagasse." *Chemical Engineering Journal* 420: 127708. doi:10.1016/j.cej.2020.127708.

Vandenberghe, Luciana Porto de Souza, Priscilla Zwiercheczewski de Oliveira, Gustavo Amaro Bittencourt, Ariane Fátima Murawski de Mello, Zulma Sarmiento Vásquez, Susan Grace Karp, and Carlos Ricardo Soccol. 2021. "The 2G and 3G Bioplastics: An Overview." *Biotechnology Research and Innovation* 5 (1): e2021004. doi:10.4322/biori.202104.

Vigier, Karine De Oliveira, Gregory Chatel, and François Jérôme. 2015. "Contribution of Deep Eutectic Solvents for Biomass Processing: Opportunities, Challenges, and Limitations." *ChemCatChem* 7 (8): 1250–1260. doi:10.1002/cctc.201500134.

Vuong, Paton, Daniel J Lim, Daniel V Murphy, Michael J Wise, Andrew S Whiteley, and Parwinder Kaur. 2021. "Developing Bioprospecting Strategies for Bioplastics Through the Large-Scale Mining of Microbial Genomes." *Frontiers in Microbiology* 12: 697309. doi:10.3389/fmicb.2021.697309.

Xia, Qinqin, Chaoji Chen, Yonggang Yao, Jianguo Li, Shuaiming He, Yubing Zhou, Teng Li, Xuejun Pan, Yuan Yao, and Liangbing Hu. 2021. "A Strong, Biodegradable and Recyclable Lignocellulosic Bioplastic." *Nature Sustainability* 4 (7): 627–635. doi:10.1038/s41893-021-00702-w.

Yarnold, Jennifer, Hakan Karan, Melanie Oey, and Ben Hankamer. 2019. "Microalgal Aquafeeds As Part of a Circular Bioeconomy." *Trends in Plant Science* 24 (10): 959–970. doi:10.1016/j.tplants.2019.06.005.

Yin, Fen, Dongna Li, Xiaojun Ma, and Chong Zhang. 2019. "Pretreatment of Lignocellulosic Feedstock to Produce Fermentable Sugars for Poly(3-Hydroxybutyrate-Co-3-Hydroxyvalerate) Production Using Activated Sludge." *Bioresource Technology* 290: 121773. doi:10.1016/j.biortech.2019.121773.

Zhou, Shengde, T B Causey, A Hasona, K T Shanmugam, and L O Ingram. 2003. "Production of Optically Pure d-Lactic Acid in Mineral Salts Medium by Metabolically Engineered *Escherichia coli* W3110." *Applied and Environmental Microbiology* 69 (1): 399–407. doi:10.1128/AEM.69.1.399-407.2003.

4 Second-Generation Bioplastics from Waste Vegetable Oils

Krishna Gautam, Shreya Dwivedi, Pallavi Gupta, Sunita Varjani, and Vivek Kumar Gaur

CONTENTS

4.1 INTRODUCTION

Increasing industrialization and urbanization will result in the depletion of fossil fuels, encouraging research into alternative sources. Currently, there are 7.9 billion people around the world, but that number is projected to rise to 8.6 billion by 2030, 9.8 billion in 2050 and 11.2 billion in 2100, as per a report from the United Nations (Bayer et al., 2014). Population growth is accompanied by an increase in the manufacture of synthetic plastics, which are non-biodegradable and last for decades in the environment (Bayer et al., 2014). Over 380 million tonnes (MT) of plastic are produced each year, and that number is rising at a 4% annual rate (Rosenboom et al., 2022). This poses a serious threat to environmental and human health. Bioplastics are a groundbreaking technology, being bio-based polymers and biodegradable, therefore lessening the harmful impacts of synthetic plastics on the environment and ecology (Sidek et al., 2019). Utilizing leftovers from the production and consumption of vegetable oils as substrates for bioplastic production can help reduce the harmful effects on the environment (da Silva et al., 2022; Sidek et al., 2019).

Bioplastics are synthetic polymers made with the aid of microorganisms from both food and non-food sources, such as vegetable waste oil, pea and corn starch. Based on the types of bioresources employed in its manufacturing, it has been divided into three generations: first generation (1G), plant-based food or animal feed; second generation (2G), non-food crops; and third generation (3G), algae as source (Degli Esposti et al., 2021). Due to their ease of biodegrading and composting, 2G bioplastics are referred to as sustainable alternatives to synthetic or traditional plastics (De Corato, 2021). For the extraction of edible oils, a variety of seeds are used, and the by-products of oil extraction from seeds, known as seed oil cakes (SOCs), account for around half of the weight of the original seeds. SOCs are seen to be great choices for raw materials to be used in a biorefinery for

DOI: 10.1201/9781003344018-4

the creation of high-value-added products in accordance with circular economy paradigms because they are rich in proteins, fibers and secondary metabolites (Mirpoor et al., 2021). The most well-known and common kind of bioplastic is polylactic acid (PLA); however, from an industrial perspective polyhydroxyalkanoates (PHAs) are increasingly gaining more market interest in recent years. Bio-polyesters are garnering significant interest as technical-grade polymers due to a unique combination of features: (1) potential replacement for industrial thermoplastics such as polyvinyl chloride (PVC), polyethylene (PE), polyethylene terephthalate (PET) and polypropylene (PP); (2) derived from renewable sources, being bio-based; (3) ability to biodegrade in both anaerobic and aerobic environments, as well as in aquatic habitats; and (4) variety in structural composition and bio-compatibility with tissues and cells (Tortajada et al., 2013).

Natural polymers from agricultural sources present an intriguing alternative for edible and biodegradable plastics because the agricultural sector generates a large number of by-products along with biological macromolecules, such as proteins and carbohydrates, that are able to mimic the polymeric matrices derived from crude oil. This chapter gives an overview of the properties of bioplastics, in particular 2G bioplastics and used vegetable oil as a source for production and the application of bioplastics. Microbial biorefineries and engineering techniques are included in the list of strategies for enhancing bioplastics production and commercialization. Additionally, we address the current state of the bioplastic market on the global stage and its potential future perspectives in terms of both production and potential uses.

4.2 PROPERTIES OF SECOND-GENERATION BIOPLASTICS

Based on the type of bio-resource processed, classifications between the various types of bioplastics have been developed. Second-generation bioplastics came into existence as 1G bioplastic production was posing environmental hazards like scarcity of water and loss of biodiversity due to accelerated cutting down of trees, which in turn resulted in affecting social stability of the population. In the year 2015, European Renewable Energy Directive specified and amended that the substrate preference should be given to non-food feedstocks (European Parliament and of the Council, 2015; European Parliament and Council of the European Union, 2009). Bio-based PET, PLA, PHA and polybutylene succinate (PBS) are examples of new bioplastics (Patel et al., 2018). Replacement of synthetic plastics with bioplastics is also being preferred as they emit much less carbon dioxide during their incineration and hence contribute to the mitigation of the greenhouse effect. Moreover, they are structurally diverse, biologically similar to tissues and cells due to biological origin and easily degradable in aerobic or anaerobic environmental conditions (Chen, 2009). In the global plastics market as of 2014, 10% of the synthetic plastic production was replaced with PLA bioplastic as it is sustainable and it has comparable properties with polystyrene (PS; Bioplastics European, 2014). Likewise, production of PHA and poly(glycolic acid) (PGA) has been taken into consideration because of their synonymous properties with PET and low-density polyethylene (LDPE); presently their contribution is only 1.4% and <1% in the biomaterial market, respectively, but their production is estimated to increase fourfold by the year 2023 due to excellent mechanical and biodegradation potential (Bioplastics European, 2014; Pillai and Sharma, 2010). In addition, due to highly variable monomer composition, the polymeric properties of bioplastics can be enhanced. Meanwhile, as shown in Figure 4.1, bioplastics derived from 2G biowaste have various unique features and applications that are comparable to synthetic plastics (Nandakumar et al., 2021). Studies suggests that due to involvement of non-food crops in the production of bioplastics, they do not compete with the social requirement of food, and being generated from waste feedstocks, they are inexpensive and reported to drop the processing cost by 30% (Tesfaye et al., 2017).

Irrespective of their global acceptance as an alternate polymer, their industrial stock growth is still a challenge (De Corato, 2021). As an example, the feedstock from winery waste can be recycled into biopolymers as they have large amounts of carbon (De Corato, 2021). However, the recycling can be hampered due to interference of antimicrobial tannins and antioxidants like anthocyanins, polyphenols and flavonoids that mediates the accumulation of PHAs in the cells that constrains cell growth

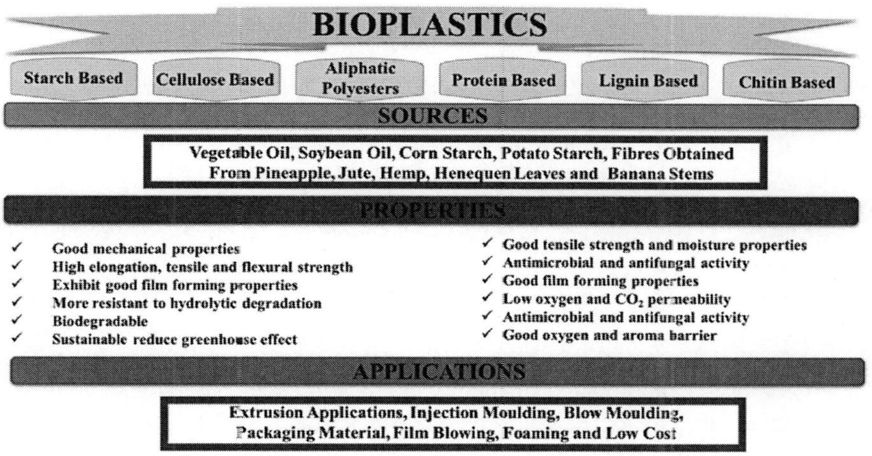

FIGURE 4.1 Various types of bioplastics with raw sources, properties and potential applications.

(De Corato, 2020). It has been reported that the production of biopolymers can reach up to 1200 tons per year by using a carbon dioxide extraction technique for the extraction of natural antioxidants from grape pomace, production of PHAs and its storage in the cells of a *Cupriavidus necator* strain (De Corato, 2021). Evidently, a progressive relationship was reported between the quantity of pomace used and net production of bioplastics. In wine residues, the high concentration of tartaric acid and carbon content have the potential to be converted into succinic acid and PHAs (De Corato, 2021). There are several other examples of feedstocks reported for the production of 2G bioplastic productions including lignocellulosic biomass (wasted agricultural straw, industrial rice residues, wood cellulose) and non-food vegetable oils (castor beans and other oil residues; Vandenberghe et al., 2021).

In terms of applicability of 2G bioplastics, they are extensively used in the packaging industry, molding industry and in generation of fibers in which short-chain PHAs and their co-polymers are major contributors (Chen, 2009). However, some medium-chain PHAs are also known to be favorable contenders as they are derivatives of long-chain PHAs and hence have alteration in crystalline structure, water resistance capacity, elasticity, low oxygen permeability and biodegradation potential (Tortajada et al., 2013). Generally, PHAs with short chains, or scl-PHAs, often have monomers with 3 to 5 carbon atoms (e.g., 3-hydroxybutyrate [3HB] and 3-hydroxyvalerate). PHAs with long chains (lcl-PHAs) have monomers that contain more than 14 carbon atoms (e.g., 3-hydroxyhexadecanoate). In contrast, medium-chain-length PHAs (mcl-PHAs) range from 6 to 14 carbon atoms (e.g., 3-hydroxyhexanoate [3HHx] and 3-hydroxyheptanoate [3HHo]; Reddy et al., 2022). Since they can be compressed and molded to create bio-resorbable and compostable packaging materials, they have potential medicinal applications. They can also be used in the paint industry due to their binding property and in the food industry as coating and packaging material and in the production of bio-rubbers that are biodegradable, being also used for the coating of fruits and vegetables which increase their shelf life and generate a profitable market (De Corato, 2020). In 2019, it was reported that the food packaging industry alone contributed to 52% of the global bioplastic market, which is around 1.26 million tons (De Corato, 2021). However, other sectors also contribute the noticeable percentage to the global bioplastic market including the construction industry (4%), agricultural sector (7%), automotive industry (7%), coating and adhesive industry (7%), consumer goods sector (10%) and textile industry (10%; De Corato, 2021).

4.3 PROPERTIES OF VEGETABLE OIL WASTE FOR BIOPLASTICS PRODUCTION

Production of bioplastics from vegetable oil waste (VOW) originating from household or industrial practices is one the most innovative strategies, as VOW is a spectacular carbon source and due to this property, it can be used directly without any pre-processing (Khatami et al., 2021). VOW generated from sunflower oil, frying corn oil, rapeseed oil and palm oil have been proven promising agents for production of bioplastics with the application of bacterial strains like *Pseudomonas* and *Cupriavidus necator* (Table 4.1; Obruca et al., 2014). It is known that some groups of bacteria produce various types of biopolymers like polyesters, polysaccharides, polyamides and polyphosphates. Among these PHAs are one of the naturally occurring biopolymers that possess analogous traits to those synthetic polymers produced artificially (Yang et al., 2010). It has been studied that rapeseed oil waste can sustainably converted into homopolymer polyhydroxybutyrate (PHB) bioplastic by using a gram-negative soil bacterial strain of *C. necator* up to 1.2 g/L (Verlinden et al., 2011). Similarly, Obruca et al. (2010) proved that rapeseed oil waste produces about 80% cell dry weight of PHA using *C. necator.* By biological or chemical processing VOW can be used to produce bioplastics by the process of acidification (to prevent the formation of pathogens and other unwanted microorganisms), microbial fermentation (to facilitate the conversion of complex organic molecules into simpler ones by the action of microbes) and hydrolysis (in which the addition of the elements of water allows enzymes to enhance the cleaving of molecular bonds; Krishnamurthy and Amritkumar, 2019). Some vegetable oils like crude palm oil and soybean oil have been tested to link two or more polymers with a covalent bonding and hence are known as cross-linking agents (Krishnamurthy and Amritkumar, 2019). Presence of mono- and triglycerides in the crude palm oil waste also act as appropriate plasticizers. Stability and resistance toward high temperature or heat can also be improved by the incorporation of crude palm oil (Makhtar et al., 2013). Globally, the processing of edible vegetables generates a huge proportion of inedible or waste feedstock (Bayer et al., 2014). This waste biomass from food oil waste, effluents from palm oil mill and palm kernel shell have required characteristics for the manufacturing of bioplastics through microbial fermentation and they are rich in cellulosic content, which is a sustainable polymer with crystalline structure ideal for formation of strong fibers (Bayer et al., 2014; Lam et al., 2017; Sathitsuksanoh et al., 2013).

TABLE 4.1
Microbial Biorefineries Strategies for Bioplastic Production Using Oil Waste

Type of Biopolymer	Substrate Utilized	Microbial Strain	Yield/ Accumulation	Production Scale	References
P(3HB-co-3HHx)	Coconut Oil	*Aeromonas hydrophila*	15–45	–	(Qiu et al. 2005)
PHA	Waste cooking oil	*Bacillus thermoamylovorans*	74.4% (w/w) of DCW	Shake flask	(Sangkharak et al. 2021)
PHA	Waste cooking oil	*Halomonas hydrothermalis*	2.26 g/L	Fermentation	(Pernicova et al. 2019)
PHA	Waste cooking oil	*Pseudomonas aeruginosa* 42A2	29.4%	Aerobic fermentation	(Fernández et al. 2005)
PHA	Waste vegetable oil	*Pseudomonas putida* KT2440	1.91 g/L	Batch bioreactor	(Borrero-de Acuña et al. 2019)
PHA	Waste cooking oil	*Pseudomonas chlororaphis* 555	13.87 g/L	Pulse-fed batch fermentation	(Ruiz et al. 2019)
P(3HB-co-mcl-PHA)	Jatropha oil, coconut oil	*Ralstonia eutropha/ Cupriavidus necator*	20%–80%	–	(Samadhiya et al. 2021)
PHA	Effluent from palm oil mill	*Bacillus licheniformis* M2-12	62.97% of DCW	5 L fermenter	(Talan et al. 2022)
PHB	waste palm oil	*C. necator* H16	83%	–	(Talan et al. 2022)
PHA	waste palm oil	*C. necator* H16	6.8%	–	(Talan et al. 2022)

As already mentioned, cotton, soybean, groundnut, rapeseed and sunflower are major oil producing plants and at global level, 50% or more of the total bioplastic market was occupied by soybean oil cake and the rest by cotton (10%) and rapeseed (10%; Bayer et al., 2014). These SOCs are used in the production of bioplastics that are used for food packaging and other materials (Jang et al., 2011). Particularly, it has been reported that cotton seed has certain proteins that possess excellent bio-film producing characteristics when combined with glycerol, which is a known plasticizer, and the resultant bioplastic was noticed to have enhanced thermal and mechanical properties which can be potentially used in agricultural practices (Yue et al., 2012). Reportedly, sesame oil seeds contain protein which is an extensive source for the production of bioplastics as they are thermally stable, less soluble in water and possess good ductile strength; several studies have been conducted lately to develop edible biofilms from sesame seed proteins (Sharma and Singh, 2016).

4.4 STRATEGIES FOR PRODUCTION OF BIOPLASTICS

4.4.1 Microbial Biorefineries

Considering the current plastic pollution and increasing demand for various plastic products highlights the use of biorefineries framework for the microbial nourishment (Samadhiya et al., 2022). The biorefinery aims to maximize or optimize the environmental, economic and social benefits by utilizing all synergies for effective and sustainable production. Microbial biorefineries also serve as a platform for the conversion of biomass into polymers. Biorefineries are the main strategies aligning with the bio-based circular bioeconomy that breaks the take-make-dispose cycle (Leong et al., 2021). Microbial biorefineries are being used to produce biopolymers that are of great interest by offering waste management solutions. The biorefineries approach to produce bioplastics not only promotes the relationship between energy and the surroundings but also protects the ecological community by reducing the carbon footprint. Several types of environmentally friendly bioplastics like PHAs can be produced using various bioprocesses that rely on the use of renewable resources. Additionally, various forms of oil waste can be used for the development of microbial bioplastics, which can be produced directly by bioprocesses like fermentation, as shown in Figure 4.2. Moreover, the 3G biorefineries strategy addresses the green conversion of renewable sources of energy into a range of profitable bio-based products such as biofuels, bioplastics and biogas in order

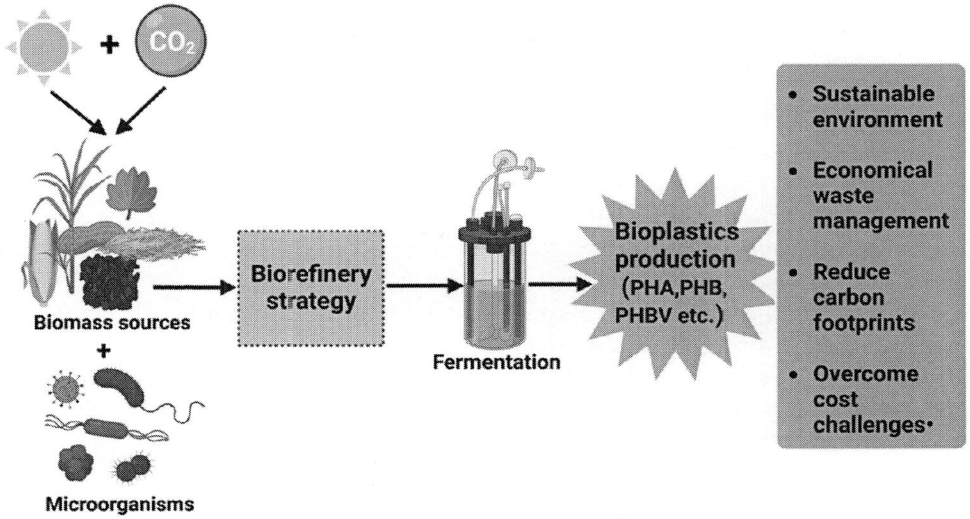

FIGURE 4.2 Figure depicting biorefinery strategies for the production of bioplastics.

to make microalgal biomass an economically feasible feedstock for replacement of conventional bioplastics (Samadhiya et al., 2022). The biorefinery strategy also assists in overcoming the sustainability and cost challenges associated with the commercialization of PHA.

For easing the load of scaling up, the custom-built consortium of photosynthetic bacteria and microalgae will achieve not only excessive bioplastics (PHAs) but also the precursors of biomass molecules. In brief, promising eco-friendly bioplastic components can be obtained from bacterial consortium and microalgae (*Bacillus subtilis* and *Ralstonia eutropha*) as a replacement for conventional fossil-based plastics, eventually unraveling the approach of eliminating plastic pollution along with the burden on renewable resources (Samadhiya et al., 2022). The food waste biorefineries strategy is considered one of the sustainable pathways for the production of PHAs. Fernández et al. (2005) reported the bioprocess by utilizing the waste cooking oil as a substrate that resulted in the production of 29.4% PHAs using *Pseudomonas aeruginosa* 42A2. Similarly, Sangkharak et al. (2021) studied the microbial bioplastic production by *Bacillus thermoamylovorans* utilizing a waste cooking oil (WCO) as a carbon substrate and reported the yield of PHA up to 74.4% (w/w). The WCOs, such as domestic and industrial waste cooking oils, can successfully be employed for biopolymer production. As the WCO generated from these sources does not need prior pre-treatment, it can be efficiently used as a substrate for the production of PHAs using an appropriate microbial strain in fermenter processing (Talan et al., 2022). A study applied *Cupriavidus necator* H16 to produce PHA from waste palm oil and tallow and yielded 6.8% and 83% dry microbial biomass and PHB accumulation, respectively (Taniguchi et al., 2003). The utilization of tallow as a carbon substrate produced the copolymer poly(3-hydroxybutyrate-co-3-hydroxyvalerate) (PHBV). Another study used various WCOs (rapeseed oil, palm oil, sunflower oil) as PHA production substrates for *C. necator* strain H16, yielding biomasses of 10.8, 11.9 and 10.9 g/L, respectively, and PHB contents of 52.4%, 67.9% and 58%, respectively (Talan et al., 2022). Chaudhry et al. (2011) reported 35.63% PHA production by utilizing corn oil with *Pseudomonas* species. Similarly, other researchers also studied the production of PHA by utilizing waste vegetable oil with wild and recombinant strains of *Pseudomonas putida* KT440 that resulted in the production of 1.91 g/L PHA (Talan et al., 2022). Thus, to address waste disposal problems on human health and environment, waste generated in biorefineries must be disposed of in a biodegradable and sustainable manner. As a result, biological valorization of excessive waste and side streams through various bioprocesses for bioplastics production and further useful by-products significantly promotes a sustainable circular bioeconomy concept by characterizing a low carbon economy through greenhouse gas depletion.

4.4.2 Engineering Approaches to Produce Bioplastics

4.4.2.1 Metabolic Engineering

Conventional metabolic engineering along with technology of evolutionary engineering and synthetic biology is known as "system metabolic engineering." The advancements in microbial strain development that enhances the biorefinery efficiency by reducing time and production cost depends on traditional metabolic engineering approaches. Various microorganisms including *Corynebacterium glutamicum, Escherichia coli* and *Cupriavidus necator* have been used as platform microbial cell factories by profitably applying systems metabolic engineering for development (Choi et al., 2020). Therefore, the goals of metabolic engineering can be achieved by increasing product formation, speeding up the process, saving energy, reducing by-product production and developing species with resistance to environmental stress. Extending the substrate range, improving yield and productivity, eliminating by-products, improving process efficiency, improving cellular characteristics, and expanding the range of products to include heterologous protein production are the various strategies for metabolic engineering (Kumar and Prasad, 2011). The bioprocess, which was powered by highly specific enzymes and metabolic pathways, was able to transform crude biomass into usable precursors, effectively acting as an interface between biomass and polymer trade demands and facilitating the growth of biopolymers (Matsumoto and Taguchi, 2013). Researchers were able to fully comprehend the various biological processes taking place in PHA-accumulating species by conducting systems-level research on signaling, metabolic and regulatory networks. Thus the awareness gained will contribute toward novel approaches for better production of PHAs, such as tailor-made PHAs with demanded molecular mass and monomer composition (Hiraishi and Taguchi, 2013). Because of the widespread use of microorganisms in several industries, they have been used to produce a broad variety of valuable products in food and pharmaceutical industries and agriculture processes (Kumar and Prasad, 2011). In contrast to petroleum-based plastics, a variety of microbial species were evolved for the production of economical PHAs. Microbes that are not natural producers of PHAs can be engineered, such as *E. coli*, by introducing the *phaCAB* operon from a native producer such as *C. necator* (Choi et al., 2020). In another study, yeast strains were engineered with the *phaABCRe* operon, which contains the genes of the *Ralstonia eutropha* PHA biosynthetic pathway: *phbA*, encoding β-ketothiolase; *phbB*, encoding NADPH-linked acetoacetyl-CoA reductase; and *phbC*, encoding PHA synthase as demonstrated in Figure 4.3. The resulting recombinant yeast strain was able to produce PHB, which can be used to synthesize biodegradable bioplastics (Kumar and Prasad, 2011). Similarly, the genes that encode PHB biosynthetic pathways of *R. eutropha* were transferred into microbial strain *Saccharomyces cerevisiae* for the heterologous production of bioplastics polymers (Kumar and Prasad, 2011). Several researchers were able to develop one-step fermentative production processes for artificial microbial polyesters (which could not be produced by native microbes but could be produced by engineered microbes based on the biosynthesis of PHA) through intense strain development focusing on PhaC engineering, systems metabolic engineering and heterologous metabolic pathways (Choi et al., 2020). Poly (3HB co-3HHx) was recognized in two different species of *Aeromonas* using coconut oil, olive oil and trilaurin as a carbon source where the fraction of 3HHx depends on the different concentrations of carbon source and culture conditions. PhaC (the key enzyme involved in the biosynthesis of PHA) has high substrate specificity for 3HHx-CoA and 3HB-CoA, which is accountable to produce 3HB-CoA and

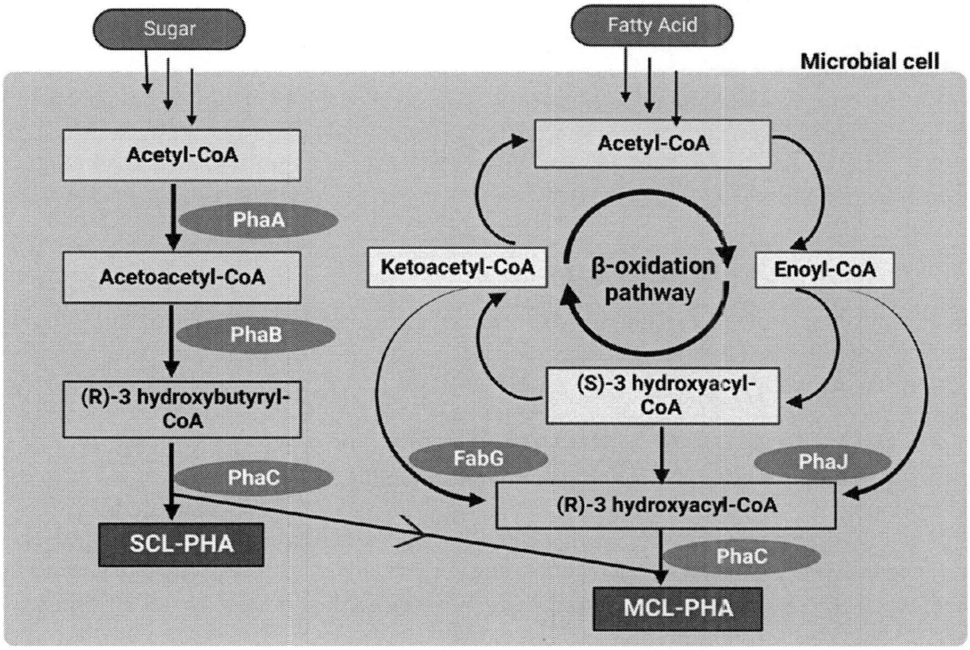

FIGURE 4.3 Pathways for PHA production.

3HHx-CoA from the correspondent enoyl-CoAs that have been identified through the molecular characterization of *Aeromonas caviae* biosynthesis genes (Choi et al., 2020). Therefore the progress of metabolic engineering approach facilitated the recombinant bacteria to synthesize PHA efficiently. For example, several genetically engineered strains such as *C. necator*, *E. coli* ΔarcA, *E. coli* GCSC6576 (pSYL107), *E. coli* CGSC 4401 (pJC4), *E. coli* CGSC 4401 (pJC4) and *E. coli* CGSC 4401 (pJC4) has been reported for the production of up to 168 g/L PHB (Favaro et al., 2019).

4.4.2.2 Protein Engineering

Protein engineering has become a powerful biotechnological toolbox that has gained popularity as a method for enhancing enzymatic activities, bolstering enzyme stabilities, and broadening product spectrum in natural product biosynthesis (Li et al., 2020). There are several issues such as cost-effectiveness and physiochemical properties that hindered the production and widespread use of PHAs; therefore the protein engineering approach of *phaC* that plays a major role in PHA biosynthesis was successfully studied for the enhancing the efficiency of PHA production and the properties of resulted polymers (Hiraishi and Taguchi, 2013). PHA synthase catalyzes the polymerization of coenzyme A (CoA) thioesters of HAs (hydroxyalkanoic acids) into polymers of PHA, which results in the releasing of the CoA moiety. The *phaC* gene, clustered with other genes (*phaA, phaB, phaG* and *phaJ*, encoded by β-ketothiolase, acetoacetyl-CoA PHA synthase, 3-hydroxyacyl-acyl carrier protein-coenzyme A transferase and enoyl-CoA hydratase, respectively), has been successfully implicated in PHA production (Zou et al., 2017). Structurally *PhaC* has two terminals: N-terminal and C-terminal catalytic domains. According to research, the N-terminal domain of *PhaC* is an essential key for its expression, substrate specificities, and activity. It also influences the molecular mass of the PHA produced (Neoh et al., 2022). Four different classes of natural *PhaC* were reported; Class I PHA synthases and class II PHA synthases that are single subunit enzymes discovered from *R. eutropha* are the widely studied PHA synthases that are involved in the production of medium-chain-length PHA; Class II PHA synthases discovered from *Pseudomonas aeruginosa* also encompass single-unit enzymes and have been involved in the synthesis of medium-chain-length PHA such as polyhydroxyhexanoate (PHHx) and polyhydroxyoctanoate (PHO); Class III PHA synthases discovered from *Allochromatium vinosum* requires two subunits to be functional and has been utilized in in vivo and in vitro production of short-chain-length PHAs involving P(3HB-co-3HV); Class IV PHA synthases were found in *Bacillus megaterium* consisting of two subunit enzymes that are involved in the synthesis of 2HP, 3HHx, 3HO, 4HB and 6HHx precursors (Neoh et al., 2022). Researchers also studied the protein engineering approach for the optimization of plastic degrading enzymes by improving and enhancing enzyme stability and catalytic efficiency (Zhu et al., 2022).

4.5 GLOBAL MARKET OF BIOPLASTICS

The market for bioplastics is expanding as a result of consumer awareness of sustainable plastic alternatives and widespread attempts to minimize the use of traditional plastics that are not biodegradable. Traditional plastics, which are mostly made of petroleum, take decades to disintegrate and spend a lot of time in landfills. When biodegradable plastics are thrown away or reincorporated into the environment, they degrade more quickly. Additionally, compared to conventional plastics, biodegradable polymers decompose significantly more quickly because of the actions of microorganisms. Biopolymer output peaked at 2.11 MT in 2018 and is expected to increase to 2.62 MT by 2023. Bioplastics still account for less than 1% of all plastic manufacturing despite this tremendous industry expansion (Niaounakis, 2019). Comparing the production of bioplastic to that of all other forms of plastic, it is shown that just 1% of the 320 MT produced by other forms of plastics are bioplastics (Mozaffari and Kholdebarin, 2019). The market for bioplastics has been expanding recently as a result of increased demand. By 2025, the amount of the worldwide bioplastics and biopolymers market is anticipated to increase from USD 10.5 billion to USD 27.9 billion (Degli Esposti et al.,

2021). The expansion of the worldwide packaging end-use business is the main factor driving this strong growth. Although higher scales might lower pricing, increase the need, and offer incentives for expanding the use of bioplastics technology and investing in recycling, infrastructure remains a high-risk industry with the major issue being unclear demand due to high pricing and unknown end of life treatment (Rosenboom et al., 2022). Unfortunately, the primary factors limiting the use of bioplastics are their more expensive manufacture and often less desirable mechanical qualities as compared to petro-plastics (Dilkes-Hoffman et al., 2018). The production of bioplastics was mostly concentrated in Asia (45%), with Europe producing 25% of the total, although this is anticipated to increase as a result of the European Commission's commitment to the development of a circular economy (Brizga et al., 2020). According to USDA data, about 600 MT of oilseeds were produced globally in 2018–2019, mostly for the manufacture of edible oils, although SOCs, a by-product of oil extraction, often make up roughly half of the weight of the original seed (Bayer et al., 2014). These SOCs, which are leftovers from the production of vegetable oils, are known to have significant concentrations of proteins, polysaccharides and fiber that may be recovered and may serve as a renewable source for cutting-edge bio-based goods. As a result, SOCs could be a desirable feedstock for the construction of biorefineries that produce both edible and non-edible oils. Additionally, these vegetable SOCs may be utilized to make bioplastics.

4.6 FUTURE PERSPECTIVES

The negative effects on the environment brought on by the vast amount of non-biodegradable trash are encouraging research into the production of novel biodegradable materials that can be produced from natural resources including biomass, plants and microorganisms. Future bioplastics innovations might boost production efficiency, create new uses for the material and open up new business prospects.

1. The government restrictions and initiatives against single-use plastic are a significant driver of the bioplastic market expansion. An important factor in the market growth is the spontaneous biodegradability of bioplastics such as PHA in environmental settings. Additional opportunities will be made possible by newly available raw resources and the expansion of the Asia-Pacific area (Degli Esposti et al., 2021).
2. A renewable, plentiful and more morally sound feedstock alternative to 1G biomass is lignocellulosic agriculture and other biowastes. However, in order to provide polymer building blocks in a financially viable and environmentally responsible way, biorefinery processes must become more efficient and follow green chemistry guidelines (such as employing safe chemicals and lowering energy consumption).
3. The production of bioplastics is made possible by the biotechnology of microorganisms because it has considerable commercial potential for a variety of industries, including agriculture, medicine, pharmaceuticals and veterinary care. Therefore a new set of guidelines and standards for bioplastics should be developed for their manufacturing, use and waste disposal globally. As a result, labeling laws need to be improved to take into account how much energy, raw materials and pollutants are used in manufacturing and using a product (Sidek et al., 2019).
4. Sustainability cannot be assumed simply by using renewable resources. Sustainability is mostly dependent on a material's process of manufacturing, ultimate usage and recycling capabilities, but less so on its constituent parts. However, with advancements in technology, bioplastics might help various businesses that rely on plastic transition to a circular economy.
5. International cooperation, continuous research and current technical developments are all necessary for the commercialization and demonstration of bioplastics. Bioplastic materials must be constructed on an integrated ecologically friendly basis in order to improve the sustainability of materials and processes throughout the course of their lifetime (Sidek et al., 2019).

6. PHA is still more expensive to produce as compared to synthetic polymers. The main reasons for the high cost are the slow microbial growth and the poor PHA generation from basic ingredients.
7. Overall, the culture process parameters and feeding ratio of different precursors need to be carefully optimized. And particular focus will be placed on elements that influence molecular weights, such as the concentration, activity and simultaneous degradation of biopolymer synthase.

4.7 SUMMARY AND CONCLUSIONS

An increasing number of scientists and researchers are working with bioplastics, an environmentally friendly alternative manufactured from renewable biomass including vegetable fats and oils, algae, seaweed, and maize or potato starch. One of these, used cooking oil, was discovered to be a possible raw material for the development of bioplastics. Being bio-sourced, biodegradable and biocompatible, the polymer derived from such sources has wide attraction. To improve microorganism efficiency in biomass consumption, bioplastic polymerization (particularly for PHAs) and biological depolymerization for recycling, microbial biorefinery and metabolic engineering have been found to be promising tools. PHA are seen to be the most promising bioplastics now being developed, justifying deployment and greater usage in many application sectors, particularly high added-value sectors like the biomedical and cosmetics industries. Yet even with advanced technologies, bioplastics have the ability to accelerate many plastic-intensive sectors toward a circular economy.

4.7.1 Acknowledgment

The authors received no financial support for the publication of this article.

4.7.2 Conflict of Interest Statement

The authors declare no competing financial or personal interests.

REFERENCES

Bayer, I.S., Guzman-Puyol, S., Heredia-Guerrero, J.A., Ceseracciu, L., Pignatelli, F., Ruffilli, R., Cingolani, R., Athanassiou, A., 2014. Direct transformation of edible vegetable waste into bioplastics. Macromolecules 47(15), 5135–5143. https://doi.org/10.1021/ma5008557

Bioplastics European, 2014. Bioplastics facts and figures. Institute for Bioplastics and Biocomposite, Nova-Institute, Berlin.

Borrero-de Acuña, J.M., Aravena-Carrasco, C., Gutierrez-Urrutia, I., Duchens, D., Poblete-Castro, I., 2019. Enhanced synthesis of medium-chain-length poly (3-hydroxyalkanoates) by inactivating the tricarboxylate transport system of Pseudomonas putida KT2440 and process development using waste vegetable oil. Process Biochemistry 77, 23–30. https://doi.org/10.1016/j.procbio.2018.10.012

Brizga, J., Hubacek, K., Feng, K., 2020. The unintended side effects of bioplastics: Carbon, land, and water footprints. One Earth 3(1), 45–53. https://doi.org/10.1016/j.oneear.2020.06.016

Chaudhry, W.N., Jamil, N., Ali, I., Ayaz, M.H., Hasnain, S., 2011. Screening for polyhydroxyalkanoate (PHA)-producing bacterial strains and comparison of PHA production from various inexpensive carbon sources. Annals of Microbiology 61(3), 623–629. https://doi.org/10.1007/s13213-010-0181-6

Chen, G.Q., 2009. A microbial polyhydroxyalkanoates (PHA) based bio- and materials industry. Chemical Society Reviews 38(8), 2434–2446. https://doi.org/10.1039/b812677c

Choi, S.Y., Rhie, M.N., Kim, H.T., Joo, J.C., Cho, I.J., Son, J., Jo, S.Y., Sohn, Y.J., Baritugo, K.-A., Pyo, J., 2020. Metabolic engineering for the synthesis of polyesters: A 100-year journey from polyhydroxyalkanoates to non-natural microbial polyesters. Metabolic Engineering 58, 47–81.

da Silva, C.A., Dos Santos, R.N., Oliveira, G.G., Ferreira, T.P. de S., de Souza, N.L.G.D., Soares, A.S., de Melo, J.F., Colares, C.J.G., de Souza, U.J.B., de Araújo-Filho, R.N., Aguiar, R.W. de S., Dos Santos, G.R., Gabe-,

E.E., Campos, F.S., 2022. Biodiesel and bioplastic production from waste-cooking-oil transesterification: An environmentally friendly approach. Energies 15(3), 1073. https://doi.org/10.3390/en15031073
De Corato, U., 2020. Improving the shelf-life and quality of fresh and minimally-processed fruits and vegetables for a modern food industry: A comprehensive critical review from the traditional technologies into the most promising advancements. Critical Reviews in Food Science and Nutrition 60(6), 940–975. https://doi.org/10.1080/10408398.2018.1553025
De Corato, U., 2021. Bioplastics from winemaking by-products in the buildings sector: A feasibility study on the main opportunities, barriers and challenges. Circular Economy and Sustainability 1(4), 1313–1333. https://doi.org/10.1007/s43615-021-00048-7
Degli Esposti, M., Morselli, D., Fava, F., Bertin, L., Cavani, F., Viaggi, D., Fabbri, P., 2021. The role of biotechnology in the transition from plastics to bioplastics: An opportunity to reconnect global growth with sustainability. FEBS Open Bio 11(4), 967–983. https://doi.org/10.1002/2211-5463.13119
Dilkes-Hoffman, L.S., Pratt, S., Lant, P.A., Laycock, B., 2018. The role of biodegradable plastic in solving plastic solid waste accumulation. Plastics to Energy: Fuel, Chemicals, and Sustainability Implications 469–505. https://doi.org/10.1016/B978-0-12-813140-4.00019-4
European Parliament and of the Council, 2015. Directive 2015/652 of the European Parliament and of the Council. Official Journal of the European Union. https://eur-lex.europa.eu/legal-content/EN/TXT/?uri=celex%3A32015L0652
European Parliament, Council of the European Union, 2009. Directive 2009/48/EC of the European Parliament and of the Council of 18 June 2009 on the safety of toys. Official Journal of the European Union. https://eur-lex.europa.eu/LexUriServ/LexUriServ.do?uri=OJ:L:2009:170:0001:0037:EN:PDF
Favaro, L., Basaglia, M., Casella, S., 2019. Improving polyhydroxyalkanoate production from inexpensive carbon sources by genetic approaches: A review. Biofuels, Bioproducts and Biorefining 13(1), 208–227. https://doi.org/10.1002/bbb.1944
Fernández, D., Rodríguez, E., Bassas, M., Viñas, M., Solanas, A.M., Llorens, J., Marqués, A.M., Manresa, A., 2005. Agro-industrial oily wastes as substrates for PHA production by the new strain *Pseudomonas aeruginosa* NCIB 40045: Effect of culture conditions. Biochemical Engineering Journal 26(2–3), 159–167. https://doi.org/10.1016/j.bej.2005.04.022
Hiraishi, T., Taguchi, S., 2013. Protein engineering of enzymes involved in bioplastic metabolism. Protein engineering—technology and application, InTech, Rijeka, Croatia, 133–165.
Jang, S.A., Shin, Y.J., Song, K. Bin, 2011. Effect of rapeseed protein-gelatin film containing grapefruit seed extract on "Maehyang" strawberry quality. International Journal of Food Science and Technology 46(3), 620–625. https://doi.org/10.1111/j.1365-2621.2010.02530.x
Khatami, K., Perez-Zabaleta, M., Owusu-Agyeman, I., Cetecioglu, Z., 2021. Waste to bioplastics: How close are we to sustainable polyhydroxyalkanoates production? Waste Management 119, 374–388. https://doi.org/10.1016/j.wasman.2020.10.008
Krishnamurthy, A., Amritkumar, P., 2019. Synthesis and characterization of eco-friendly bioplastic from low-cost plant resources SN Applied Sciences 1(11), 1432. https://doi.org/10.1007/s42452-019-1460-x
Kumar, R.R., Prasad, S., 2011. Metabolic engineering of bacteria. Indian Journal of Microbiology 51, 403–409.
Lam, W., Wang, Y., Chan, P.L., Chan, S.W., Tsang, Y.F., Chua, H., Yu, P.H.F., 2017. Production of polyhydroxyalkanoates (PHA) using sludge from different wastewater treatment processes and the potential for medical and pharmaceutical applications. Environmental Technology (United Kingdom) 38(13–14), 1779–1791. https://doi.org/10.1080/09593330.2017.1316316
Leong, H.Y., Chang, C.K., Khoo, K.S., Chew, K.W., Chia, S.R., Lim, J.W., Chang, J.S., Show, P.L., 2021. Waste biorefinery towards a sustainable circular bioeconomy: a solution to global issues. Biotechnology for Biofuels 14(1), 1–15. https://doi.org/10.1186/s13068-021-01939-5
Li, C., Zhang, R., Wang, J., Wilson, L.M., Yan, Y., 2020. Protein engineering for improving and diversifying natural product biosynthesis. Trends in Biotechnology 38, 729–744.
Makhtar, N.S.M., Rais, M.F.M., Rodhi, M.N.M., Bujang, N., Musa, M., Hamid, K.H.K., 2013. *Tacca leontopetaloides* starch: New sources starch for biodegradable plastic. Procedia Engineering 68, 385–391. https://doi.org/10.1016/j.proeng.2013.12.196
Matsumoto, K., Taguchi, S., 2013. Enzyme and metabolic engineering for the production of novel biopolymers: Crossover of biological and chemical processes. Current Opinion in Biotechnology 24, 1054–1060.
Mirpoor, S.F., Giosafatto, C.V.L., Porta, R., 2021. Biorefining of seed oil cakes as industrial co-streams for production of innovative bioplastics. A review. Trends in Food Science and Technology 109, 259–270. https://doi.org/10.1016/j.tifs.2021.01.014

Mozaffari, N., Kholdebarin, A., 2019. A review: Investigation of plastics effect on the environment. Bioplastic Global Market Share and Its Future Perspectives. Technogenic and Ecological Safety 5(1), 47–54. https://doi.org/10.5281/zenodo.2600664

Nandakumar, A., Chuah, J.A., Sudesh, K., 2021. Bioplastics: A boon or bane? Renewable and Sustainable Energy Reviews 147, 111237. https://doi.org/10.1016/j.rser.2021.111237

Neoh, S.Z., Chek, M.F., Tan, H.T., Linares-Pastén, J.A., Nandakumar, A., Hakoshima, T., Sudesh, K., 2022. Polyhydroxyalkanoate synthase (PhaC): The key enzyme for biopolyester synthesis. Current Research in Biotechnology 4, 87–101.

Niaounakis, M., 2019. Recycling of biopolymers—The patent perspective. European Polymer Journal 114, 464–475. https://doi.org/10.1016/j.eurpolymj.2019.02.027

Obruca, S., Marova, I., Snajdar, O., Mravcova, L., Svoboda, Z., 2010. Production of poly(3-hydroxybutyrate-co-3-hydroxyvalerate) by *Cupriavidus necator* from waste rapeseed oil using propanol as a precursor of 3-hydroxyvalerate. Biotechnology Letters 32, 1925–1932. https://doi.org/10.1007/s10529-010-0376-8

Obruca, S., Petrik, S., Benesova, P., Svoboda, Z., Eremka, L., Marova, I., 2014. Utilization of oil extracted from spent coffee grounds for sustainable production of polyhydroxyalkanoates. Applied Microbiology and Biotechnology 98, 5883–5890. https://doi.org/10.1007/s00253-014-5653-3

Patel, M.K., Bechu, A., Villegas, J.D., Bergez-Lacoste, M., Yeung, K., Murphy, R., Woods, J., Mwabonje, O.N., Ni, Y., Patel, A.D., Gallagher, J., Bryant, D., 2018. Second-generation bio-based plastics are becoming a reality—Non-renewable energy and greenhouse gas (GHG) balance of succinic acid-based plastic end products made from lignocellulosic biomass. Biofuels, Bioproducts and Biorefining 12(3), 426–441. https://doi.org/10.1002/bbb.1849

Pernicova, I., Kucera, D., Nebesarova, J., Kalina, M., Novackova, I., Koller, M., Obruca, S., 2019. Production of polyhydroxyalkanoates on waste frying oil employing selected Halomonas strains. Bioresource Technology 292, 122028. https://doi.org/10.1016/j.biortech.2019.122028Pillai, C.K.S., Sharma, C.P., 2010. Review paper: Absorbable polymeric surgical sutures: Chemistry, production, properties, biodegradability, and performance. Journal of Biomaterials Applications 25(4), 291–366. https://doi.org/10.1177/0885328210384890

Qiu, Y.Z., Han, J., Guo, J.J., Chen, G.Q., 2005. Production of poly (3-hydroxybutyrate-co-3-hydroxyhexanoate) from gluconate and glucose by recombinant Aeromonas hydrophila and Pseudomonas putida. Biotechnology Letters 27(18), 1381–1386. https://doi.org/10.1016/S1369-703X(03)00036-6

Reddy, V.U.N., Ramanaiah, S. V, Reddy, M.V., Chang, Y.-C., 2022. Review of the developments of bacterial medium-chain-length polyhydroxyalkanoates (mcl-PHAs). Bioengineering 9, 225.

Rosenboom, J.G., Langer, R., Traverso, G., 2022. Bioplastics for a circular economy. Nature Reviews Materials 7(2), 117–137. https://doi.org/10.1038/s41578-021-00407-8

Ruiz, C., Kenny, S.T., Narancic, T., Babu, R., O'Connor, K., 2019. Conversion of waste cooking oil into medium chain polyhydroxyalkanoates in a high cell density fermentation. Journal of Biotechnology 306, 9–15. https://doi.org/10.1016/j.jbiotec.2019.08.020

Samadhiya, K., Sangtani, R., Nogueira, R., Bala, K., 2022. Insightful advancement and opportunities for microbial bioplastic production. Frontiers in Microbiology 12, 674864. https://doi.org/10.3389/fmicb.2021.674864

Sangkharak, K., Khaithongkaeo, P., Chuaikhunupakarn, T., Choonut, A., Prasertsan, P., 2021. The production of polyhydroxyalkanoate from waste cooking oil and its application in biofuel production. Biomass Conversion and Biorefinery 11, 1651–1664. https://doi.org/10.1007/s13399-020-00657-6

Sathitsuksanoh, N., Xu, B., Zhao, B., Zhang, Y.H.P., 2013. Overcoming biomass recalcitrance by combining genetically modified switchgrass and cellulose solvent-based lignocellulose pretreatment. PloS One 8(9), e73523. https://doi.org/10.1371/journal.pone.0073523

Sharma, L., Singh, C., 2016. Sesame protein based edible films: Development and characterization. Food Hydrocolloids 61, 139–147. https://doi.org/10.1016/j.foodhyd.2016.05.007

Sidek, I.S., Draman, S.F.S., Abdullah, S.R.S., Anuar, N., 2019. Current development on bioplastics and its future prospects: An introductory review. INWASCON Technology Magazine 1, 03–08. https://doi.org/10.26480/itechmag.01.2019.03.08

Talan, A., Pokhrel, S., Tyagi, R.D., Drogui, P., 2022. Biorefinery strategies for microbial bioplastics production: Sustainable pathway towards Circular Bioeconomy. Bioresource Technology Reports 17, 100875. https://doi.org/10.1016/j.biteb.2021.100875

Taniguchi, I., Kagotani, K. and Kimura, Y., 2003. Microbial production of poly (hydroxyalkanoate)s from waste edible oils. Green Chemistry 5(5), 545–548. https://doi.org/10.1039/B304800B

Tesfaye, T., Sithole, B., Ramjugernath, D., 2017. Valorisation of chicken feathers: A review on recycling and recovery route—current status and future prospects. Clean Technologies and Environmental Policy 19, 2363–2378. https://doi.org/10.1007/s10098-017-1443-9

Tortajada, M., da Silva, L.F., Prieto, M.A., 2013. Second-generation functionalized medium-chain-length polyhydroxyalkanoates: The gateway to high-value bioplastic applications. International Microbiology 16(1), 1–15. https://doi.org/10.2436/20.1501.01.175

Vandenberghe, L.P. de S., Oliveira, P.Z. de, Bittencourt, G.A., Mello, A.F.M. de, Vásquez, Z.S., Karp, S.G., Soccol, C.R., 2021. The 2G and 3G bioplastics: An overview. Biotechnology Research and Innovation 5(1), e2021004. https://doi.org/10.4322/biori.202104

Verlinden, R.A.J., Hill, D.J., Kenward, M.A., Williams, C.D., Piotrowska-Seget, Z., Radecka, I.K., 2011. Production of polyhydroxyalkanoates from waste frying oil by *Cupriavidus necator.* AMB Express 1, 1–8. https://doi.org/10.1186/2191-0855-1-11

Yang, Y.H., Brigham, C.J., Budde, C.F., Boccazzi, P., Willis, L.B., Hassan, M.A., Yusof, Z.A.M., Rha, C., Sinskey, A.J., 2010. Optimization of growth media components for polyhydroxyalkanoate (PHA) production from organic acids by *Ralstonia eutropha*. Applied Microbiology and Biotechnology 87, 2037–2045. https://doi.org/10.1007/s00253-010-2699-8

Yue, H.B., Cui, Y.D., Shuttleworth, P.S., Clark, J.H., 2012. Preparation and characterisation of bioplastics made from cottonseed protein. Green Chemistry 14(7), 2009–2016. https://doi.org/10.1039/c2gc35509d

Zhu, B., Wang, D., Wei, N., 2022. Enzyme discovery and engineering for sustainable plastic recycling. Trends in Biotechnology 40, 22–37.

Zou, H., Shi, M., Zhang, T., Li, Lei, Li, Liangzhi, Xian, M., 2017. Natural and engineered polyhydroxyalkanoate (PHA) synthase: Key enzyme in biopolyester production. Applied Microbiology and Biotechnology 101, 7417–7426.

5 Cellulose-Based Bioplastics

Rekha Unni, Reshmy R, Aravind Madhavan, Parameswaran Binod, Ashok Pandey, Mukesh Kumar Awasthi, and Raveendran Sindhu

CONTENTS

5.1 INTRODUCTION

Developing bioplastics-based materials from renewable energy resources is an active field of study that is gaining traction among scientific communities. For the first generation of bio-based plastics, biomass resources (cellulose and starch), fatty acids, and biodegradable wastes were used to synthesize the monomers, which are the plastic's basic constituents. Along with lignin and hemicellulose, cellulose constitutes 30%–50% of the total biomass in lignocellulosic materials. Because of its availability and environmentally beneficial qualities, including renewability, biocompatibility, and biodegradability, it is the best option to replace synthetic plastics. Cellulose is a promising feedstock for the production of chemicals and monomers (Rinaldi & Sch, 2009).

The linear homopolysaccharide known as cellulose is composed of glucose units that are connected end to end by a β-1,4-glycosidic bond between carbons C1 and C4 (Figure 5.1). The repeating component of polymer is two anhydro-d-glucopyranose units (AGU), which is similar to cellobiose. Strong intramolecular and intermolecular hydrogen bonding creates the linear cellulose chains, which have three hydroxyl groups per AGU, to aggregate into larger semi-crystalline units called microfibrils (Nechyporchuk et al., 2016). It is fibrillated and has a range of sizes, shapes, surface chemistry, and physical characteristics. $(C_6H_{10}O_5)n$ is the chemical formula for cellulose, where n denotes the quantity or degree of polymerization of repeating sugar units (DP).

DOI: 10.1201/9781003344018-5

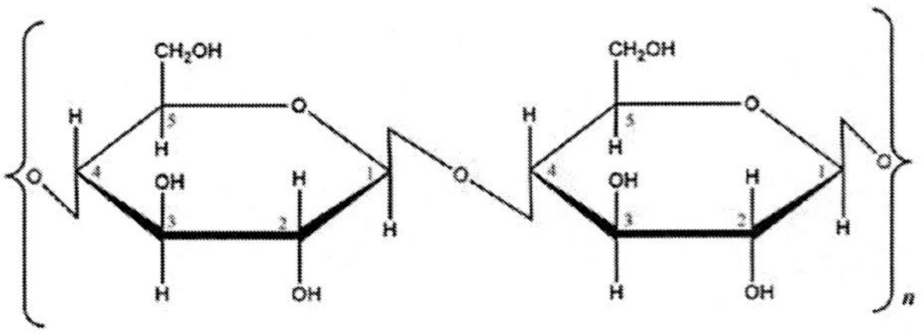

FIGURE 5.1 Structure of cellulose $(C_6H_{10}O_5)_n$.

The current chapter discusses the primary methods of producing cellulose-based plastic polymers and focuses on the most significant recent developments in this area. With respect to current trends, this chapter is divided into five sections. The creation of plastic polymers based on monomers obtained from cellulose is critically examined in the first section. The utilization of unmodified cellulose fibers and their derivatives in the field of bioplastics has made recent developments, which are briefly discussed in the following section. The final section investigates the progress made to date in the use of cellulose at the nanometer range in plastic nanocomposites.

5.2 THE EVOLUTION OF BIOPLASTICS MADE FROM DIFFERENT TYPES OF CELLULOSE

Cellulose-based plastics have been used for a long time in a variety of industries. The first and second approaches were initially used to create cellulosic plastic products based on cellulose fibers, cellulose derivatives, and cellulose-derived monomers (Figure 5.2). Cellulose was initially regenerated by chemical modification without depolymerization by using a well-known viscose technique for producing cellophane and rayon. Later, derivatization was abandoned in favor of the cuoxam and lyocell techniques for creating cellulose plastic goods. As an alternative, cellulose is able to be depolymerized catalytically to yield monomeric glucose, which serves as the raw material for the manufacture of a large range of monomers through catalytic or biotechnological conversion. These cellulose-derived monomers have been employed in the manufacture of sustainable bioplastics (Rose & Palkovits, 2011). The use of cellulose derivatives as thermoplastic bulk materials was first industrialized in the 19th century. The traditional viscose technique, which produces hazardous by-products, was used to create regenerated cellulose-based bioplastic sheets (CS_2 and H_2S). The most common cellulose derivatives used in industry are cellulose acetate; cellulose esters for shaping, impregnation, and films; renewed cellulose for fibers; and cellulose ethers, which are extensively used in paints, medicine, food, hygiene products, and building material. However, the processability and utilization of this type of cellulose in biodegradable plastic films were limited until solvent solutions for dissolving cellulose became available. Industry has recently used functional polymeric materials such as composites and hybrid

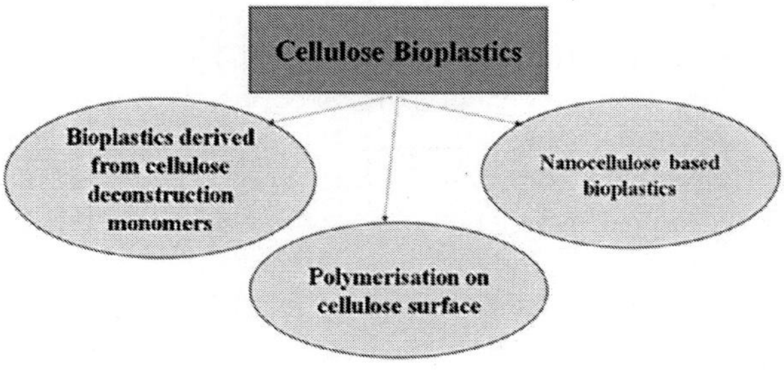

FIGURE 5.2 The main routes used for cellulose bioplastic preparation.

composites, as well as new and improved polymeric thermoplastic film components (Rani et al., 2014). The third production technique gave rise to the subsequent generation of cellulosic bioplastic materials, which includes nanoscale cellulose fibers. By incorporating nanocellulose into different polymeric materials, this generation of cellulosic bioplastic materials was created. Figure 5.2 represents the main routes used for cellulose bioplastic preparation.

5.3 DEPOLYMERIZATION OF CELLULOSE

Two methods for depolymerizing cellulose to glucose are β-1,4-glycosidic link cleavage and catalyst-free hydrolysis on enzymatic hydrolysis in supercritical water (Karam et al., 2018). In more recent times, heterogeneous or homogeneous reactions with acids have produced depolymerized cellulose. The availability of bulk cellulose–derived monomers for the manufacture of plastics is greatly expanded by such processes.

A platform chemical generated from cellulosic biomass with tremendous promise and versatility is lactic acid. A novel homo-fermentative, facultative anaerobe called *Enterococcus faecalis* is used to commercially generate it by fermenting sugars at high concentrations. Up to 98% high-purity lactic acid is produced from glucose. It can be used as a renewable feedstock and active precursor for the synthesis of a variety of bioplastics. Lactic acid can be directly polycondensed or converted into poly(lactic acid) (PLA) using ring-opening polymerization. The most studied and utilized biodegradable and renewable thermoplastic polyester is called PLA, and it has the potential to replace traditional petrochemical-based polymers. Due to its mechanical qualities, which are comparable to those of poly(ethylene terephthalate) (PET) and polypropylene (PP), PLA has been employed as a substitute for some petroleum-based polymers in business-related applications. Its use has been constrained in a variety of applications due to several of its characteristics, including rigidness and a small thermal distortion temperature. Brittleness was a problem for PLA, but recent advances in the creation of blends, composites, and nanocomposites have solved it and expanded the application possibilities. The replacement of traditional petrochemical-based plastics now used in food packaging with fully bio-based PLA/polyhydroxybutyrate (PHB) blend substances shows very encouraging futures. Due to its nucleation effect, high crystalline PHB can be added to the PLA matrix via melt blending to boost the crystallinity of the PLA. Additionally, this inclusion produced materials with improved mechanical resistance and barrier performance for high potential application in food packaging (Arrieta et al., 2017; Boufarguine et al., 2013; Hassan & Balakrishnan, 2013).

The acid-catalyzed dehydration of glucose produces 5-hydroxy methyl furfural (5-HMF), a precursor for the production of furan-based monomers that could be used as potential medicinal, polymer, and bioplastic substitutes. Another alluring cellulose-based monomer is itaconic acid, an unsaturated dicarboxylic acid. It is produced when fungus digest glucose obtained from cellulose. This cellulose-based monomer produces novel polymers, including polyesters and polyamides based on poly(itaconic acid), which are yet further alternatives to petrochemical products (Patil & Kim, 2016). Polyethylene (PE), polyesters, and PP are examples of conventional polymers sourced from petroleum that are also produced by decomposing cellulose into different monomers. The chemical, physical, and mechanical characteristics of these polyolefins are identical to those of their petrochemical equivalents.

The technological obstacle, however, prevents the mass manufacture of these bioplastics from depolymerization of cellulose are cellulose separation from lignocelluloses. Important cellulose extraction methods include TEMPO oxidation, enzyme hydrolysis, and alkaline-acid hydrolysis. Since the TEMPO catalyst is toxic, cellulose cannot be used for the desired application, and enzyme hydrolysis cannot remove all non-cellulosic contents (Yang et al., 2019).

5.4 BIOPLASTICS MADE OF NANOCELLULOSE AND NANOPOLYMERS

Polymer nanocomposites have piqued the interest of researchers because of their opportunities for low-cost industrial output, as well as their low risks and environmental friendliness. Cellulosic nanomaterials have the potential to be the next generation of sustainable energy reinforcements used in the creation of high-performance bioplastics due to their advantages such as nano scale, high aspect ratios, low densities, low production costs, and superior biodegradability. Furthermore, the related nanocomposites have excellent mechanical properties.

The term "nanocellulose" describes cellulose molecules with at least one nanoscale dimension (1–100 nm). The prime reason for this increase in interest in nanosized materials is the ability to reduce the size of cellulose fibers and produce exceedingly homogeneous materials with superior mechanical qualities. It is regarded as a biodegradable plastic because of its bioavailability. Nanocellulose is classified into three types based on its synthesis method and the circumstances that determine its size, composition, and properties: nanocrystalline cellulose, nanofibrillar cellulose or microfibrillar cellulose, and bacterial cellulose.

5.4.1 Nanocrystalline Cellulose (CNC)

Typically, acid hydrolysis of cellulosic components mixed with water is used to produce cellulose nanocrystals (CNCs). Strong sulfuric acid dissolves the amorphous regions of cellulose but not the crystalline regions. Despite the fact that this method produces a rigid, rod-like CNC with nearly 90% purity, the sulfate groups remain attached to the surface of the fibers as impurities. CNCs typically have dimensions ranging from 200 to 500 nm and from 3 to 35 nm (Lin & Dufresne, 2014).

5.4.2 Nanofibrillar Cellulose (CNF)

A cellulosic source is often subjected to a succession of high-shear mechanical processes in order to create cellulose nanofibers that have both crystalline and amorphous areas. Turbak et al. (1983) reported nanofibrillar cellulose for the first time. Grinding, high-pressure homogenization, and microfluidization are the three most popular high-shear mechanical processes used to transform cellulose fibers into nanoform (Turbak et al., 1983).

5.4.3 Bacterial Cellulose (BC)

Microbial cellulose is another name for bacterial cellulose (BC). It is frequently produced as a separate molecule from bacteria (such as *Acetobacterxylinum*) and does not require further processing to remove contaminants such as lignin, pectin, and hemicellulose. The glucose chains are fed into the bacterial body and expelled through tiny holes in the cell wall during BC biosynthesis. When glucose is combined with the cell wall, BC nanofibers take on the shape of ribbons. This ribbon-like web-shaped structure generates a distinct nanofiber system 20–100 nm long (Raghavendran et al., 2020).

The nanocellulose extracted from various sources is subjected to surface modification using various techniques such as esterification, etherification, oxidation, silylation, and polymer grafting. Even at low concentrations, surface-modified highly crystalline nanocomposites offer superior mechanical, thermal, and barrier capabilities to the untreated polymer matrix when employed to reinforce polymer matrices. Because various facilities for the mass manufacturing of CNCs have been disclosed in recent years, it also offers cheap raw material for eco-friendly packaging components, including food packaging. These bioplastics benefits allow them to compete head to head with popular petroleum-based plastics like PE, PP, polystyrene, and poly(vinyl chloride) (Prakobna et al., 2015).

Bottom-up assembly of nanocellulose can provide a highly porous, mechanically robust substrate with a high specific surface area for a variety of visiting nanomaterials. To incorporate these nanomaterials into the nanocellulose substrate, three different techniques can be used: direct deposition onto the nanocellulose surface, direct addition through nanocellulose dispersion, and synthesis of the guest nanomaterial in nanocellulose-based products such as BC membranes. The resulting nanocellulose-based nanocomposites combine the advantages of the guest nanomaterial and the substrate while also exhibiting synergistic properties. They also have a diverse range of potential applications, including antimicrobial filters, pollution biosensors, and sustainable energy equipment (Yu et al., 2015).

5.5 CELLULOSE DERIVATIVES AS BIOPLASTICS

Numerous semi-synthetic cellulose derivatives primarily use cellulose as their base material. Some cellulose derivatives along with their synthesis, properties, and applications are described as follows.

5.5.1 Cellulose Esters

The three co-esters of cellulose acetate with propionate and butyrate are the most significant cellulose esters. The main significant cellulose ester among them is cellulose acetate. It was applied as a coating to airplane fabric after being first used for photographic film. It is a thermoplastic, meaning that when heated, it will melt and become softer (Gilbert, 2017). Three steps can be thought of in the homogenous acetylation process to prepare the acetate: pretreatment of cellulose, acetylation, and hydrolysis.

Acetic anhydride and glacial acetic acid are combined with the cellulose fibers, and sulfuric acid is used as a catalyst. Triacetate of cellulose is the end product. Water is then added to slightly hydrolyze the triacetate to cellulose acetate in the following step. Figure 5.3 depicts a flowchart of the cellulose esterification process.

Cellulose acetate is a highly transparent, resistant, and malleable material. Chemical resistance to hydrocarbons, vegetable oils, and organic and inorganic acids is high. Plasticizers are frequently added to or mixed with cellulose esters like butyrate acetate and propionate acetate, which have better flexibility, toughness, and moisture resistance, to make them even more flexible.

Textiles (fibers and strings for high-quality fabrics), plastic films such as optoelectronic film for LCD technology and anti-fog eyewear, and consumables such as cellulose-based filters, glass boxes, and labels are just a few examples.

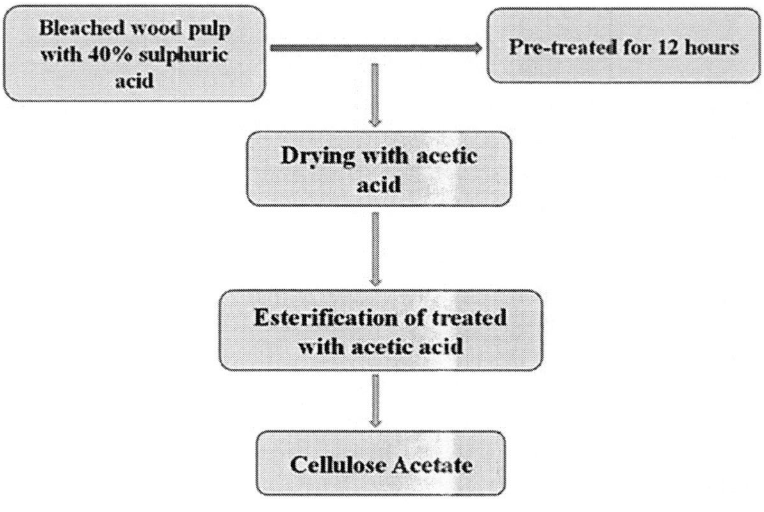

FIGURE 5.3 A flow chart showing the steps involved in esterification of cellulose.

5.5.2 Cellulose Nitrates

The polynitrate ester of the natural polysaccharide cellulose is called cellulose nitrate. The most significant and sole inorganic ester of cellulose that is commercially available is this semi-synthetic polymer, which was created more than 150 years ago. It has been a significant contributor to numerous advancements in the industrial arts and sciences over the years due to its distinctive physical features and affordable cost. The first significant advancement in explosives since the invention of black powder, cellulose nitrate was initially utilized in the production of military explosives, where it earned the nickname "gun cotton." Celluloid, which was created when it was discovered that cellulose nitrate could be stabilized with camphor (in a ratio of 4:1), marked the beginning of the development of engineering plastics (Selwitz, 1988).

The cellulose solution, the acids added to nitric acid to help with the nitration process, and the conditions of the reactions themselves, such as temperature, pressure, or different chemicals added to stabilize or catalyze the reaction, are all important in the synthesis of nitrocellulose. All of these elements have an impact on the nitrocellulose's shape, structure, reactivity, and usefulness both during and after production. The general method of preparation involved dissolving a small amount of fine cotton wool in a solution of nitric and sulfuric acids. After 2 minutes, the cotton was peeled in order to determine the esterification amount and remove all acid residue before being washed in cold water. Then, it was gently dried at a temperature of around 40°C.

Braconnot, a French chemist and pharmacist, created the first nitrocellulose in 1832. By treating cotton or wood pulp (also known as crude cellulose) with strong nitric acid (85%, v/v), he created an ignitable and unstable solid. With a low nitrated cellulose content (a maximum nitrogen level of 4%–5%), this process created the heterogeneous, unstable compound known as xyloidine. The precursor to nitrocellulose was found by Braconnot, but it was Schönbein who developed nitrocellulose as a stable product 14 years later. The reaction between cotton and a solution of nitric acid and sulfuric acid was used to produce it, and the German-Swiss chemist who created it patented it. Currently, this technique is utilized to produce commercial nitrocellulose with a few minor changes. Until Vieille created his renowned "B gunpowder" in 1884, there had been almost 40 years without any significant improvements to the nitrocellulose mixture. It was the first gunpowder without smoke. He figured out how to dissolve nitrocellulose as a colloid in alcohol-ether solutions, making nitrocellulose the primary ingredient in gunpowder (Wisniak, 2011).

In terms of how nitrocellulose is produced, it has already been mentioned that employing concentrated nitric acid alone results in a heterogeneous, volatile, and low-nitrogen nitrocellulose. Prior to the addition of 98% nitric acid, the synthesis process was enhanced by the inclusion of nitric acid vapors as a pretreatment. In this instance, the vapor phase aided in the production of stable homogeneous nitrocellulose with a higher degree of nitration (i.e., nitrogen content of 13.6%). A durable product with a significant level of nitration (maximum nitrogen concentration of 13.9%) was synthesized by the reaction of cellulose with a solution of nitric and sulfuric acid. It has been shown that 1:1 to 1:3 is the ideal ratio for these two acids (nitric acid:sulfuric acid).

Nitrocellulose with a nitrogen concentration of up to 13.7% was created using 1:1 to 1:3 mixtures of nitric and phosphoric acids. This method produced a highly stable nitrocellulose, however phosphoric acid corrodes iron and steel, which is problematic for the production of nitrocellulose. Presently the two most common methods for creating nitrocellulose are the batch-type mechanical dipping technique and the continuous nitration processing method. Both processes rely on the etherification of cellulose's hydroxyl groups with nitric acid in a solution of nitric and sulfuric acids (Torre, 2012).

The nitrogen content of nitrocellulose determines the uses or applications of nitrocellulose. Nitrocellulose with low nitrogen content (12%) is used to make films, inks, lacquers, and paints. However, nitrocellulose with a high level of nitrogen (>12%) is used to make explosives. Heavily nitrated nitrocellulose is a component of several explosives, including dynamite and propellants, with average molar weights ranging from 20 to 250 kDa. Strong explosives known as "dynamites," which are mostly employed for civil purposes, are made up of inert, flammable, and destructive components (Plank, 2004). Explosive components include ammonium nitrate and nitroglycol (such as calcium carbonate). Gunpowders, which are utilized to propel projectiles quickly, are divided into four categories based on their chemical makeup: black gunpowders, homogenous gunpowders, compounded gunpowders, and high-explosive gunpowders. Nitrocellulose serves as a frequent active ingredient in homogeneous gunpowders, sometimes referred to as colloidal or nitrocellulose-based gunpowders. Three different types of homogeneous gunpowders can be distinguished based on the number of active ingredients: (1) single-based gunpowders, which are primarily made of nitrocellulose; (2) double-based gunpowders, which are made of nitrocellulose and one additional explosive (dinitrotoluene, nitroglycerin, or dinitroethyleneglycol); and (3) triple-base gunpowders, which are made of nitrocellulose and two additional explosives (nitroglycerin, dinitroethylene glycol, or dinitrotoluene; López-López et al., 2010).

5.5.3 Cellulose Ethers

The most important cellulose ethers are ethyl cellulose, methyl cellulose, hydroxy ethyl cellulose, and hydroxy methyl cellulose. Ethyl cellulose is produced by first treating the cellulose with a sodium hydroxide solution and then treating it with an alkyl halide. In an alkaline medium, epoxides or chloroacetate react with cellulose to produce hydroxy ethyl and hydroxypropyl cellulose instead of alkyl halide (Miller & Krochta, 1997). Alkyl sulfate can also be used to treat alkali cellulose. For instance, treating methyl sulfate is a typical step in the production of methyl cellulose (Figure 5.4).

The majority of cellulose ethers are water soluble and UV stable (Arca et al., 2018).

They are used as emulsifiers, thickeners, lubricating oils, emulsifying agents, rheology modifiers, and film formers in a variety of industries, including food, medical products, home healthcare goods, hydrocarbon chemicals, building, printing, sealants, battery packs, and fabrics (Taylor et al., 2004). Carboxymethyl cellulose is the most widely used commercial cellulose ether in the world (over half of the total cellulose ether consumption). By volume, methyl cellulose and its derivatives are the second most commonly used materials (Aminabhavi & Patil, 1998; Gardner et al., 2008).

5.5.4 Carboxymethyl Cellulose

A nontoxic, biodegradable, and biocompatible polymer made from cellulose is called carboxymethyl cellulose (CMC). It is sodium carboxymethyl cellulose, commonly known as cellulose gum, sodium cellulose glycolate, or water-soluble cellulose ether. It is employed in a variety of industrial applications including food technology, pharmaceuticals, paper, cosmetics, paints, thickeners, emulsion stabilizers, and water-retention agents because of its colloidal, binding, thickening, absorbing, stabilizing, and film-forming capabilities. Commercially, CMC is made by reacting cellulose with sodium monochloroacetate in a solution comprising alcohol, water, and NaOH that has an excess

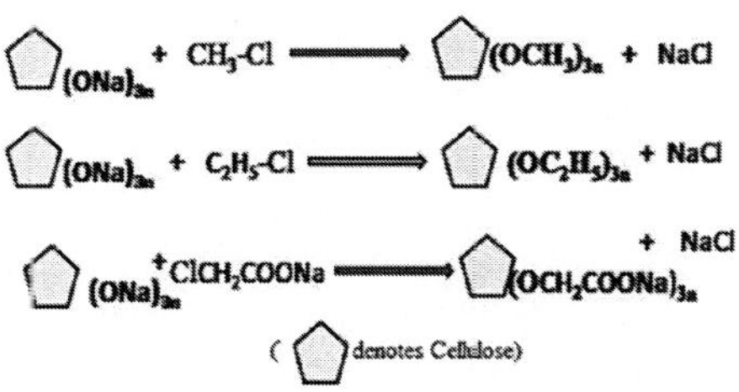

FIGURE 5.4 Different methods of production of cellulose ethers.

of alcohol. In addition to homogeneous carboxymethylation, rotating drum technique, fluidized bed technique, sheet carboxymethylation, Werner-Pfleiderer type mixers, and a solventless approach utilizing a twin screw press and a paddle reactor, there are several ways to make cellulose ethers (Wenliang Zhang, 2022).

It is a heterogeneous response when cellulose is carboxymethylated. The diffusion rates of the chemicals NaOH and $ClCH_2COONa$ within the cellulose particles determine its pace. As a result, the cellulose particle's state of aggregation is crucial. The particles' crystalline structure will be sufficiently damaged if they become looser, which will speed up the reaction. The diffusion of the reagents into the particles is hampered by the high degree of crystalline aggregation present in natural and synthetic celluloses, which is typically over 70%. When doing the carboxymethylation for cellulose, the NaOH and $ClCH_2COONa$ will remain in the solvent rather than diffuse into the crystalline aggregation if the crystalline aggregation could not be properly eliminated (Zhang & Li, 2014).

Carboxymethyl cellulose serves as a gelling agent and viscosity modifier in a variety of industries, including the production of drugs and processed foods. It serves a variety of purposes in different products, including lowering the freezing point of ice cream, lowering the fat content of food items, and encapsulating medications for ingesting and chemical release. Additionally, it is employed in the manufacturing of drilling mud for oil wells and the thickening of textile dyes.

5.6 APPLICATION OF CELLULOSE-DERIVED BIOPLASTICS

5.6.1 Food Packaging

Materials that are used to handle, protect, transport, or temporarily confine objects that are typically thrown away as garbage after use are referred to as packaging. Packaging is an essential step in the food processing business for maintaining food quality and safety, offering food protection, and displaying food. Approximately 65% of the world's bioplastic production in 2018 was accounted for by the bioplastics used in the packaging industry. Compared to cellulose fibers, nanoscale cellulosic fibers have a greater specific surface area, and they are also better able to form hydrogen bonds, which helps them to build a dense network that is tough for molecules to get through. It is advantageous for barrier applications to have this feature since it helps to block the flow of oxygen, which is crucial for the food packaging sector. Bacterial nanofibrils (BNF) are extremely long nanofibers with a diameter of 20–100 nm that are made up of very pure cellulose nanofiber networks created from low-molecular-weight sugars and alcohols by bacteria. It has become clear that CMF, CMC, CNF, CNC, and BNF are essential ingredients in the creation of cellulose-based food packaging (Jiang & Ngai, 2022).

5.6.2 Biomedicine

Assessing a material's biocompatibility and confirming how it interacts with cells are key requirements for biomedical applications, especially when the material must remain in touch with living tissue and must not have any cytotoxic or other adverse consequences. Cellulose has distinct advantages over synthetic biopolymers, including biodegradability, biocompatibility, low production costs, availability, sustainability, non-toxicity, and superior mechanical qualities. Because of their unique ability to be completely resorbed in predetermined time periods ranging from months to a few years, these properties provide potential as bioresorbable polymers that play an increasingly important role in biomedical applications. BC is a biomaterial with enormous potential in dental and oral applications. Recently, cost-effective and user-friendly functional biopolymeric-based materials have been used as a viable tool for developing, repairing, and regenerating functional tissues and organs in the human body. The use of cellulosic composites has been proposed in the development of scaffold constructs that can be implanted in patients to replace failing or malfunctioning organs. Furthermore, the inclusion of the appropriate reinforcement material for tissue-engineered biocomposite scaffolds is a significant factor in improving its characteristics and long-term biocompatibility.

The cellulose derivatives cellulose acetate, hydroxyethyl cellulose, hydroxypropyl cellulose, cellulose sulfate, carboxymethyl cellulose, methyl cellulose, and ethyl cellulose are the most often utilized ones for tissue engineering. Being a non-ionic, water-soluble polymer with a glucose bond, hydroxyethyl cellulose is a good choice for tissue engineering applications. Cell viability is increased and cell proliferation is significantly stimulated by hydroxyethyl cellulose. In addition, hydroxyethyl cellulose greatly increases cell proliferation when present in high concentrations. Nanofibrillar cellulose is an ideal matrix for wound healing due to its high volume–to–surface area ratio, moisture capability, and high porosity. Its structure also allows it to mimic the architecture of extracellular matrix or tissue/organs. Nanofibrillar cellulose hydrogel is a novel material for controlling excessive wound contraction in vivo and in vitro.

CNFs have a significant potential in biomedicine as a carrier for controlled drug delivery due to their suitable flexibility, elasticity, low density, low toxicity, and relatively reactive surface, which can be used for grafting specific groups, in addition to being renewable and cheap. CNFs can stabilize oil-water and air-water interfaces due to their rheological, barrier, and physicochemical properties. Furthermore, because CNFs have a high surface area per unit mass, nanoparticles can be stabilized, increasing the likelihood of positive molecular interactions with poorly water-soluble drugs. A novel material, cellulose acetate phthalate, is the most efficient solution for pH-controlled drug release. One of the most significant functions of cellulose acetate phthalate is microencapsulation in an aqueous or an organic medium. Electrospun cellulose acetate phthalate fibers aid in HIV resistance. These fibers are nontoxic to vaginal epithelial cells and vaginal lactobacilli even after dissolution. These fibers have the potential to be used to stack anti-HIV drugs and to inhibit HIV infection during sexual contact. Hydroxypropyl methyl cellulose is one of the most common hydrophilic biodegradable polymers used in controlled-release formulations and approved by the FDA. Injectable chitosan/glycerophosphate thermosensitive solutions containing vancomycin-loaded hydroxypropyl methyl cellulose microparticles are produced for the local treatment of osteomyelitis. A porous and spongy hydroxypropyl methylcellulose hydrogel structure allows for a long-term release profile in vitro, making it an excellent candidate for use in sustained antibiotic delivery (Seddiqi et al., 2021; Siddiqi et al., 2018).

5.6.3 Construction

Recently, cellulose in various forms is utilized as a significant material cements, paint, laminations, and so forth in the construction field to improve quality. In the construction sector, mainly cellulose in the form of their ether derivatives is used. The main functionalities provided better properties are methyl and carboxymethyl groups. Cellulose materials are incorporated in construction sector to provide viscosity and water retention. The main cellulose additive used to promote water retention in dry-mix binders and grouts is methyl cellulose (MC). Tile adhesives, plastering with a gypsum or lime base, and joint filler for wall boards are a few major uses. The most significant group of thickeners in this market are nonionic cellulosic ethers, which are universal thickeners for the paint industry. Cellulose ethers nowadays play an emerging role as adhesives with adequate viscosity to prevent tile sagging. Other types of cellulose derivatives used in water-borne dispersion paints are methyl hydroxyethyl cellulose (MHEC), methyl hydroxypropyl cellulose (MHPC), and hydroxyethyl cellulose (HEC). They give the paint structural viscosity and the water retention that is crucial for capillary adhesion forces on walls or under layments. Improved spreadability and increased abrasion resistance are further advantages. The incorporation of cellulose in concrete facilitates to the superior aspect ratio, and tightly bound fibers effectively stabilize crack spots, preventing cracks from forming. The ultra-high-performing fracture maintenance of concrete could be enhanced by the addition of cellulose fibers.

5.6.4 Wastewater Remediation

Nanocellulose biosorbants can function as a membrane, filter, or scaffold depending on the pore geometry and flow mode. The effectiveness of NC membranes or filters is impacted by NC's high aspect ratio, which renders it selective in absorbing toxins from dirty water. Natural NC hydrophilicity and its well-managed surface qualities have the capacity to absorb heavy metals and also assist in removing bio-fouling, a significant issue in membrane technology (Voisin et al., 2017). NC can be utilized effectively for removing cadmium, chromium, silver, and various industrial dyes. Chitpong and Husson investigated the removal efficiency of cadmium in traces using ion exchange membranes from wastewater (Chitpong & Husson, 2016). Chitpong and coworkers later designed a CNF-based membrane filter for ultra-filtration of metal ions such as Fe^{2+}/Fe^{3+}, Ag^{+}, Cu^{2+}, and Cd^{2+} (Chitpong & Husson, 2017).

5.6.5 Textiles

There are numerous applications in the field of medical textiles that have been formed as a result of particular qualities such as antistatic behavior, moisture content, low levels of contaminants, high mechanical capabilities, and liquid and cellulose fiber adsorption. One of the most intriguing materials for antibacterial functionalization is cellulose fibers. In fact, it is believed that cellulose fiber surface modification is the greatest method for achieving traditional textile durability for medical usage (Risti et al., 2011). Due to its molecular makeup and substantial active surface area, cellulose fibers can serve as an excellent matrix for the creation of biocompatible, bioactive, and intelligent materials. When cellulose reacts with methylol-5, 5-dimethylhydantoin and hypochlorite produce chloramines that have an antibacterial effect on the fiber's surface. The printing paste can effectively spread spirooxazine- and Tinuvin 144–containing ethanol cellulose nanocomposites. Nanoparticles have the ability to keep their photochromic capabilities while being printed at high temperatures without compromising the physical qualities of paste printing. Additionally, they are utilized in textile sizing, paints, detergents, and paper additives that increase paper strength.

5.7 CONCLUSION

For the manufacture of bioplastics via three main methods, various types of cellulose offer attractive resources. The first method entails breaking down cellulose into monomers that can be polymerized. The second and third methods involve adding different types of cellulose to polymer composites as additives or matrices, such as natural cellulose, cellulose derivatives, and nanocellulose. Utilizing cellulose effectively will restrict the consumption of finite fossil fuels. Our society will benefit from having access to the right polymeric materials thanks to the emphasized production pathways for cellulose-based polymers. This will be useful in creating sustainable polymeric goods to satisfy social, environmental, and economic needs. Cellulosic biopolymers can currently replace petroleum-based polymers; however, there are obstacles that must be overcome in the next years. These difficulties include developing quick, inexpensive, and time-saving procedures for all phases of the manufacturing of cellulosic biopolymers, such as cellulose separation from lignocelluloses, synthesis of suitable forms of cellulose, and polymerization. Solutions to such difficult issues require a variety of methods based on process engineering, bioengineering, and chemistry. Researchers will benefit from this rigorous analysis as they plan, choose, and create diverse cellulose shapes for a variety of polymeric material activities.

REFERENCES

Aminabhavi, T. M., & Patil, G. V. (1998). Designed monomers and polymers a review on the sustained release of cardiovascular drugs through hydroxypropyl methylcellulose and sodium carboxymethylcellulose polymers. *Designed Monomers and Polymers*, *1*(10), 347–372.

Arca, H. C., Mosquera-Giraldo, L. I., Bi, V., Xu, D., Taylor, L. S., & Edgar, K. J. (2018). Pharmaceutical applications of cellulose ethers and cellulose ether esters. *Biomacromolecules*, *9*(6), 1–81. https://doi.org/10.1021/acs.biomac.8b00517

Arrieta, M. P., Dolores Samper, M., Aldas, M., & López, J. (2017). On the use of PLA-PHB blends for sustainable food packaging applications. *Materials*, *1008*(10), 1–26. https://doi.org/10.3390/ma10091008

Boufarguine, M., Guinault, A., Miquelard-Garnier, G., & Sollogoub, C. (2013). PLA/PHBV films with improved mechanical and gas barrier properties. *Macromolecular Materials and Engineering*, *298*(10), 1065–1073.

Chitpong, N., & Husson, S. M. (2016). Nanofiber ion-exchange membranes for the rapid uptake and recovery of heavy metals from water. *Membranes*, *6*(4), 59–75. https://doi.org/10.3390/membranes6040059

Chitpong, N., & Husson, S. M. (2017). Polyacid functionalized cellulose nanofiber membranes for removal of heavy metals from impaired waters. *Journal of Membrane Science*, *523*, 418–429. https://doi.org/10.1016/j.memsci.2016.10.020

Gardner, D. J., Oporto, G. S., & Mills, R. (2008). Adhesion and surface issues in cellulose and nanocellulose. *Journal of Adhesion Science and Technology*, *22*(11), 545–567. https://doi.org/10.1163/156856108X295509

Gilbert, M. (2017). Chapter 22—cellulose plastics. In *Brydson's plastics materials*. Elsevier Ltd. https://doi.org/10.1016/B978-0-323-35824-8.00022-0

Hassan, A., & Balakrishnan, H. (2013). Polylactic acid based blends, composites and nanocomposites. *Advances in Natural Polymers*, *6*(9), 361–396. https://doi.org/10.1007/978-3-642-20940-6

Jiang, Z., & Ngai, T. (2022). Recent advances in chemically modified cellulose and its derivatives for food packaging applications: A review. *Polymers*, *14*(10), 1533.

Karam, A., Amaniampong, P. N., Fernández, J. M. G., Oldani, C., Marinkovic, S., Estrine, B., Vigier, K. D. O., & Jérôme, F. (2018). Mechanocatalytic depolymerization of cellulose with perfluorinated sulfonic acid ionomers. *Frontiers in Chemistry*, *6*, 1–9. https://doi.org/10.3389/fchem.2018.00074

Lin, N., & Dufresne, A. (2014). Nanocellulose in biomedicine: Current status and future prospect. *European Polymer Journal*, *59*, 302–325. https://doi.org/10.1016/j.eurpolymj.2014.07.025

López-López, M., Ángeles, M., De, F., Sáiz, J., Luis, J., Vega, A., Torre, M., & Garcia-Ruiz, C. (2010). New protocol for the isolation of nitrocellulose from gunpowders: Utility in their identification. *Talanta*, *81*(3), 1742–1749. https://doi.org/10.1016/j.talanta.2010.03.033

Miller, K. S., & Krochta, J. M. (1997). Oxygen and aroma barrier properties of edible films: A review. *Trends in Food Science and Technology*, *81*, 1–10.

Nechyporchuk, O., Belgacem, M. N., & Pignon, F. (2016). Current progress in rheology of cellulose nanofibril suspensions. *Biomacromolecules*, *17*(7), 2311–2320. https://doi.org/10.1021/acs.biomac.6b00668

Patil, M. P., & Kim, G. (2016). Eco-friendly approach for nanoparticles synthesis and mechanism behind antibacterial activity of silver and anticancer activity of gold nanoparticles. *Applied Microbiology and Biotechnology*, *5*(11), 1–14. https://doi.org/10.1007/s00253-016-8012-8

Plank, J. (2004). Applications of biopolymers and other biotechnological products in building materials. *Applied Microbiology and Biotechnology*, *66*(7), 1–9.

Prakobna, K., Galland, S., & Berglund, L. A. (2015). High-performance and moisture-stable cellulose-starch nanocomposites based on bioinspired core-shell nano fibers. *Biomacromolecules*, *16*(3), 904–912. https://doi.org/10.1021/bm5018194

Raghavendran, V., Asare, E., & Roy, I. (2020). Bacterial cellulose: Biosynthesis, production, and applications. In *Advances in microbial physiology* (Vol. 77). Elsevier Ltd. https://doi.org/10.1016/bs.ampbs.2020.07.002

Rani, M. S. A., Rudhziah, S., Ahmad, A., & Sabirin Mohamed, M. (2014). Biopolymer electrolyte based on derivatives of cellulose from Kenaf Bast fiber. *Polymers*, *6*(9), 2371–2385.

Rinaldi, R., & Sch, F. (2009). Design of solid catalysts for the conversion of biomass. *Energy & Environmental Science*, *2*(5), 610–626. https://doi.org/10.1039/b902668a

Risti, T., Fras, L., Novak, M., Kralj, M., Sonjak, S., Cimerman, G., & Strnad, S. (2011). Antimicrobial efficiency of functionalized cellulose fibres as potential medical textiles. In *Science against microbial pathogens: communicating current research and technological advances*, A. Méndez-Vilas (Ed.), pp. 36–51.

Rose, M., & Palkovits, R. (2011). Cellulose-based sustainable polymers: State of the art and future trends. *Macromolecular Rapid Communications*, *32*(5), 1299–1311. https://doi.org/10.1002/marc.201100230

Seddiqi, H., Oliaei, E., Honarkar, H., Jin, J., Geonzon, L. C., Bacabac, R. G., & Klein-Nulend, J. (2021). Cellulose and its derivatives: Towards biomedical applications. *Cellulose*, *28*(5), 1891–1931.

Selwitz, C. (1988). *Cellulose nitrate in conservation*. Getty Publications, 1–78.

Siddiqi, K. S., Ur Rahman, A., Tajuddin, & Husen, A. (2018). Properties of zinc oxide nanoparticles and their activity against microbes. *Nanoscale Research Letters*, *13*, 141. https://doi.org/10.1186/s11671-018-2532-3

Taylor, P., Cha, D. S. U., & Chinnan, M. S. (2004). Biopolymer-based antimicrobial packaging: A review biopolymer-based antimicrobial packaging: A review. *Critical Reviews in Food Science and Nutrition*, *44*(1), 223–237.

Torre, M. (2012). Nitrocellulose in propellants: Characteristics and thermal properties. *Advances in Materials Science Research*, *7*, 1–21.

Turbak, A. F., Snyder, F. W., & Sandberg, K. R. (1983). Microfibrillated cellulose, a cellulose product: Properties, uses and commercial potential. *Journal of Applied Polymer Science: Applied Polymer Symposium*, *37*(6), 815–827.

Voisin, H., Bergström, L., Liu, P., & Mathew, A. P. (2017). Nanocellulose-based materials for water purification. *Nanomaterials*, *7*, 1–18. https://doi.org/10.3390/nano7030057

Wenliang Zhang, L. (2022). Synthesis and applications of carboxymethyl cellulose hydrogels. *Gels*, *8*(6), 529–539.

Wisniak, J. (2011). The development of dynamite. From Braconnot to Nobel. *Educación Química*, *19*(1), 7[illegible]. https://doi.org/10.22201/fq.18708404e.2008.1.25765

Yang, X., Reid, M. S., Olsén, P., & Berglund, L. A. (2019). Eco-friendly cellulose nanofibrils designed by nature: Effects from preserving native state. *ACS Nano*, *14*(12), 724–735.

Yu, H. Y., Chen, G. Y., Wang, Y. B., & Yao, J. M. (2015). A facile one-pot route for preparing cellulose nanocrystal/zinc oxide nanohybrids with high antibacterial and photocatalytic activity. *Cellulose*, *22*(1), 261–273. https://doi.org/10.1007/s10570-014-0491-0

Zhang, Y., & Li, L. (2014). Reaction mechanism of carboxymethyl starch-based wood adhesive. *Computer Modelling and Technologies*, *18*(11), 1150–1155.

6 Second- and Third-Generation Sources for Bioplastics

Production and Innovations in Applications

Jeyaprakash Dharmaraja, Retnam Krishna Priya, Sutha Shobana, Sundaram Arvindnarayan, L. Bartolucci, E. De Maina, P. Mele, V. Mulone, and Gopalakrishnan Kumar

CONTENTS

DOI: 10.1201/9781003344018-6

6.1 INTRODUCTION

In almost all the countries, enormous plastic wastes are generated from their essential products, such as PE (polyethylene), PET (polyethylene terephthalate), PVC (polyvinyl chloride), PS (polystyrene), PP (polypropylene), PBS (polybutylene succinate), PBA (polybutylene adipate), and PCL (polycaprolactone). For instance, in 2019 the universal plastics production was 368 million tonnes; in 2008 it was merely 245 million tonnes. Recently, the estimation of worldwide plastics utilization and production is about1.6 million tonnes per day, increased due to the COVID-19 pandemic outbreak via medical equipment use (masks, needles, gloves, etc.; Benson et al., 2021). Although the worldwide manufacturing of plastics somehow reduced in 2020, the demand has been expanded by 40%, along with a medical and personal protective equipment waste generation increase by about 370% (Gorrasi et al., 2021; Tofa et al., 2019; Chia et al., 2021; The Guardian, 2020; Mu et al., 2022; Razeghi et al., 2021; Nanda and Berruti, 2021a, 2021c; Othman et al., 2021). Ecologically alternative procedures of reusing wastes derived from the plastic materials could be analyzed in place of their landfill and ocean dumping (Fawzy et al., 2020; Suresh et al., 2021; Nanda et al., 2021; Nanda and Berruti, 2021b). The continuous growth of production and utilization of petrochemically derived oil-based plastic products, coupled with their cost, durability, strength-to-weight ratio, their contribution to the ease of everyday life, and their recalcitrant nature, have led to a considerable amplification of plastic wastes in the form of municipal solid wastes.

At present, bioplastics play a crucial role, being solicited in a widespread range of products, including trade, packing, and domestic items. They have made life easier and comfortable with less cost. The growths of population, fiscal development, requirement for supplies, and variations in lifestyle have remarkably been increasing the stipulation and fabrication of products that require plastics. Second-generation bioplastic products (2G) are derived from lignocelluloses, being both technically and economically attractive due to their special features (e.g., light weight, firmness, suppleness, cost-effectiveness, ease of obtainability, large production). Synthetic plastics have numerous industrial and household applications. In the packing industry, apart from their own positive characteristics, alternative materials to substitute their use are hard to pursue, as their residual constituents mixture both macro and micro non-biodegradable molecular contaminants (Wang et al., 2019). Third-generation bioplastics (3G) are derived from alternative carbon sources that are not the lignocellulosic residues from 2G, such as algae oils, and have been applied in a wide range of products in both the medical and biological fields (Akhtartavan et al., 2019). The devices used for drug delivery systems and tissue engineering for biomedical purposes have been successfully made by such biodegradable plastics (Daniel et al., 2020). The algae-derived biofilm membranes present pores ranging from 6 to 30 cm diameter, and the nanocellulosic materials with their composites have effectively been used in bioplastics research. Recently, innovative developments are being investigated, which mainly includes 3D printing (Filiz et al., 2021; Mahendiran et al., 2021). On account of their bio-compatibility, algae-derived PHAs (polyhydroxyalkanoates) are utilized for medical applications, akin to post-surgical ulcer therapy, wound-healing dressings, cancer detection, cardiac valves, synthetic blood arteries, bone-tissue engineering, and so forth (Iolanda et al., 2021; Behera et al., 2022; Wu et al., 2022). The need to find a substitute for sustainable alternative for oil-based plastics requires well-developed physicochemical, biological, and biodegradable polymers, which have made them also attractive in biomedical applications. Moreover, small and large plastics particles could function as the supports to transmit pathogenic microbes, consequently enhancing diseases for living beings together with the airborne and waterborne ones. In this chapter, the sources, production process, characteristics, and innovations of bioplastics from both 2G and 3G biomass materials are briefly highlighted, with the addition of environmental impacts, novel applications, economy, and productivity.

6.2 BIOPLASTICS

In Europe, in order to compensate 60% of the growing overall plastic demand, the production of bioplastics products from renewable resources has attracted much attention. They are classified into two categories: biodegradable and non-biodegradable ones. Unlike oil-based plastics, the biodegradable bioplastics like PLAs (polylactic acids), PHAs, PET, Bio-PE (polyethylene), and Bio-PBS (polybutylene succinate) can be recycled/incinerated (Mekonnen et al., 2013; Ranganadhareddy et al., 2022). Bioplastics can be considered as feasibly environmentally friendly and economic solutions to solve many ecological issues created by petrochemically derived fibers, fulfilling goals of sustainable development in the spheres of health care, industrialization, safety, consumption, production patterns, and climate-associated hazards. The various ecological issues caused by conventional plastics can be reduced with the use of bio-based fibers, as they represent less pollution through their biodegradability, less consumption of finite resources through their generation from bio-renewable substrates, and correct disposal of agroindustrial residues and wastes through their use during the production process. The environmental sustainability promotes better utilization and restores terrestrial ecosystems, managing forests, combating desertification and reducing both land degradation and biodiversity loss.

6.2.1 Sources of Bioplastics

Some important sources of bioplastics are 2G biomass, which is mainly composed of cellulose, hemicellulose, and lignin, as well as 3G derived celluloses, starch, proteins, and lipids. Such source-based bioplastics are interesting due to their biodegradable, high durability, and strong and tough nature (Ranganadhareddy et al., 2022; Qasim et al., 2021). In this section, the characteristics of bioplastics based on celluloses, starches, proteins and lipids will be discussed.

6.2.1.1 Cellulose-Based Bioplastics

Cellulose is a natural polysaccharide that can be derived from agricultural crops, forestry, algae, and even bacterial biomass materials (Nanda et al., 2013, 2014). The bioplastics made by cellulose present low density and low cost (Polman et al., 2021; Ranganadhareddy et al., 2022). The existence of weak hydrogen bonds in cellulosic bioplastics decreases their mechanical and flexibility strength. In addition, the current research suggests combining the celluloses with other polysaccharides like pectin and chitosan for improving their characteristics, predominantly over stability, flexibility, and transparency (Yaradoddi et al., 2020). As well, the celluloses are water sensitive and present weak interfacial linkages and thermal stability, consequently making them less well attractive, although this can be overcome by means of some pretreatments and chemical functionalization of celluloses (Polman et al., 2021).

6.2.1.2 Starch-Based Biopolymers

Currently, starch-based biopolymers are considered among the most promising bioplastics due to their renewable and biodegradable characteristics, availability, cost-effectiveness, and the possibility of making edible films. When compared to petrochemically derived plastics, such kinds seem to be analogous in certain properties, mainly mechanical and transparency (Shahabi- Ghahfarrokhi et al., 2019). The most important resources of starch are frequently food crops like cassava, corn, rice, potatoes, and wheat, with a highlight for corn crops with 80% of the global starch production (Diyana et al., 2021). Thereby, the extracted starch can suitably be utilized to produce biochemicals, biofuels and some other value added biomaterials (Nanda et al., 2015). Under extreme conditions like extrusion, pressure, and temperature, starch-based bioplastics result in the formation of odorant derivatives, which should be analyzed and significantly removed. Some of the unpleasant odorant compounds produced while composing food packaging biofilms are eugenols, 2-nonenals, trimethylamines, and vinyl amyl ketones. Recently, starch-based biopolymer films have widely been blended to algae-derived and vegetable-derived essential oils like cinnamon, clove, ginger, and peppermint, not only to avoid such kind of issues but also to improve their antimicrobial characteristics.

6.2.1.3 Protein- and Lipid-Based Biopolymers

Proteins are hetero-polymers of amino acid monomeric units connected through peptide linkages. One of the advantages of protein-based biofilms are the better mechanical properties when compared to both the cellulose and starch derived ones. The micro and nano structure of the protein is naturally fibrous and is arranged via covalent linkages, hydrogen bonds, and ionic forces of attraction (Mohamed et al., 2020). Generally, lipids extracted from plants, insects, and animals essentially include fatty acids, fatty alcohols, glycerides, phospholipids, and terpenes. Due to the hydrophobic and non-polar nature of lipids, they can delay the moisture crusade within food particles inside the packaging, further enhancing the same with antimicrobial properties due to the existing vital terpenoid-like oils.

6.3 PRODUCTION AND CHARACTERISTICS OF BIOPLASTIC PIONEERS

The PHAs (polyhydroxyalkanoates), among them their most frequent biopolymer PHBs (polyhydroxybutyrate), can be produced from many bioresources with the use of specific microorganisms. Examples of their production include the use of sugarcane molasses as substrate for *Bacillus cereus* and *Staphylococcus epidermidis*, agroindustrial residues for *Bacillus* sp., date syrup for *Pseudodonghicolaxiamenensis*, oil palm leaf non-food sugars with biodiesel industry by-products and spent cooking oil for *Cupriavidusnecator*, wheat straw lignocellulosic hydrolysates for *Burkholderiasacchari*, wheat bran hydrolysate for *Ralstoniaeutropha*, and bakery wastehydrolysate for *Halomonasboliviensis*. In a secondary process, algae oil produced by the genera *Chlorella*, *Spirulina*, and others can also be used as raw material for bioplastic production (Sabathiniet al., 2018). The PHAs can be procured via three significant pathways (Figure6.1). Other examples of bioplastics include PLAs, polyamide 11/nylon 11 (PAMs 11), and polyhydroxyurethanes (PHUs), which will briefly be discussed together with PHAs in this section.

6.3.1 Polyhydroxyalkanoates (PHAs)

Large-scale biocompatible and biodegradable PHA production can be achieved by using, for example, economical cellulosic carbon substrates, wastewater sources, and activated sludge (Rodriguez-Perez et al., 2018). Biomass-derived thermoplastic-polyester composed of PHA monomeric [(R)-3-hydroxyalkanoic acids] units can biologically be produced by different gram-positive and gram-negative bacteria (Keshavarz and Roy, 2010). Different alkyl (R–: methyl to tridecyl) groups are composed of hydroxyalkanoate units (Mannina et al., 2020; Yu et al., 2020). PHA can be produced by a large variety of microorganisms and carbon sources, as shown in Table 6.1, as long as the substrates present suitable contents of nitrogen, oxygen, phosphorous, microelements, and a proper pH (Kumar et al., 2020). It can also be fabricated using lipids and complex sugars as a form of storing up energy in the form of carbon contents (Tarrahi et al., 2020). They may suitably be synthesized by certain standard methods like melting processes. PHAs can generate bio-based materials with superior mechanical properties, depending on the chemical composition of their monomeric units and the variance of surface composition. For instance, PHA-derived bioplastics are more stable to UV rays and organic and inorganic solvents than the PLA-derived ones (Tarrahi et al., 2020). The halogenated solvents, such as dichloromethane, dichloroethane, and chloroform, can be used as solvent for PHAs compounds. Compared to conventional petrochemical plastics, these bioplastics have been drawn much attention owing to their biodegradable and non-toxic nature (Ray and Kalia, 2017; Koller et al., 2017). PHA-derived bioplastics have been recognized for their vibrant applications in industrial and bio-medicinal fields, mainly for making biodegradable antibacterial and anticancer implants, biocontrol agents, drug carriers, memory enhancers, and engineering tissue (Ray and Kalia, 2017; Choi et al. 2020).

6.3.2 Poly-3-Hydroxybutyrate (PHBs)

PHBs are among the most known PHAs, presenting high biodegradability and biocompatibility. These bioplastics pioneers can also be obtained through microbial fermentation of renewable carbon sources as shown in Table 6.1 (Mlalila et al., 2018). Their synthetic path includes condensation process of acetoacetyl-coenzyme A in the presence of an enzymatic catalystβ-ketothiolase to generate acetoacetyl-CoA. On reduction, 3-hydroxybutyryl-CoA (Figure 6.1) is consequently obtained. Last, catalysis by PHB synthase of previously formed coenzyme namely, 3-hydroxybutyryl-CoA yields PHBs (Sirohi et al., 2021a, 2021b). The final yield of the fermentation process can be enhanced by different bacterial cultures, and recent research indicates the possibility of utilizing waste biomass materials to produce PHBs as an economic alternative. They are able to resist UV rays, their glass transition temperature is about 2°C with 60% crystallinity, and when their tensile strength is 43 MPa they present a high melting temperature (177°C; Sirohi et al., 2021a, 2021b). They resist UV rays, and their glass transition temperature is about 2°C with 60% crystallinity. The main reason for low global production is associated to the high poly-3-hydroxybutyrate production cost. Its production path has acquired commercial importance through the use of actinomycetes, algae, archaea, bacteria, and marine prokaryotes, since their fermentation is cost-effective and their product is easy to recover. Among them, marine microbial species usually reserve energy in the form of intracellular poly-β-hydroxybutyrate granules. During the production, cellular freezing and dehydration are common downstream processes (Ranganadhareddy et al., 2022). Furthermore, a high salt concentration decreases the possible contamination by other microorganisms.

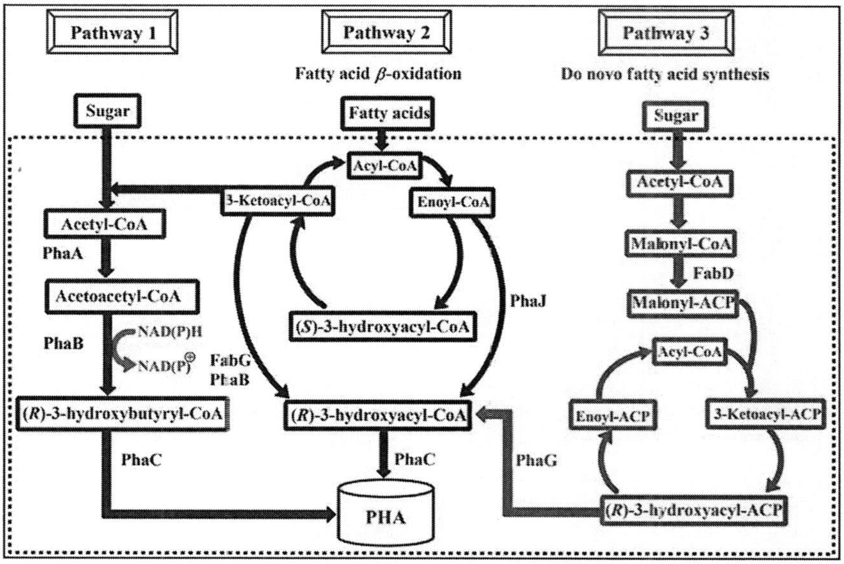

FIGURE 6.1 Three metabolic pathways for PHA production.

Abbreviations: FabG = 3-ketoacyl acyl carrier protein (ACP) reductase; PhaA= β-ketothiolase; PhaB= acetoacetyl coenzyme A(CoA) reductase; PhaC= PHA synthase; PhaG= acyl-ACP-CoA trans acylase; PhaJ= enoyl-C ketoacylacyl carrier protein (ACP) reductase.

Source: Modified from Coppola et al. (2021).

TABLE 6.1
Summary of Algae-Based 3G Bioplastic Production

Biomass Species	Product	Process Description
Spirulinaplatensis	*S. platensis*–PVA blend film	6.0% MAH and 30.0% glycerol; tensile strength at 27.7–28.3 kgf/cm² and 59.2%–66.0% elongation
	S. platensis–wheat gluten blend thermoplastic	30.0% microalgae: wheat gluten, compression mold at 120°C, 40 bar, 10 min, glycerol or 1,4-butanediol as plasticizers
Spirulina sp.	*Spirulina*–polybutylene succinate (PBS) composites	Melt blending at 130°C, 6 min, with 6% MAH-grafted PBS and 15%–50% *Spirulina* biomass
Chlorella	*Chlorella*–PVA blend films	Ultrasonic *Chlorella* treatment, solvent casting at 80°C, with glycerol and citric acid; 15.3 kgf/cm² tensile strength and 99.6% elongation
Chlorella vulgaris	*Chlorella*–PVA blend film	Compression mold at 120°C between *Chlorella*: glycerol and MAH-grafted PVA; 42.3 kgf/cm² tensile strength and 13.0% elongation
Scenedesmus sp.	PHB	Glucose, N, P, Fe, and salinity concentration optimized by Taguchi design; 0.8%–29.9% (w:w) dry weight final accumulation
Chlorogloeafritschii	PHB	5.0% (v:v) inoculum in 150 mLBG–11 medium, pH 7.5, 32°C, 100 μmol/m²/s light, CO_2 supply at 160 rpm; 51% (w:w) substrate conversion yield
Botryococcusbraunii	PHB	60% of sewage wastewater at pH 7.5, 40°C; 247 mg/L of PHB
Spirulina sp. LEB 18	PHB	Adapted culture with 8.4 g/LNaHCO³, 0.05 g/L NaNO3, and 0.5 g/LKH$_2$PO$_4$; 30.7% (w:w) dry weight
Chlorella sp. and *Scenedesmus* sp.	PHB, lipids, and bioethanol	Hybrid pretreatment and fermentation with wastewater treatment inoculum; 0.4 g PHB/g dry cell weight; 76.2% sugar utilization
Chlorella pyrenoidosa	PHB and biodiesel	27% (w:w) dry weight after 14 days under optimized conditions

Source: Modified from Vandenberghe et al. (2021)].

6.3.3 Polylactic Acid (PLAs)

PLAs are other renewable, biocompatible, biodegradable, and thermoplastic polymers formed by lactides or lactic acid monomeric units, which can be obtained from renewable agricultural materials as well. The PLAs obtained from renewable agricultural materials presents some advantages, such as less energy-intensive production, low carbon footprint, easy lactic acid recycling via hydrolytic process, lactide monomers production from lactic acid, biodegradable characteristics as packaging material, reduction of landfills burden, and release of CO_2 in biodegradation. Those advantages improve agricultural-based economy and environmentally friendly approaches (Jamshidian et al., 2010). When compared to petrochemically derived plastics such as PET and PS, they present non-toxicity and better mechanical strength (Karamanlioglu et al., 2017). In addition to a low cost and facilitated production, there are no toxic intermediates or by-products and their final polymer shows enhanced thermal plasticity, elasticity, and resistance toward aroma and flavor adhesion (Jamshidian et al., 2010). Recently, the research has been turned on inferior thermal and mechanical assisted stereo-multifaceted complex PLAs (i.e., Neo-PLAs). Beyond, the stereo form of PLAs is composed of poly(L-lactides), and their enantiomeric poly(D-lactides) form those display 230°C as high melting temperature (Tutoni and Becker, 2020).

6.3.4 Polyamide 11/Nylon 11 (PAM 11)

PAM 11 bioplastics are classified under the family of nylon and are originated by the polymerization of 11-aminoundecanoic acid. Their major precursors are vegetable oils such as castor oil, a bio-based renewable resource. This kind of bioplastic possesses high chemical, mechanical, and thermal stability, wherein the lignocellulosic fibers reinforce their polymer, thereby making an extremely attractive alternative from the industrial and eco-friendly point of view (Dechet et al., 2019). In recent decades, PAM 11 impacts a special attention due to its ductile strength and low ecological impact, as well as its less energy-intensive application and lower carbon footprint.

6.3.5 Polyhydroxyurethanes (PHUs)

PHUs are favorable sustainable materials for the production of urethanes and polyurethanes (PUs), as there is a growing concern in order to restrict the use of toxic isocyanates. Recent studies reveal that the utilization of CO_2 for the production of PHUs effectively demonstrates an alternative to reduce isocyanate dependency, as the production path of PHUs also eliminates CO_2 (Nair et al., 2019). The hybrid PHUs can effectively replace PUs, as the former area result of cyclic carbonates and amines polymerization and, consequently, avoid the utilization of isocyanates and its environmental pollution (Cornille et al., 2017). Therefore PHUs have been received as promising compounds in the bioplastics research. Synthetically, production pathways of PHUs mainly involve different routes: (1) aziridine co-polymerization with CO_2, (2) polyaddition of bicyclic carbonate to a diamine, and (3) bicarbonate trans-urethanization with a diol. The hybrid non-isocyanate PHUs can be produced from the co-polymerization of PHUs with different kinds of polymers like epoxies/acrylates in amines and cyclic carbonates (Suryawanshi et al., 2019; Ecochard and Caillol, 2020) via polyaddition, ring-opening polymerization, trans-urethanization, and rearrangement and are the common mechanistic path routes (Suryawanshi et al., 2019). The sustainable, bio-based, iso-cyanate–freed PHUs (Miao et al., 2014) can specifically be produced from vegetable oils like linseed oil, castor oil, cashew nut shell oil, and others. The applications and physical properties of the polymer can be modified according to their linkages. For instance, the polyols could be blended to some organic fillers like celluloses for increasing their stiffness (Suryawanshi et al., 2019), the hydroxyl groups could be increased to higher densities for improving mechanical properties and adhesive applications (Cornille et al., 2017), and the polymer could be blended with acrylates and methacrylates for coating applications. Additionally, the high thermal stability of PHUs can also indicate applications as thermo-setting and thermo-plastics, and their rigidity and flexibility make them suitable for manufacturing foams, latex, or hydro-gels (Ecochard and Caillol, 2020; Mahendiran et al., 2021).

6.4 INNOVATIONS IN BIOPLASTICS PRODUCTION SOURCES

Bioplastics represent one of the best solutions to the extreme production of hydrocarbon-based petrochemically derived plastics, since their mode of preparation and application cause lesser impacts to the environment. In this section, innovations in bioplastics carbon sources or raw materials are discussed as examples of environmental friendly strategies.

6.4.1 Bioplastics Made from Thymidine and CO_2

The Centre of Sustainable Chemical Technologies (CSCT) is developing another biodegradable form of bioplastics with the use of thymidine sugar and carbon dioxide at low pressures and room temperature (Gregory et al., 2016).

6.4.2 Bioplastics Made from Corn and Shellfish

The University of Otago developed food packaging bioplastics from by-products derived from corn and shellfish industries. This kind of research can eventually reduce oil dependence for the production of biodegradable plastics, thereby reducing waste-dumping into the environment.

6.4.3 Bioplastics Made from Banana Peels

A Turkish teenager named Elif Bilgin won the Science in Action Award for developing a bioplastic production process using banana peels. She suggested a chemical process for conversion of the banana peel into resistant bioplastics. This idea could offer a solution for reducing both waste materials and oil dependence for production of plastics.

6.4.4 Avocado-Based Biodegradable Straws

There is a recent research that uses avocado to produce straws used for drinking beverages. They are made from avocado pits, present natural color, and are one of the most sustainable and eco-friendly straws available today. The product is named "Hardcore Avocado Straws," and according to the advertising, they do not crack, warp, or dissolve in a beverage, and are totally biodegradable and compostable. Unlike plastic straws, these bio-based straws are derived from plants and contain no chemicals or toxic materials, being suitable in terms of durability and strength.

6.4.5 Bioplastics Made from Shrimp

Shrimps are classified as arthropods, living animals that possess a hard exoskeleton made primarily of chitin biopolymer, and their shells are usually disposed of. However, waste shells can be utilized to make bioplastics. Angeline Arora from the Sydney Girls High School has devised a grinding technique for the shells to separate chitin in its purest form. This technique involves a process of soaking the shells in acids and then submitting them to alkaline treatment to remove the recalcitrant contaminants. This content can be sequentially mixed with fibroin, a protein from spider silk, and pressed into thin sheets of bioplastics.

6.4.6 Bioplastics Made from Seaweed

Currently, most bioplastics are derived from corn, sugarcane, vegetable oils, or starchy crops, many of them being edible crops. However, non-food substitutes for producing bioplastics can also be utilized. Seaweed is an interesting non-food alternative, as it is plentiful, requires low space for cultivation, and does not require fertilizers (Behera et al., 2022).

6.5 NOVEL APPLICATIONS OF BIOPLASTICS

Bioplastics can significantly reduce the current necessity on fossil resources, as fossil fuel use dangerously depletes the atmosphere. Therefore there is a current responsibility to develop emerging techniques for the production and application of bioplastics. In general, bioplastics are formed from renewable resources such as corn oil, sugar beet, sugarcane bagasse, grass plants, and algae. An entire variety of routine products is urgently being mass-produced with bioplastics, such as packaging and food-service products, biomedical applications, and even consumer electronics and automotive components. Bio-based polymers have been at the vanguard of investigation for biomedical applications over the last 50 years because of their biocompatibility and biodegradability (Prasanth et al., 2021; Ranganadhareddy et al., 2022). It is worth stating that bioplastics have been tested in a large variety of applications in the fields of packaging, textiles, consumer goods, agriculture, horticulture, and motorized and other value-added industries. However, as yet they are not so frequent and are less useful in coatings, electronics industry, adhesives, and civil engineering sectors. Around2020, their major application was in packaging industries, representing 47% or roughly 1 million tons of total bioplastics produced.

6.5.1 Applications of PHA-Derived Bioplastics

Because of the biodegradable nature of PHA-derived bioplastics, they play significant role in recent medications (Ulery et al., 2011). Modifications of the polymer can also contribute to medical and healing applications (Ray and Kalia, 2017). For instance, PHAs can be used as a raw material for producing nano-particles to target the repaired tissues in tissue engineering and to create tablet shells very effectively (Nigmatullin et al., 2015). However, the processing of the polymer and the device design are equally critical for successful tissue engineering applications (Wu et al., 2022). The synthesis of PHAs using microalgae is occurring during cultivation of microbes in nutrient-deficient circumstances (Ranganadhareddy et al., 2022). The metabolic route involves a wide spectrum of bacterial species to synthesize PHAs as well as fatty acids, and is mainly linked to a prevalent precursor acetyl-CoA. Since PHAs are permeable to oxygen and consist of moisture content, possessing mechanical efficiency, they have been utilized effectively for food packaging; also they provide both water and oxygen shielding (Ranganadhareddy et al., 2022).

6.5.2 Applications of PHB-Derived Bioplastics

In the medical field, PHBs-derived bioplastics can be used in the production of medical implants, cardiac valves, bandage materials, and medicinal encapsulation. In the agricultural industry, it can be used for manure encapsulation and for slow release of fertilizers. In the food industry, it can be used for food packing, encapsulating food supplements, and health care foods. For instance, the eugenol-incorporated PHBs present antifungal characteristics, utilized in industries as food wrappers. Generally, it involves the production of expendable and trivial items consumed day today (Ranganadha et al., 2020). Abiotic stress assisted by *Synechocystis* sp. has shown an increase of nearly 81% of PHBs. They could be utilized as an effectual antimicrobial agent for aquaculture. Moreover, they prevent pathogen contamination in biofilms, thereby inhibiting infection. They possess antifouling characteristics so that they can be blended to other metals and are applicable to ship hulls for the protection from unwanted settling of marine creatures.

6.5.3 Applications of PLA-Derived Bioplastics

Similar to PHAs, PLAs can also be used as bio-based green wrappers for renewed nutrition food materials and for soil administration and production of crops. However, because of their hardness and resistance, they can be used in major segments like building construction, transportation, furnishing equipment, electronic machines, cloths, and fiber strands. In addition, the aesthetic demands for PLAs composites are driven by 3D printing, conveyance, and fabric industries, gaining attention due to their properties in multifaceted uses (Dang et al., 2022). PLAs can effectively extend the durability of microalgae-derived plastics, and thereby can improve the mechanical characteristics. The two main microalgae species, *Chlorella* and *Spirulina*, are essentially suitable for the production of bioplastics, owing to a very high composition of proteins (51%–58% and 46%–63%, respectively; Chong et al., 2022).

6.5.4 Applications of PAM 11–Derived Bioplastics

As analogized to other polyamides, this kind of PAM 11 has a high melting point of about 200°C, which indicates an active reinforced material for naturally occurring fibers (Oliver-Ortega, 2019).

It is hardly perishable and can be reprocessed alongside with its composites, making it pertinent for industrial yields with maximum durability. For instance, PAM 11 is used to make pipelines for biogas fields, tubes for water flow, electrical wires and clips, metal covertures, and astronautics and motor vehicle mechanical pieces. In daily life, it can be used as cloth in fabric production fields, toothbrush bristles, brogans, and the strings of badminton rackets (Nanni and Messori, 2020). PAM 11 can also be used to make surgically oriented molds, specifically for lap linkages; repairing femurs, spinal cords, and mandibular repairing; and 3D printing of artificial teeth (Filiz et al., 2021).

6.5.5 Applications of PHU-Derived Bioplastics

PHU-derived bioplastics possess improved mechanical (Cornille et al., 2017) and adhesion properties. Consequently, their blended form with acrylates and methacrylates can be used for making veneer pertinence. They can be applied in both thermo-setting and thermo-plastics industries owing to their high thermal stability property. Besides, their rigorous and flexible nature make them appropriate for making froths and latex/hydrogels (Ecochardand Caillol, 2020; Mahendiran et al., 2021).

6.5.6 Applications of Cellulose-Based Bioplastics

Cellulose-based bioplastics are gaining much consideration due to their biodegradability, high stability, strength, and hardness (Okolie et al., 2021; Qasim et al., 2021). They have low density, low cost, and a smooth nature (Polman et al., 2021). Yet existing hydrogen bonds can reduce the flexibility and mechanical nature of these bioplastics. Investigation into cellulose-based biopolymers recommends that transparency can be enhanced by combining the celluloses with some other polysaccharides like pectin or chitosan (Yaradoddi et al., 2020). On the other hand, the polymers present water fragility, absence of interstitial grip, and low thermal stability, thus reducing their popularity (Polman et al., 2021; Siqueira et al., 2021). Current investigations revealed the usage of cellulose-based bioplastics as biodegradable fillers in the fabrication of polymers improves their expendable life. Aguilar et al. (2019) investigated that bioplastics of magnetic nature can be produced based on cellulose using cotton and sugarcane bagasse as substrates and enhancing their features for both optical and magnetic purposes. Hossain et al. (2018) used nano-cellulose–based bioplastics, derived from leaves of corn, for developing an edible nano-coating material functional for medication and pill coating. Azminand Nor (2020) established cellulose-based biofilms, produced from the waste husks of sugarcane bagasse and cocoa pods, by combining one-fourth of fiber and three-fourths of celluloses to create a moisture-resistant food packaging film (Plastemart et al., 2021; Yaradoddi et al., 2020). Additionally, cellulose based biopolymers and their offshoots are widely used in food packaging industries (Liu, 2021; Salim et al., 2022). Shaghaleh et al. (2018) noted that cellulosic dendrimers have wide applications in the accessible petroleum-based polymers. In addition, they can be used as cellulosic well-designed polymeric materials for a range of uses, from chemical science and power storage equipment to biomedical applications.

6.5.7 Applications of Starch-Based Bioplastics

Essential oils made from the starch-based bioplastics (Bps) lead to usage in cosmetic and pharmacological industries (Helal et al., 2019). Souza et al. (2013) found that the essential oils, derived from clove and cinnamon, can be blended with cassava starch–based films, which can be used for suitable wrapping and bakery goods. The authors also distinguished an incredible antimicrobial activity to penicillium commune and other fungi. Utami et al. (2019) made tapioca films from cinnamon-based essential oils, which also intended that as the concentration of the essential oil is increased, the tensile strength of the film slowly decreased. Gao et al. (2020) noted that such bioplastics merged with 0.5% soybean oil can retain good tensile strength of about 5.2 MPa. Recent research made on starch-based biofilms showed that coupling with plasticizers, fibers and other vital plants as well as algae oils can display better biochemical and mechanical properties.

6.5.8 Applications of Protein-Based Bioplastics

Due to their abundant availability, biodegradable, good nutritious value, and enlargement competence characteristics, these protein-based bioplastics are typically used as wrapping materials. As they can create edible films, they present healthier and power-driven features compared to the polysaccharides (Mohamed et al., 2020; Roja et al., 2019). The artificially made elastomers, for instance, can be combined with keratin to improve subsequent equipment mechanical properties and thermal stability. Saps from soybean protein can produce compounds analogous to carbon black–filled elastomers (Garrison et al., 2016).

6.5.9 Applications of Lipid-Based Bioplastics

In general, these bio-based biopolymers are applied in commercial films and coverings with benefits such as glossy appearance, moisture resistance, and low production cost (Mohamed et al., 2020; Cruz et al., 2014, 2022). These bio-based films prepared from fats and oils are highly transparent and elastic, and are used for packing fruits, meat products, vegetables, and other food (Rodrigues et al., 2016). As lipids are hydrophobic, they slow down the moisture action within food particles, reducing the presence of microorganisms. Further antimicrobial properties can be enhanced due to the presence of essential oils in coverings and films.

6.6 ECONOMY IN THE PRODUCTION OF BIOPLASTICS

Research into the production of biodegradable and compatible polymeric has been determined by vision of policies that have their origins in the Brundtland Report of 1987, which provides a platform in favor of a more sustainable society (Brundtland, 1987). These polymeric components are significant fractions of the emerging portfolio toward renewable raw materials development to convey economic, environmental, and social benefits. As plant-derived biodegradable bioplastics, such naturally occurring materials can be accumulated by bacteria as a response to limitation of the inorganic nutrients, generally with the existence of carbon-rich feedstock. Early research chiefly explains the accumulation of bioplastics precursors like PHAs and PLAs and their technical development for industrial-scale production. They can be produced by materials such as corn-derived glucose on the industrial scale, reaching, for example, both PHBs in the original form and certain copolymers such as PHBVs (Obruca et al., 2014a). In recent years, research has been attempting to convert domestic, agricultural, and industrial wastes into bioplastics precursors, as they can become platform resources for a bio-based society (Obruca et al., 2014b). Wastes come from creation and the end of life of products and services generated for society, and the amount generated by the society is directly linked to economic growth. Usually, wastes have low or negative value because of the burden of disposal, no perceived use, or unknown technology to extract value from it. At the same time, wastes are underutilized resources given the current drivers that motivate a new green economy and increasing resource efficiency, and the current management practices and source-control strategies are more and more liable to change for the success of this purpose. The outcome of various bioplastics production processes depends on substrate costs and development of its precursors. In Europe, for instance, huge amounts of surplus whey are available from the dairy industry, providing mainly lactose for PLA production (Cywar et al., 2022). The outcomes of various substrate costs and the yield on the bioplastics production cost depend on the development of its precursors. Bioplastics production began with Imperial Chemical Industries (ICI) in 1975. From *Alcaligeneseutrophus* (H16), ICI started manufacturing of Biopol as early as 1982, and Cargill

Dow polymers have been marketed as EcoPla polymeric resins in Japan. Novamont's starch-based bioplastics material, namely Mater-Bi, has also been marketed there by Nippon Gohsei, and has been used in transport packaging for electrical goods, agricultural mulch film, and in composting trials (Obruca et al., 2014a, 2014b). Mitsubishi and Nippon Shokubai use the trade names LUNARE ZT and Lunare SE of Japan. Lacea is a kind of bioplastics procured from the beet, corn, cane, and tapioca resources and has analogs in both petrochemically derived polystyrene/polythene. In Japan, Daicel Chemical Industries has developed biodegradable blends of two dissimilar kinds of material, namely polycaprolactone and acetyl cellulose resin, under the brand name of Celgreen (Obruca et al., 2014a, 2014b).

6.7 CONCLUDING REMARKS

In recent years, the renewable, biodegradable, and sustainable inexpensive resources for the development of bioplastics are highly desired as an alternative for crude oil–based plastics. Bioplastics, which are derived from bio-resources such as carbohydrates, proteins, and lipids, are favorable choices for a greener future due to their biocompatible and favorable mechanical features. The research field of using wastes and residues as raw materials for the production of functional bioplastics is relatively unexplored and is growing each year. This comes in agreement with the construction of a sustainable economy, replacing fossil carbon as the chemical industry's raw material for bioplastics precursors such as PHAs, PLAs, PAM 11, PHUs, and cellulose. The biodegradability of intracellular components of bioplastics precursors results in the conversion of organic materials to simpler inorganic molecules, such as water and carbon dioxide or organic methane through the biological processing mediated by microbial enzymes. Many efforts to modify the feedstock to construct industrially certain functional polymers by traditional biotechnical approaches attained only very inadequate accomplishment, suggesting that effective biomass conversion necessitates the synergistic action of the complex networks. Such kinds of commercial and eco-friendly bioplastics can be synthesized to decrease bio-wastes and protect natural resources, manage and utilize the natural resources efficiently and sustainably for zero-waste or circular bioeconomy, and reduce waste generation substantially through prevention, reduction, recycling, and reuse.

REFERENCES

N.M. Aguilar, F. Arteaga-Cardona, M.E.A. Reyes, J.J. Gervacio-Arciniega, U. Salazar-Kuri, Magnetic bioplastics based on isolated cellulose from cotton and sugarcane bagasse, *Mater. Chem Phys.*, 238 (2019) 121921.

S. Akhtartavan, M. Karimi, K. Karimian, N. Azarpira, M. Khatami, H. Heli, Evaluation of a self-nanoemulsifying docetaxel delivery system, *Biomed. Pharmacother.*, 109 (2019) 2427–2433.

S.N.H.M. Azmin, M.S.M. Nor, Development and characterization of food packaging bioplasticfilm from cocoa pod husk cellulose incorporated with sugarcane bagasse fibre, *J. Bioresour. Bioprod.*, 5 (2020) 248–255.

S. Behera, M. Priyadarshanee, S. Das, Polyhydroxyalkanoates, the bioplastics of microbial origin: properties, biochemical synthesis, and their applications, *Chemosphere*, 294 (2022) 133723.

N.U. Benson, D.E. Bassey, T. Palanisami, COVID pollution: impact of COVID-19 pandemic on global plastic waste footprint, *Heliyon*, 7 (2021) e06343.

G.H. Brundtland, *Report of the World Commission on Environment and Development: Our Common Future*, Oxford University Press, United Nations World Commission on Environment and Development (1987).

R.W. Chia, J.Y. Lee, H. Kim, J. Jang, Microplastic pollution in soil and groundwater: a review, *Environ. Chem. Lett.*, 19 (2021) 4211–4224.

S.Y. Choi, I.J. Cho, Y. Lee, Y.J. Kim, K.J. Kim, S.Y. Lee, Microbial polyhydroxyalkanoates and nonnatural polyesters, *Adv. Mater.*, 32 (2020) 1907138.

J.W.R. Chong, X. Tan, K.S. Khoo, H.S. Ng, W. Jonglertjunya, G.Y. Yew, P.L. Show, *Environ. Res.*, 206 (2022) 112620.

G. Coppola, M.T. Gaudio, C.G. Lopresto, V. Calabro, S. Curcio, S. Chakraborty, Bioplastic from renewable biomass: a facile solution for a greener environment, *Earth Syst. Environ.*, 5 (2021) 231–251.

A. Cornille, M. Blain, R. Auvergne, B. Andrioletti, B. Boutevin, S. Caillol, A study of cyclic carbonate aminolysis at room temperature: effect of cyclic carbonate structures and solvents on polyhydroxyurethane synthesis, *Polym. Chem.*, 8 (2017) 592–604.

R.A. Cruz, A. Oehmen, M.A. Reis, The impact of biomass withdrawal strategy on the biomass selection and polyhydroxyalkanoates accumulation of mixed microbial cultures, *New Biotechnol.*, 66 (2022) 8–15.

M.V. Cruz, A. Paiva, P. Lisboa, F. Freitas, V. Alves, P. Simões, S. Barreiros, M.A.M. Reis, Production of polyhydroxyalkanoates from spent coffee grounds oil obtained by supercritical fluid extraction technology, *Bioresour. Technol.*, 157 (2014) 360–363.

R.M. Cywar, N.A. Rorrer, C.B. Hoyt, T.G. Beckham, E.Y.X. Chen, Bio-based polymers with performance-advantaged properties, *Nat. Rev. Mater.*, 7 (2022) 83–103.

B.T. Dang, X.T, Bui, D.P. Tran, H.H. Ngo, L.D. Nghiem, P.T. Nguyen, H.H. Nguyen, C. Lin, K.Y.A. Lin, S. Varjani, Current application of algae derivatives for bioplastic production: a review, *Bioresour. Technol.*, 347 (2022) 126698.

S. Daniel, L. Amanda, L, Adriana, D. Mario, Approaching the environmental problem of microplastics: importance of WWTP treatments, *Sci. Total Environ.*, 740 (2020) 140016.

M.A. Dechet, A. Goblirsch, S. Romeis, M. Zhao, F.J. Lanyi, J. Kaschta, J. Schmidt, Production of polyamide 11 microparticles for additive manufacturing by liquid-liquid phase separation and precipitation, *Chem. Eng. Sci.*, 197 (2019) 11–25.

Z.N. Diyana, R. Jumaidin, M.Z. Selamat, I. Ghazali, N. Julmohammad, N. Huda, R.A. Ilyas, Physical properties of thermoplastic starch derived from natural resources and its blends: a review, *Polymers*, 13 (2021) 1396.

Y. Ecochard, S. Caillol, Hybrid polyhydroxyurethanes: how to overcome limitations and reach cutting edge properties?, *Eur. Polym. J.*, 137 (2020) 109915.

S. Fawzy, A.I. Osman, J. Doran, D.W. Rooney, Strategies for mitigation of climate change: a review, *Environ. Chem. Lett.*, 18 (2020) 2069–2094.

W. Gao, W. Wu, P. Liu, H. Hou, X. Li, B. Cui, Preparation and evaluation of hydrophobic biodegradable films made from corn/octenylsuccinated starch incorporated with different concentrations of soybean oil, *Int. J. Biol.*, 142 (2020) 376–383.

T.F. Garrison, A. Murawski, R.L. Quirino, Bio-based polymers with potential for biodegradability, *Polymers*, 8 (2016) 262.

G. Gorrasi, A. Sorrentino, E. Lichtfouse, Back to plastic pollution in COVID times, *Environ. Chem. Lett.*, 19 (2021) 1–4.

G.L. Gregory, L.M. Jenisch, B. Charles, G. Kociok-Kohn, A. Buchard, Polymers from sugars and CO_2: Synthesis and polymerization of a D-mannose-based cyclic carbonate, *Macromolecules*, 49(19) (2016) 7165–7169.

I.M. Helal, A. El-Bessoumy, E. Al-Bataineh, M.R. Joseph, P. Rajagopalan, H.C. Chandramoorthy, S.B.H. Ahmed, Antimicrobial efficiency of essential oils from traditional medicinal plants of Asir region, Saudi Arabia, over drug resistant isolates, *BioMed. Res. Int.*, 2019 (2019) 8928306.

A.S. Hossain, M.M. Uddin, V.N. Veettil, M. Fawzi, Nano-cellulose based nano-coating biomaterial dataset using corn leaf biomass: an innovative biodegradable plant biomaterial, *Data Brief*, 17 (2018) 162–168.

C. Iolanda, P. Claudia, I. Rachele, C. Angela, M.M. Corsaro, S. Giovanni, P. Cinzia, The power of two: An artificial microbial consortium for the conversion of inulin into polyhydroxyalkanoates, *Int. J. Biol. Macromol.*, 189 (2021) 494–502.

M. Jamshidian, E.A. Tehrany, M. Imran, M. Jacquot, S. Desobry, Poly-lactic acid: production, applications, nanocomposites, and release studies, *Comp. Rev. Food Sci. Food Saf.*, 9 (2010) 552–571.

M. Karamanlioglu, R. Preziosi, G.D. Robson, Abiotic and biotic environmental degradation of the bioplastic polymer poly(lactic acid): a review, *Polym. Degrad. Stabil.*, 137 (2017) 122–130.

T. Keshavarz, I. Roy, Polyhydroxyalkanoates: bioplastics with a green agenda, *Curr. Opin. Microbiol.*, 13 (2010) 321–326.

M. Koller, L. Maršálek, M.M. Sousa Dias, G. Braunegg, Producing microbial polyhydroxyalkanoate (PHA) biopolyesters in a sustainable manner, *New Biotechnol.*, 37 (2017) 24–38.

M. Kumar, R. Rathour, R. Singh, Y. Sun, A. Pandey, E. Gnansounou, K.Y.A. Lin, D.C.W. Tsang, I.S. Thakur, Bacterial polyhydroxyalkanoates: opportunities, challenges, and prospects, *J. Clean. Prod.*, 263 (2020) 121500.

Y. Liu, S. Ahmed, D.E. Sameen, Y. Wang, R. Lu, J. Dai, S. Li, W. Qin, A review of cellulose and its derivatives in bio polymer based for food packaging applications, *Trends Food Sci. Technol.*, 112 (2021) 532–546.

B. Mahendiran, S. Muthusamy, S. Sampath, S.N. Jaisankar, K.C. Popat, R. Selvakumar, G.S. Krishnakumar, Recent trends in natural polysaccharide based bioinks for multiscale 3D printing in tissue regeneration: a review, *Int. J. Biol. Macromol.*, 183 (2021) 564–588.

G. Mannina, D. Presti, G. Montiel-Jarillo, J. Carrera, M.E., Suárez-Ojeda, recovery of polyhydroxyalkanoates (PHAs) from wastewater: a review, *Bioresour. Technol.*, 297 (2020) 122478.

T. Mekonnen, P. Mussone, H. Khalil, D. Bressler, Progress in biobased plastics and plasticizing modifications, *J. Mater. Chem. A*, 1 (2013) 13379–13398.

S. Miao, P. Wang, Z. Su, S. Zhang, Vegetable-oil-based polymers as future polymeric biomaterials, *Acta Biomater.*, 10 (2014) 1692–1704.

N. Mlalila, A. Hilonga, H. Swai, F. Devlieghere, P. Ragaert, Antimicrobial packaging based on starch, poly (3-hydroxybutyrate) and poly (lactic-co-glycolide) materials and application challenges, *Trends Food Sci. Technol.*, 74 (2018) 1–11.

S.A. Mohamed, M. El-Sakhawy, M.A.M. El-Sakhawy, Polysaccharides, protein and lipid-based natural edible films in food packaging: a review, *Carbohydr. Polym.*, 238 (2020) 116178.

X. Mu, S. Qi, J. Liu, L. Yuan, Y. Huang, J. Xue, L. Qian, C. Wang, Y. Li, Toxicity and behavioral response of zebrafish exposed to combined microplastic and bisphenol analogues, *Environ. Chem. Lett.*, 20 (2022) 41–48.

A.S. Nair, S. Cherian, N. Balachandran, U.G. Panicker, S.K.K. Sankaranarayanan, Hybrid poly(hydroxy urethane)s: folded-sheet morphology and thermoreversible adhesion, *ACS Omega*, 4 (2019) 13042–13051

S. Nanda, R. Azargohar, A.K. Dalai, J.A. Kozinski, An assessment on the sustainability of lignocellulosic biomass for biorefining, *Renew. Sustain. Energy Rev.*, 50 (2015) 925–941.

S. Nanda, F. Berruti, A technical review of bioenergy and resource recovery from municipal solid waste, *J. Hazard. Mater.*, 403 (2021a) 123970.

S. Nanda, F. Berruti, Municipal solid waste management and landfilling technologies: a review, *Environ. Chem. Lett.*, 19 (2021b) 1433–1456.

S. Nanda, F. Berruti, Thermochemical conversion of plastic waste to fuels: a review, *Environ. Chem. Lett.*, 19 (2021c) 123–148.

S. Nanda, J. Mohammad, S.N. Reddy, J.A. Kozinski, A.K. Dalai, Pathways of lignocellulosic biomass conversion to renewable fuels, *Biomass Convers. Bioref.*, 4 (2014) 157–191.

S. Nanda, P. Mohanty, K.K. Pant, S. Naik, J.A. Kozinski, A.K. Dalai, Characterization of North American lignocellulosic biomass and biochars in terms of their candidacy for alternate renewable fuels, *Bioenergy Res.*, 6 (2013) 663–677.

S. Nanda, J.A. Okolie, R. Patel, F. Pattnaik, Z. Fang, A.K. Dalai, J.A. Kozinski, S. Naik, Catalytic hydrothermal co-gasification of canola meal and low-density polyethylene using mixed metal oxides for hydrogen production, *Int. J. Hydrog. Energy*, 47 (100) (2021) 42084–42098. https://doi.org/10.1016/j.ijhydene.2021.08.179.

A. Nanni, M. Messori, Thermo-mechanical properties and creep modelling of wine lees filled polyamide 11 (PA11) and polybutylene succinate (PBS) bio-composites, *Comp. Sci. Technol.*, 188 (2020) 107974.

R. Nigmatullin, P. Thomas, B. Lukasiewicz, H. Puthussary, I. Roy, Polyhydroxyalkanoates, a family of natural polymers, and their applications in drug delivery, *J. Chem. Technol. Biotechnol.*, 90 (2015) 1209–1221.

S. Obruca, P. Benesova, S. Petrik, J. Oborna, R. Prikryl, I. Marova, Production of polyhydroxyalkanoates using hydrolysate of spent coffee grounds, *Process Biochem.*, 49 (2014a) 1409–1414.

S. Obruca, S. Petrik, P. Benesova, Z. Svoboda, L. Eremka, I. Marova, Utilization of oil extracted from spent coffee grounds for sustainable production of polyhydroxyalkanoates, *Appl. Microbiol. Biotechnol.*, 98 (2014b) 5883–5890.

J.A. Okolie, S. Nanda, A.K. Dalai, J.A. Kozinski, Chemistry and specialty industrial applications of lignocellulosic biomass, *Waste Biomass Valor.*, 12 (2021) 2145–2169.

H. Oliver-Ortega, F. Julian, F.X. Espinach, Q. Tarrés, M. Ardanuy, P. Mutjé, Research on the use of lignocellulosic fibers reinforced bio-polyamide 11 with composites for automotive parts: car door handle case study, *J. Clean. Prod.*, 226 (2019) 64–73.

A.R. Othman, H.A. Hasan, M.H. Muhamad, N.I. Ismail, S.R.S. Abdullah, Microbial degradation of microplastics by enzymatic processes: a review, *Environ. Chem. Lett.*, 19 (2021) 3057–3073.

Plastemart (2021) Cellulose-based plastics address need for more sustainable raw materials from food, healthcare, coatings and construction. www.plastemart.com/plastic-technical-articles/cellulose-based-plastics-address-need-for-more-sustainableraw-materials-from-food-healthcare-coatings-and-construction/2356 (Accessed 2 August 2021).

E.M. Polman, G.J.M. Gruter, J.R. Parsons, A. Tietema, Comparison of the aerobic biodegradation of biopolymers and the corresponding bioplastics: a review, *Sci. Total Environ.*, 753 (2021) 141953.

S.M. Prasanth, P.S. Kumar, S. Harish, M. Rishikesh, S. Nanda, D.V.N. Vo, Application of biomass derived products in mid-size automotive industries: a review, *Chemosphere*, 280 (2021) 130723.

U. Qasim, A.I. Osman, A.H. Al-Muhtaseb, C. Farrell, M. Al-Abri, M. Ali, D.V.N. Vo, F. Jamil, D.W. Rooney, Renewable cellulosic nanocomposites for food packaging to avoid fossil fuel plastic pollution: a review, *Environ. Chem. Lett.*, 19 (2021) 613–641.

R.A. Ranganadha, K. Vidyaprabhakar, T.C. Venkateswarulu, S. Krupanidhi, M. Nazneen Bobby, P.K. Abraham, P. Sudhakar, P. Vijetha, Statistical optimization of Polyhydroxybutyrate (PHB) production by novel *Acinetobacternosocomialis* RR20 strain using response surface methodology, *Curr. Trends Biotechnol. Pharm.*, 14(1) (2020) 62–69.

A. Ranganadhareddy, P. Vijetha, C. Chandrsekhar, Bioplastic production from microalgae and their applications—a critical review, *J. Biochem. Technol.*, 13(2) (2022) 13–18.

S. Ray, V.C. Kalia, Biomedical applications of polyhydroxyalkanoates, *Indian J. Microbiol.*, 57 (2017) 261–269.

N. Razeghi, A.H. Hamidian, C. Wu, Y. Zhang, M. Yang, Microplastic sampling techniques in freshwaters and sediments: a review, *Environ. Chem. Lett.*, 19 (2021) 4225–4252.

D.C. Rodrigues, A.P. Cunha, E.S. Brito, H.M. Azeredo, M.I. Gallão, Mesquite seed gum and palm fruit oil emulsion edible films: Influence of oil content and sonication, *Food Hydrocoll.*, 56 (2016) 227–235.

S. Rodriguez-Perez, A. Serrano, A.A. Pantion, B. Alonso-Farinas, Challenges of scaling-up PHA production from waste streams. A review, *J. Environ. Manag.*, 205 (2018) 215–230.

K. Roja, S.D. Ruben, A. Susaimanickam, M. Thangavel, Extraction and characterization of polyhydroxyalkanoates from marine green alga and cyanobacteria, *Biocatal. Agric. Biotechnol.*, 22 (2019) 101358.

H.A. Sabathini, L. Windiani, D. Ismail, M. Gozan, Mechanical physical properties of *Chlorella*-PVA based bioplastic with ultrasonic homogenizer, *E3S Web of Conferences*, 67 (2018) 03046.

M.H. Salim, Y. Abdellaoui, A.A. Benhamou, E.-H. Ablouh, M. El Achaby, Z. Kassab, Influence of cellulosic nano crystals from pea pod waste on mechanical, thermal, biodegradability and barrier properties of chitosan based films, *Cellulose*, 29 (2022) 5117–5135.

H. Shaghaleh, X. Xu, S. Wang, Current Progress in Production of biopolymeric materials based on cellulose, cellulose nano fibers and cellulose derivatives, *RSC Adv.*, 8(2) (2018) 825–842.

I. Shahabi-Ghahfarrokhi, V. Goudarzi, A. Babaei-Ghazvini, Production of starch based biopolymer by green photochemical reaction at different UV region as a food packaging material: Physicochemical characterization, *Int. J. Biol. Macromol.*, 122 (2019) 201–209.

L.D.V. Siqueira, C.I.L.F. Arias, B.C. Maniglia, C.C. Tadini, Starch-based biodegradable plastics: methods of production, challenges and future perspectives. *Curr. Opin. Food Sci.*, 38 (2021) 122–130.

R. Sirohi, V.K. Gaur, A.K. Pandey, S.J. Sim, S. Kumar, Harnessing fruit waste for poly-3-hydroxybutyrate production: a review, *Bioresour. Technol.*, 326 (2021a) 124734.

R. Sirohi, J.P. Pandey, A. Tarafdar, A. Agarwal, S.K. Chaudhuri, R. Sindhu, An environmentally sustainable green process for the utilization of damaged wheat grains for poly-3-hydroxybutyrate production, *Environ. Technol. Innov.*, 21 (2021b) 101271.

A.C. Souza, G.E.O. Goto, J.A. Mainardi, A.C.V. Coelho, C.C. Tadini, Cassava starch composite films incorporated with cinnamon essential oil: antimicrobial activity, microstructure, mechanical and barrier properties, *LWT—Food Sci. Technol.*, 54 (2013) 346–352.

A. Suresh, A. Alagusundaram, P.S Kumar, D.V.N. Vo, F.C. Christopher, B. Balaji, V. Viswanathan, S. Sankar, Microwave pyrolysis of coal, biomass and plastic waste: a review, *Environ. Chem. Lett.*, 19 (2021) 3609–3629.

Y. Suryawanshi, P. Sanap, V. Wani, Advances in the synthesis of non-isocyanate polyurethanes, *Polym. Bull.*, 76 (2019) 3233–3246.

R. Tarrahi, Z. Fathi, M.O. Seydibeyoğlu, E. Doustkhah, A. Khataee, Polyhydroxyalkanoates (PHA): from production to nanoarchitecture, *Int. J. Biol. Macromol.*, 146 (2020) 596–619.

The Guardian, Microplastics revealed in the placentas of unborn babies (2020). www.theguardian.com/environment/2020/dec/22/microplastics-revealed-in-placentas-unborn-babies.

T.S. Tofa, K.L. Kunjali, S. Paul, J. Dutta, Visible light photocatalytic degradation of microplastic residues with zinc oxide nanorods, *Environ. Chem. Lett.*, 17 (2019) 1341–1346.

G. Tutoni, M.L. Becker, Underexplored stereocomplex polymeric scaffolds with improved thermal and mechanical properties, *Macromolecules* 53 (2020) 10303–10314.

B.D. Ulery, L.S. Nair, C.T. Laurencin, Biomedical applications of biodegradable polymers, *J. Polym. Sci. B: Polym. Phys.*, 49 (2011) 832–864.

R. Utami, L.U. Khasanah, G.J. Manuhara, Z.K. Ayuningrum, Effects of cinnamon bark essential oil (*Cinnamomumburmannii*) on characteristics of edible film and quality of fresh beef, *Pertanika J Trop Agric Sci.*, 42(4) (2019) 1173–1184.

L.P.S. Vandenberghe, P.Z. Oliveira, G.A. Bittencourt, A.F.M. Mello, Z.S. Vásquez, S.G. Karp, C.R. Soccol, The 2G and 3G bioplastics: an overview, *Biotechnol. Res. Innov.*, 5(1) (2021) e2021004 (1–10).

Z. Wang, Y. Qin, W. Li, W. Yang, Q. Meng, J. Yang, Microplastic contamination in freshwater: first observation in Lake Ulansuhai, Yellow River Basin, China, *Environ. Chem. Lett.*, 17 (2019) 1821–1830.

N. Wu, Y. Tu, G. Fan, J. Ding, J. Luo, W. Wang, C. Zhang, C. Yuan, H. Zhang, P. Chen, S. Tan, H. Xiao, Enhanced photodynamic therapy/photothermo therapy for nasopharyngeal carcinoma via a tumour microenvironment-responsive self-oxygenated drug delivery system, *Asian J. Pharm. Sci.*, 17(2) (2022) 253–267.

J.S. Yaradoddi, N.R. Banapurmath, S.V. Ganachari, M.E.M. Soudagar, N.M. Mubarak, S. Hallad, S. Hugar, H. Fayaz, Biodegradable carboxymethyl cellulose based material for sustainable packaging application, *Sci. Rep.*, 10 (2020) 21960.

L.P. Yu, X. Yan, X. Zhang, X.B. Chen, Q. Wu, X.R. Jiang, G.Q. Chen, Biosynthesis of functional polyhydroxyalkanoates by engineered *Halomonasbluephagenesis*, *Metab. Eng.*, 59 (2020) 119–130.

7 Third-Generation Bioplastics from Food Waste

Reshmy R, Eapen Philip, Deepa Thomas, Vaisakh P H, Raveendran Sindhu, Aravind Madhavan, Mukesh Kumar Awasthi, Parameswaran Binod, and Ashok Pandey

CONTENTS

7.1 INTRODUCTION

The use of plastic-based products has increased the demand for plastic on a global scale, imposing additional strain on the waste management system. Two decades ago, the globe produced half as much plastic waste as it does today. Recycling rates for plastic garbage are below 9%. Another 19% is burned in open pits, 50% ends up in landfills, and 22% manages to get past waste management systems and wind up in unauthorized dumps, open pit fires, or terrestrial or aquatic habitats. Reducing the usage of plastics derived from petroleum, which contribute to worldwide environmental contamination, is a topic of great interest. It is acceptable to anticipate the creation of an original remedy to lessen the problem (Rosenboom et al. 2022; Roy Chong et al. 2022). Bioplastics are a recently popularized substitute for traditional petrochemical plastics. Plant-derived raw materials, natural polymers, and other tiny molecules are sources that can be used to make bioplastics. Terrestrial crops like corn and potatoes, which vie with food stocks and take up a lot of land, water, and nutrients, are the source of bioplastics (Otoni et al. 2021).

The enormous volume of food waste from various sources is an additional environmental strain if it is disposed of improperly. Food waste bio-valorization into bio-products is an option under biorefinery concepts. Implementing a bio-refinery platform for food waste processing is an appealing strategy for producing valuable feedstock for bio-based plastics. The cost of producing biodegradable plastics should decrease with the implementation of such a process. Materials are synthesized from carbon-neutral resources in the fabrication of bioplastics from food waste, which is a renewable and

DOI: 10.1201/9781003344018-7

sustainable process. As a result, one of the targeted uses of food waste to improve global sustainability is the transformation of food waste into value-added chemicals (Chong et al. 2021).

When compared to the conventional approach, the development of microalgae cultivation in food manufacturing sewage has substantially decreased the cost of sewage treatment by lowering carbon emissions, energy consumption, and chemical utilization while generating microalgae materials that can be used to make affordable fertilizers and bioplastics. Thus using microalgae to make bioplastics presents a fantastic chance to improve the sustainability of plastic usage. For manufacturing of bioplastics, algae variants have been proposed, like a viable sustainable biomass resource (Rajpoot et al. 2022). Interest in microalgae has increased as a result of the recent rise in popularity of the circular bio-economy, particularly when combined with wastewater treatment, with the dual goal of enabling the manufacture of a variety of bio-based goods while bio-remediating wastewater. Numerous useful chemicals are derived from microalgae, including polysaccharides, starch, and polyhydroxyalkanoates (PHAs) through cyanobacteria. The amount of various functional groups in microalgae, like sulfate, hydroxyl, and carboxyl, can be modified or tuned to produce the desired bioplastic qualities, particularly with food, pharmaceuticals, and biomedical packaging (Sreenikethanam and Bajhaiya 2022).

7.2 MICROALGAE SOURCES AND CHARACTERISTICS

Microalgae are simple cell–structured photosynthetic microorganisms that need light, CO_2, water, and micronutrients to flourish. With more than 25,000 species previously isolated and characterized, they can be categorized as prokaryotic cyanobacteria (blue-green algae) or eukaryotic microorganisms. The major portions of the bioactive contents are composed of carbohydrates, lipids, carotenoids, and proteins. In a variety of photo bioreactors, microalgae can be grown in places that are not appropriate for agriculture. Additionally, its biomass yield is five to ten times greater than that of typical food crops. Furthermore, they have a limited generation period and exponential growth when the environment is appropriate. Microalgae could therefore serve as a feedstock for the biopolymer sector (Banerjee and Khan 2022).

Microalgae can develop in one of three ways depending on the carbon and energy sources they employ for metabolism: autotrophy, heterotrophy, or mixotrophy. The microalgae produces the necessary organic matter and energy during autotrophic growth by using CO_2 as a source of carbon and sunlight as a source of energy. Most autotrophic cultivation is done in closed photo bioreactors and open pond systems. In heterotrophic growth, organic materials are used both as a source of energy and carbon. Although glucose is the main typical carbon source, certain microalgae can also grow using glycerol. When the level of organic carbon is high, the heterotrophic first stage of the mixotrophic metabolism begins, and the autotrophic second stage begins when photosynthesis is triggered when the level of organic carbon is low. To speed up growth rates and biomass accumulation, the mixotrophic mode, which includes simultaneous CO_2 fixation and organic carbon assimilation, might be a useful alternative to photoautotrophic or heterotrophic mode (Shandilya and Pattarkine 2019).

The two main types of microalgae used to produce bioplastic are *Chlorella* sp. and *Spirulina* sp. *Spirulina* has a higher blend performance, while *Chlorella* appears to have better bioplastic behavior. Various kinds of *Chlorella* were included in biomass polymer composites using additives and polymers. Additionally, starch granules obtained from *Chlorella pyrenoidosa* and *Chlorella sorokiniana* may be used to make bioplastics (Das et al. 2018). *Spirulina* was explored for the manufacture of bioplastics, similar to *Chlorella*. For instance, using polyvinyl alcohol and salt-rich *Spirulina sp.* wastes, a bioplastic-based film was created. *Spirulina platensis* was used to create another bioplastic that has good biodegradability (Devadas et al. 2021).

Additionally, *Chlorogloeafritschii, Neochlorisoleoabundans, Phaeodactylumtricornutum, Calothrix scytonemicola, Nannocloropsisgaditana,* and *Scenedesmus almeriensis* were employed to make bioplastics. For starch production and the creation of starch-based bioplastics, ten green microalgae were examined. The strain that produced the most starch and had intriguing plasticization characteristics when heated to 120°C was *Chlamydomonasreinhardtii* 11–32A (Mathiot et al. 2019).

Microalgae can withstand and thrive on a range of fertilizer sources as well as different kinds of effluent. Utilizing wastewater would enable for considerable cost reductions in microalgae cultivation while also effectively purifying the wastewater. Wastewater streams conveniently contain both the water and nutrients necessary for production of microalgae. The fundamental goal of some microalgae growing techniques mentioned in the literature is to generate bio-components for the creation of biomaterials. The four basic categories of microalgae cultivation systems are open, closed, hybrid, and biofilm systems. In open ponds, algal biomass is open to the surroundings. They have a low cost to construct and maintain, but their capacity to effectively manage growth parameters including irradiance levels, temperature, and pollution places them at a disadvantage. The forms of open ponds include circular, lagoon, sloped thin layer, and paddlewheel-driven raceway ponds (Harmon et al. 2021). Closed cultivation systems, however, cost substantially more to set up than open ones. The examples of closed cultivation systems include plate, helical tubular, flat panel, and tubular photo bioreactors. Hybrid cultivation systems incorporate two or more bioreactors into a single operation, typically in various stages, to enhance the benefits of the both cultivation systems while reducing the disadvantages of the daughter processes, such as the large manufacturing costs connected to photo bioreactors and pollutants in the raceway ponds (Richardson et al. 2014).

Methods for growing algae in suspension are used in both open and closed algae cultivation. Systems for cultivating microalgal biofilms are an alternative. The utilization of microalgal biofilm systems can significantly reduce the cost of microalgal harvesting. In microalgal biofilm systems, the microalgae cells are connected to a surface made of material, holding them there as nutrients flow across the surface (Gross et al. 2015).

Diverse technologies are used on algal biomass to break down the cell wall and extract the desired products from the cell contents. Microwaves, supercritical fluid extraction, and ultrasound are used in physicochemical treatments. The more typical microwave frequency is 2450 MHz, which ruptures cells by causing intracellular water to boil and increasing pressure. Shock waves are produced by ultrasound technology, which causes cavitation bubbles to form and rupture cell walls. In supercritical fluid extraction, CO_2 is typically utilized to break down the cell wall and release metabolites by operating at temperatures and pressures over the critical point (Günerken et al. 2015).

Chemical procedures attempt to disrupt the microalgal cell wall and membrane by using different chemicals including solvents, chelating agents, surfactants, hypochlorites, acids, and alkalis. Solvents, for instance, dissolve the components of the cell wall, alkalis saponify the lipids in the membrane, and acids open pores that let the internal contents escape. On the other hand, enzymatic hydrolysis is a biological process with more selectivity that uses less energy. The cell wall can be destroyed by enzymes like glucosidases, glucanases, peptidases, and lipases (Mutanda et al. 2011; Günerken et al. 2015).

Understanding the metabolic and cellular processes of microalgae can be made easier by using gene engineering to boost particular biosynthetic pathways. The capacity to modify microalgae for the effective production of pertinent biofuels and bio-products is made possible by the availability of various cutting-edge genetic modification technologies (Qin et al. 2012).

7.3 CURRENT STATUS OF MICROALGAE-BASED BIOPLASTIC PRODUCTION

Commercialization of microalgae-based goods has huge market potential. For the manufacturing of bioplastics, algae compounds has been proposed like a viable reusable biomass sources. The algal compounds could be made to produce bioplastics either as an essential component, like as starches and PHAs, or merely as an addition, as sulfated polysaccharides (Karan et al. 2019). A possible approach to reducing pollution and making bioplastics would be to standardize strains, harvesting, growth conditions, and isolating microalgae in a sustainable way. Since they can be grown on nonarable land, next-generation microalgae-based bioplastic manufacture can theoretically overcome a number of difficulties with petroleum-derived polymers. This boosts our capability to transform CO_2 into feedstocks for bioplastics and increasing our ability to reverse desertification—a process that desperately needs technological assistance (Dang et al. 2022).

Many businesses have created microalgae-based bioplastics in response to the growing need for better ecofriendly and biodegradable materials to substitute petroleum-based products. The company Algix (Mississippi, US) also produces a variety of mixes with 60% PP, PBAT, and PLA and 40% algae, as well as Solaplast adaptable footwear foams, which contain an algae-ethylene vinyl acetate (EVA) mixture. The firm Algenesis Materials (California, US) converts algal oil into polyols to create polyurethane (PU)foams which are employed to create surfboard blanks and flip-flop soles. Teregroup (Italy) substituted up to 35% of its petroleum-based products by algae biomass in conjunction with Algamoil (Florida, US) to make foams, bags, and other products (Lutzu et al. 2021).

The thorough knowledge map that was produced demonstrated that microalgal bio-refinery is a scientifically active area, and that the current research attention was focused on harmoniously combining the unit operations required to make it enviro-economically viable. The market potential, regulating policy criteria, and the widespread demand to implement circular bio-economy viewpoints were emphasized to help in the real-time deployment of microalgal bio refinery.

7.4 RECENT TECHNOLOGY OF FOOD WASTE PROCESSING

Environmental protection greatly benefits from the treatment of wastewater from food production. The typical method of treating wastewater from food processing typically includes an aerobic and anaerobic digester as the primary device, then moves on to centrifugation, a clarifier, followed by granular medium purification, a slurry dryer, and chiller before releasing the treated wastewater back into the environment. There is a need for specialized equipment, which will increase investment costs. Without periodic modifications of alkalinity from microorganisms, anaerobic treatments cannot remove nitrogen and phosphate. The removal of phosphorus, nitrogen, and nitrogen-containing waste as well as the thickening of sludge are alternative ways that aerobic processes lead to significant investment and energy expenses).

Utilizing microorganisms like microalgae species for the biological treatment of wastewater is a recent development in the processing of food waste. In addition to producing bioenergy and high value-added compounds, microalgae have been recognized as an effective microorganism for wastewater remediation. Microalgae addition to food processing wastewater will result in significant decreases in annual operating costs, total capital expenditure, sludge formation, carbon emissions, and greenhouse gas mitigation, as well as the production of extra valuable by-products like inexpensive fertilizers and other by-products like bioplastics.

Microalgae-assisted food waste value addition is advantageous for fixing carbon dioxide. The microalgae-bacteria consortium describes the advantageous interactions between photoautotrophic algae and heterotrophic bacteria with regard to the exchange of oxygen and carbon dioxide. It also assists in the reduction of food wastewater contaminants prior to discharge into the ecosystem and boosts lipid efficiency in the microalgae biomass. The food waste contains significance nutrients that are ideal for the cultivation of microalgae, including lipids, carbohydrates, proteins, nitrogen, and phosphorus. Microalgae primarily need nitrogen and phosphorus, both of which are present in food waste. Employing such nutrients will increase the productivity of highly valued elements including lipids, carbohydrates, and proteins in the creation of bioplastics by promoting biomass formation and metabolic activity. Additionally, studies have shown that microalgae have the metabolic capability to efficiently reduce high concentrations of nutrients found in various food processing effluent, such as carbon, phosphorus, and nitrogen (Chong et al. 2021). Food processing wastewater contains levels of organic stuff that may be a potential source of nourishment for microorganisms, in contrast to municipal and industrial wastewater. Unlike chemical or municipal wastewater, food processing wastewater also has the advantage of being harmless.

7.5 PRODUCTION TECHNOLOGIES IN BIOPLASTIC FROM MICROALGAE

Bioplastic can be produced in various methods, including by blending synthetic polymers and bioplastic with microalgae, genetically altered strains, proteins, starch, cellulose, and PHA. Using the entire microalgae biomass, using biopolymers produced by microalgae cells, and converting microalgae into construction blocks suitable for the polymerization are the three approaches that had been investigated for produce bioplastics from microalgae (Figure 7.1) (Noreen et al. 2016).

In the first approach, extruding of complete biomass acquired following harvesting, drying, and in certain circumstances, pre-treating for separation can be employed to effectively utilize microalgal to create bioplastics (Cinar et al. 2020). This method has the benefit of decreasing downstream treatment and removing the expense of processes for removal and isolation that might enhance energy usage and, as a result, decrease the viability of the overall process. Several microalgae cells are rather small, which makes them especially ideal for the fabrication of the fibers and thin films (Fabra et al. 2017).

Utilizing biopolymers like starch, PHA, and proteins made by microalgae cells is the second strategy. Heterotrophic bacterial monocultures are currently used in industrial PHA synthesis. To shatter cells and obtain high extraction yields, energy-intensive processes and risky organic solvent are needed. They have benefit for manufacturing PHA without the use of organic materials, although typical phototrophic cultivation has significant energy expenditures, which makes it difficult to reduce the cost. As a result, it doesn't seem like the manufacture of PHA from phototrophic microalgae is now economically viable compared to the techniques used in traditional fermentation. The expenses may be lower, though, if algal biomass production and wastewater treatment are combined. Utilizing heterotrophic metabolism, choosing beneficial strains, and including nutrients in the fermentation medium can all assist reduce production costs (Tan et al. 2022).

Various polysaccharides can be produced by microalgae, based on the type and the growth circumstances. In eukaryotic green algae, the three main polysaccharides are starch, cellulose, and hemicellulose. Every microalgal cell has proteins that could be utilized to create thermoplastic biopolymers through concentrations ranging from around 10% to over 60% (Wang et al. 2016). They have good film-forming capabilities, are biocompatible, and can adhere to a variety of substrates, including wood and human tissues. They can be manufactured by electrospinning or injection

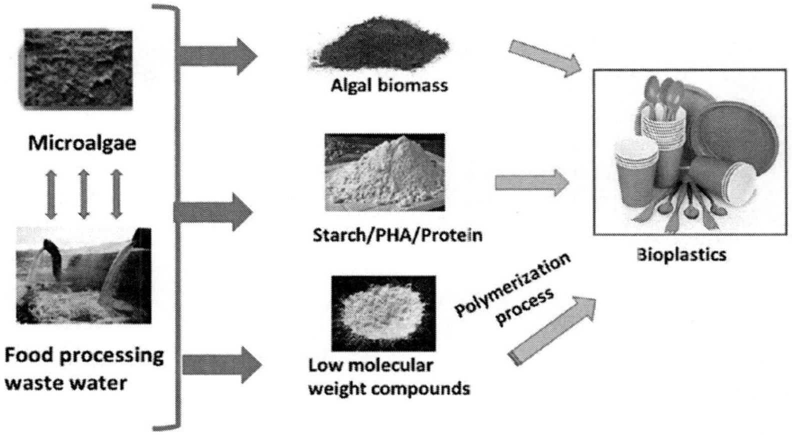

FIGURE 7.1 Approaches for the manufacture of bioplastics from microalgae.

molding. The creation of biofilms, food packaging, cosmetics, biomedical, pharmaceutical applications, and the replacement of different petroleum-based plastic goods are all possible using protein-based bioplastics (Rajpoot et al. 2022).

In addition to using the microalgae-produced polymers directly, another strategy is to use the biomass of the microalgae to make low-molecular-weight chemical compounds that can be combined to form bioplastics. Microalgae polysaccharides, for instance, can be hydrolyzed to simple sugars and subsequently transformed by fermentation into building blocks like ethanol and lactic acid (Roy Chong et al. 2022). Algae oil, or oil produced from *Phaeodactylum* sp., contains fatty acids that can be changed into diesters and then reduced to diols.

7.6 MICROALGAE-BASED BIOPLASTICS

Microalgae compounds and biomass can both be easily molded to biodegradable plastic sheets and biomaterials. Before being shaped into the desired bioplastic form, the pellets can be mixed using additives such glycerol. Bioplastics have always had the ability to take the place of traditional materials in all of the uses for plastics. It can be produced in any dimension, structure, or size and can be shaped in a wide range of methods, from fibers to thin film. Additionally, plastic beads made from algae can be employed in shooting sports, ornament-making, and fishing. However, because of the expensive price of biopolymer isolation and refinement, it is not widely explored.

7.6.1 Polyurethane

The biopolymer PU, which is readily accessible on the market, is extensively employed in elastomers, adhesives, flexible foams, and coatings. Because of its low density, excellent mechanical characteristics, and ease of fabrication, PU-based materials are extensively implemented in automotive design as parts for commercial vehicles, including bumpers, rear ends, coverings, flaps, hoods, and roof components (Abdel-Hamid et al. 2019). PU composite materials have excellent characteristics such as good flexibility, damping capability, abrasion resistance, excellent tensile strength, high weather proof and weather resilience, excellent biocompatibility, and thermal stability (Kuddus 2021).

They can be produced using oils, of which triglycerides from microalgae are a renewable source. For species to species, the fatty acid content differs. Pawar et al. reported that *Chlorella* was used by to oxidize polyols and generate them. The oil's content was 10% saturated fats and 2% undetermined fatty acids. The outcomes from microalgae oil epoxidation were positive, with conversion efficiencies that were on par with those of other vegetable oils. For the production of rigid PU foams, the epoxide ring opening of epoxidized algal oil using ethylene glycol and lactic acid was efficiently accomplished. These synthesis led to the creation of polyurethane with properties resembling those of widely accessible polyols on the market (Pawar et al. 2016). Patil et al. reported the nanocomposite coatings, blended eggshell-based, silver-loaded nanoparticles in a PU matrix with polyol from algal oil and ricinoleic acid. According to studies comparing the attributes of PU matrix composites to PU without nanoparticles, PU coverings demonstrated excellent physico mechanical characteristics (Patil et al. 2018). The microalgae oil-based PU coatings also showed antimicrobial and anticorrosive properties (Patil et al. 2019).

7.6.2 Polyhydroxyalkanoates

PHAs are primarily made up of such monomeric core component 3-hydroxyalkanoates and constitute a class of the polyesters produced intracellularly by a variety of prokaryotic species, including cyanobacteria, gram-positive and gram-negative eubacteria, and even archaea covered by harsh circumstances (Lemos Machado Abreu et al. 2017). PHAs, often referred to as poly-4-alkan-2-oxelanones, are substitute polymeric products that were derived from the naturally recyclable and biowaste sources (Ehrenhauser 2015).

PHAs are biopolyesters synthesized by a wide range of the microorganisms as intracellular additions, particularly in the context of plentiful carbon and scarce critical nutrients. They build up inside cells as transitions into the stationary stage, and that could constitute as much as 80% of its mass. These additions are indeed membrane-bound proteins and lipid aggregates. The cells can withstand oxidative damage, UV radiation, thermal shock, and osmotic instability because of their role as energy reserves. They are composed of polyhydroxyalkanoic acid monomer components, where the carboxyl group with each monomeric unit creates an ester link through hydroxyl group subsequent monomeric component. A side chain of an alkyl is created by the R-group in each monomeric unit. The different polymers and co-polymers can develop from different monomers based on the organism's substrates (Koller 2020). They have limited oxygen permeability, are insoluble, and are weatherproof. Glass-transition temperatures are between –50 and40°C, having melting temperatures ranging from 40 to 180°C. Depending upon that R-group, the temperature varies (Johnsson and Steuer 2018). The main manufacturers of PHA were cyanobacteria, which are always being researched on approaches to increase biosynthesis in the nutrient-restricted environments and through genetically engineered techniques. PHA with properties resembling those of petroleum-based plastics is known as polyhydroxybutyrate (PHB), which has also been widely commercialized (Zhang et al. 2019). It is composed of methyl groups and three carbon atom cyclic units. The metabolic pathway for producing PHB with acetyl-CoA serves as a branch site. The biosynthesis enzymes can be increased or other acetyl-CoA utilizing enzymes can be eliminated to enhance the substrate for PHB biosynthesis, which will enhance PHB manufacturing. In the core metabolism of microalgae, acetyl-CoA is a branch site. In the cells, glucose is changed into pyruvate, which can subsequently be changed into glyceraldehyde 3-phosphate and begin the Calvin cycle to make carbohydrates. The essential substrates for PHB production are acetyl-CoA, which is produced from pyruvate by the enzyme pyruvate dehydrogenase. Another method of producing acetyl coenzyme A is acetyl phosphate transformation. Cells can employ the produced acetyl-CoA in PHB production, the tri-carboxylic acid cycle, or by transforming it to acetate. In the PHB synthesis route, a group of enzymes, includes PhaA, PhaB, and PhaC/E, transform acetyl-CoA to PHB (Ganapathy et al. 2018).

7.6.3 Polylactic Acid

Polylactic acid (PLA) is an aliphatic-type thermoplastic polyester that is recyclable and compostable. It differs from many of the thermoplastic polymers because it is made from sustainable materials, especially biomass waste, by-products, and residues such as maize starch, sugarcane, and tapioca (Mekonnen et al. 2013). Lactic acid (LA) was synthesized from these materials utilizing proper biotechnology techniques as a platform chemical and a predecessor for such creation of the sustainable, biocompatible LA biopolymers. Such thermoplastic polymers, which are mostly employed for high-end purposes, are very potential in the biomedical, textile, and pharmaceuticals fields, including suture components, implantable matrix, and organ regeneration, along with specialized textile fiber uses, furniture, agricultural goods, pharmaceuticals, and drug release systems. In addition to being biodegradable and made from sustainable materials, PLA is used in diverse applications because of how simple it is to produce and behaves much under technical demand, offering it excellent characteristics at an affordable price.

Typically, microalgae biomasses are fermented to make LA and then polymerized to create PLA. One of the most effective polymers is PLA because it produces a biodegradable plastic from a smaller amount of feedstock. The amount of CO_2 emission is also lower compared to other bioplastics. It is one of the polymers that can easily switch its stereochemical composition to obtain a larger molecular mass. The various isomers can also change other properties, including amorphousness and semi crystallinity (Reddy et al. 2008). It is frequently used in food packaging application because it is an FDA-approved biopolymer. The different PLA structures can be created since monomer is chiral and occur in two optical isomeric forms. The three are poly(L-lactide), poly(D-lactide), and poly(D, L-lactide). It is usual practice to use poly(D, L-lactide) packing that contains 90% L-lactide. Because of its significant expense, a polymer with enhanced D-lactide in the mixture has a strong crystalline phase and excellent barrier and mechanical characteristics, but is not economically feasible.

7.6.4 Biopolyethylene

Biopolyethylene (bio-PE) is a polyolefin made from bio-ethylene, the naturally acceptable material that may be made from sustainable raw resources such as sugarcane and ethanol. The radiocarbon technique can be employed to distinguish among plastics and bio-PE, which is created via bio-ethylene polymerization and exhibits the same qualities and uses as traditional PE. PE is the predominant polymer used in plastic manufacture globally. It comes in a variety of kinds depending on the manufacturing process, and its characteristics can change if extra hydrocarbons such as 1-propane, 1-pentane, 1-hexane, and 1-octane are connected (Koopmans 2014).

The bio-PE has similar characteristics and features as traditional PE derived from petrochemical products, and it is synthesized using PE-specific equipment without the requirement for a distinct equipment investment. Additionally, bio-PE may be returned to the similar reusing stream because PE represents a significant benefit for the industrial sector (Mendieta et al. 2020). Since most bioproducts are hydrophilic, the moisture content influence show it functions as a gas barrier (Pandit et al. 2018). The very similar chemical compositions to petroleum-based polymers result in non-biodegradable bioplastics that possess the similar gas permeability. In the instance of bio-PE and bio-PET, lamination method was integrated with biaxial direction to enhance the O_2 and water barrier characteristics. The biaxially orientated PE or PET is very effective at preventing gas permeability because of its high degree of crystallinity and arranged packaging form. The biaxially aligned bio-PE and bio-PET films were further metalized to improve barrier characteristics (Jariyasakoolroj et al. 2020).

7.6.5 Polyethylene Terephthalate

A sizable set of polymers, including polyesters, can potentially be made using bio-based feedstocks. According to the European Bioplastic Report, the polymers polyethylene terephthalate (PET), polybutylene terephthalate (PBT), polybutylene succinate (PBS), polybutylene adipate (PBA), and the copolymers polybutylene adipate-terephthalate (PBAT), poly(butylene succinate adipate), poly(butylene succinate-co-lactate), and poly(butylene succinate terephthalate) are the most important for a shift to bio-based substances. Other important polymers include polyvinylacetate (PVAc), polyacrylates, and poly(trimethylenenaphthal). Those polymers are made out of a bio-based diol, whereas a diacid or diester may be derived out of a petrochemical or a bio-based source. The most popular polyester, PET, can be used for packaging (35%) and fiber (65%) applications. In this final application, bottles (76%), containers (11%), and films (13%), are all mentioned. PET has a significant impact on the plastics industry, however: because of its slow degradability and poor sustainability, it creates considerable environmental problems, particularly for waste treatment when used for short-term applications, like for food packaging (Guidotti et al. 2018). It was originally made available for sale in 1940, primarily for applications involving synthetic fibers and films. It was first used to make bottles in 1970 and production has only increased since then.

The bio-based PET that is produced in this manner is equivalent to petrochemical PET and may be processed via blow or injection molding and extrusion. In spite of the fact that the glycolysis method is useful for depolymerizing opaque and colored PET that cannot be recycled because of the presence of pigments, this method is not economical or environmentally sustainable due to the high energy costs associated with the high temperatures needed and the lengthy reaction times required for depolymerization (Chen and Patel 2012). The produced EG and PTA monomers could be employed once more to create PET and many other polymers. A variety of enzymes from different microorganisms could be employed as a more sustainable solution than conventional PET depolymerization techniques.

As can be seen, only a small number of materials are in significant and commercial phases, with the majority of bio-based plastics being in the research and development and pilot plant stages. Different types of technological difficulties are only conceivable in the first scenario. Since succinic acid manufacturing is theoretically feasible, there are no technical obstacles to the creation of partial bio-PBS. PBS is produced on a wide scale by esterifying succinic acid with butanol but using petrochemical precursors. The manufacture of partial bio-PET is even simpler than PBS. Scaling up is already difficult and demanding, but producing bio-PP involves a number of procedures for which there is a lack of information (Misra et al. 2011). However, even after the technology is ready for use in a plant, a number of factors can still affect how quickly it is commercialized. These include financial, technological, human resources, interactions with other industries, cooperation with other plants and agro-industrial chain companies, and regulatory factors related to fiscal policies and public policy.

7.6.6 Polyamide

A synthetic polymer is called a polyamide. It is created by joining the amino groups of two different molecules with the carboxylic acid groups of a third. It contains synthetic fibers like nylon. Because of its strength and durability, synthetic polyamide is widely used in the culinary, textile, and automotive sectors. Industrial production of aliphatic polyamides (PA) involves either step-growth polycondensation or ring-opening polymerization. Designed polymers, or PA, are utilized as elevated materials (Jiang and Loos 2016). Examples include the ring-opening polymerization of caprolactam to produce PA6 and the polycondensation of adipic acid with hexamethylenediamine to produce PA6,6. Some bio-based Pas, including PA11, PA10,10, and partially bio-based variants made from the same subunits have previously been produced.

The PA12 that is now on the market is solely derived from fossil fuels. On an industrial scale, it is made by ring-opening polymerizing lauryl lactam that has been converted through a number of steps of production beginning with butadiene. Additionally, it can be made from x-amino dodecanoic acid. In Japan, 12-amino dodecanoic acid is used by UBE to create PA12 throughout the manufacturing process. The only differences between conventional fossil-based PA12 and entirely bio-based PA12 are the inputs used and the production process; as a result, bio PA12 should be viewed as a drop-in replacement. Currently, it has been established that fatty acids sourced from sustainable sources are appropriate monomers and building blocks for the manufacture of bioplastics, including PA12. For instance, the industrial manufacture of bio-based PA11, which is offered for sale by Arkema, uses x-amino undecanoic acid, which is created from 10-undecenoic acid produced from castor oil. A pilot-scale factory for the production of bio-based ALA by the German chemical company Evonik began operating in Slovakia in 2013. The company was already employing palm kernel oil (PKO) as a base for the creation of other chemicals when the biotechnological manufacturing process began. PKO has a significant amount (more than half) of saturated C12 lauric acid (Degli Esposti et al. 2021). Most of the PKO-containing fruit palm trees are produced in Asia. Arkema is looking into an alternative method using the castor oil-derived C11 undecanoic acid methyl ester and cross-metathesis with acrylonitrile.

7.6.7 Polytrimethylene Terephthalate

One of the most significant commercial polyesters is polytrimethylene terephthalate (also known as poly-1,3-propylene terephthalate [PTT]). It is a semi-crystalline, semi-aromatic thermoplastic that is simple to thermoform, shape, and spin into fibers. Its molding qualities are equivalent to those of polybutylene terephthalate, while its thermophysical and mechanical properties are similar to those of polyethylene terephthalate (PET). For instance, it possesses great flow and surface finish, good

TABLE 7.1
Characteristics of Bioplastics from Food Waste Using Microalgae and Their Applications

Bioplastics	Characteristics	Applications	Reference
Polyurethane	Flexible, abrasion resistance	Automotive, construction	(Kuddus 2021), (Pawar 2016)
Polyhydroxyalkanoates	Biodegradable, UV resistant	Biomedical, pharmaceutical	(Koller2020), (Ehrenhauser 2015)
Polylactic acid	Nontoxic, biocompatible	Food packaging, biomedical	(Mekonnen et al. 2013), (Reddy2008)
Biopolyethylene	Durability, thermal stability	Agricultural, textile	(Mendieta 2020), (Jariyasakoolroj2020)
Polyethylene terephthalate	Lightweight, chemical resistance	Electrical insulator, food packaging,	(Chen and Patel 2012), (Misra 2011)
Polyamide	Excellent mechanical strength, thermal resistance	Textiles, medical implants	(Degli Esposti2021), (Jiang and Loos 2016)
Polytrimethylene terephthalate	Great gas barrier, elasticity	Food packaging, textile	(Rand 2021), (Sashiwa2018)

dimensional stability, and good tensile and flexural strength. Similar to PBT, it is chemically resistant to a variety of substances, including carbon tetrachloride, gasoline, oils, fats, glycols, alcohols, dilute acids, and bases (RAND 2021). However, steam and hot water have an impact on it. Triexta fibers and carpeting are the two main applications for polytrimethylene terephthalate. This fiber is used to create robust, elastic polyester textiles and yarns with excellent abrasion and wrinkle resistance. However, it is utilized far less frequently than PET and PBT.

Most engineering plastics come from nonrenewable petroleum-based feedstock, and using them has environmental advantages, such as better fuel economy and plastic that can be reused and recycle. Sustainable and renewable engineering plastics resources are establishing themselves more firmly in the market, but the cost advantage of using renewable sources of energy often restricts the use of chemicals to specialized applications of mass production, where cost is a key characteristic, if the technical qualities are comparable. In some instances, the chemical feedstock derived from renewable resources is directly competitive with its petroleum-based counterpart. The bio-based component of the poly(trimethylene terephthalate) (PTT) polymer is 1,3-propanediol (PDO), which is produced by a fermentation process on an industrial scale. In terms of technical qualities, PTT competes with PET and PBT, two other polyesters that are often employed, for instance in automotive applications. While offering a similar level of technical performance, the partially bio-based PTT with 37% renewable content significantly exceeds the PET and PBT polyesters in terms of sustainability (Sashiwa et al. 2018). Glass fiber–reinforced PTT composites have been injection molded to use as high-performance biomaterials for vehicle parts. Table 7.1 depicts the characteristics and applications of different bioplastics produced from microalgae.

7.7 CHALLENGES OF ALGAE BIOPLASTICS

Microalgae are widely known for their simplicity in production, ability to thrive in unfavorable settings, and potential to grow in effluents; however, developing marketable products from this feedstock has significant challenges at every stage along the road, from planting to commercial launch. Not all types can acclimate to a variety of cultural contexts. The culture stage may be difficult since the research strain may not be acclimated to the intended circumstances. The second challenge comes from the expensive harvesting process, which comes after cultivation. The majority of research reveals some ways to increase metabolite production, but there is a pressing need to find efficient means to extract them. Additionally, the amount of biomass and compounds generated will not be enough for industrial output. Although there is a great potential of contamination, this problem can be resolved by creating heterotrophic conditions using an external carbon source. Although it is generally recognized that genetic engineering increases metabolite production, it also presents a number of difficulties, such as the requirement for a genetic profile, the difficulty of altering genes, and the upkeep and genomic stability of mutant strains. Furthermore, modified cyanobacteria cannot be produced in open platforms because they can harm the environment. Genetic engineering can be replaced with mutagens, which can be used to create mutations at random. However, in order to locate the variant with the appropriate characteristics, substantial screening is required.

Mixed cultured with a cyanobacterial predominance are also recognized for producing significant levels of PHA, although it might be challenging to maintain this majority until purification. Maintaining high N:P ratios is seen to be a possible approach because cyanobacteria can withstand high nitrogen concentrations. Though scaling up production will undoubtedly take longer, mixed cultivations have only been utilized on the pilot scale and in laboratory. The use of pure algal bioplastics is constrained by their inferior mechanical strength when compared to currently used polymers. However, one viable solution for solving this issue is to use sustainable biomass as additives, namely compatibilizers, plasticizers, and fillers. Bioplastics are growing in the market because of their significant biodegradability. But the circumstances needed for this might not exist in landfills, where they are typically disposed of. As a result, future bioplastics that degrade under typical circumstances will need to be modified. In addition, bioplastics may release a negligible amount of greenhouse gases if improperly disposed of. On the other side, these gases can be gathered and used for different things, such the creation of biogas. Future studies on algae should concentrate on identifying solutions to each of the mentioned challenges in order to improve industrial productivity. Only about 1% of plastics are biodegradable nowadays; the remainder is made of fossil materials and poses a threat to momentous terrestrial and aquatic flora and fauna. If this situation continues, it will affect the world's species and might possibly cause some to become extinct, diminishing biodiversity. This demonstrates the negative effects of plastic use on the planet, which can be lessened by using long-lasting algal bioplastics. Because algal bioplastics are used, it serves as an example of an environmentally favorable cycle. Algae produce and accumulate a range of metabolites through the process of photosynthesis, which uses sunlight and CO_2 as raw materials. These metabolites can be enhanced by modifying the nutritional environment, employing gene editing, or generating mutations. These metabolites can be converted into biopolymers, which can then be separated from other materials and refined to create bioplastics. These bioplastics break down after being disposed of and release CO_2, which algae utilize to reproduce. A greener and safer earth will be the consequence of research into bioplastics, their advancement, and eventually their complete replacement with plastics.

7.8 CONCLUSION AND FUTURE PERSPECTIVE

Overuse of plastics contributes to pollution, which harms the planet and its native animals. Algae bioplastics have several advantages, but more study needs to be done in this field. Additionally, it is vital to make use of contemporary genetic technology to increase the synthesis of metabolites for the creation of bioplastics. Utilizing only these renewable resources contributes to the development of highly compostable bioplastic because it has been demonstrated that using different algal and cyanobacterial strains as additives increases the mechanical qualities of bioplastics to a level comparable to adding other synthetic components.

A number of genetic modification techniques can be used to target a transcription factor or a gene involved in a metabolic pathway. However, because algae are intricate eukaryotic organisms with a dearth of sequenced genomes, most investigators are constrained to examining just those organisms whose genomes have been sequenced. Algal genome sequencing needs to be independently researched in order to study algal genetics. To increase yields even more, these genetically modified organisms can also be grown in nutrient-poor environments. Algal gene editing experiments, though, are restricted to laboratory research, and these genetically altered organisms cannot be farmed for the production of industrial metabolites. The methods for increasing metabolite synthesis can be used singly or in various combinations to provide economic yields on a broad scale.

REFERENCES

Abdel-Hamid SMS, Al-Qabandi OA, Elminshawy NAS, et al. (2019) Fabrication and characterization of microcellular polyurethane sisal biocomposites. Molecules 24:1–16. https://doi.org/10.3390/molecules24244585

Banerjee S, Khan AA (2022) Role of algal-bacterial association in combined wastewater treatment and biohydrogen generation: An overview on its challenges and future. In: Integrated Environmental Technologies for Wastewater Treatment and Sustainable Development. Elsevier, pp. 509–532.

Chen GQ, Patel MK (2012) Plastics derived from biological sources: Present and future: A technical and environmental review. Chem Rev 112:2082–2099. https://doi.org/10.1021/cr200162d

Chong JWR, Khoo KS, Yew GY, et al. (2021) Advances in production of bioplastics by microalgae using food waste hydrolysate and wastewater: A review. Bioresour Technol 342.

Cinar SO, Chong ZK, Kucuker MA, et al. (2020) Bioplastic production from microalgae: A review. Int J Environ Res Public Health 17.

Dang BT, Bui XT, Tran DPH, et al. (2022) Current application of algae derivatives for bioplastic production: A review. Bioresour Technol 347:126698.

Das SK, Sathish A, Stanley J (2018) Production of biofuel and bioplastic from chlorella pyrenoidosa. In: Materials Today: Proceedings. Elsevier, pp. 16774–16781.

Degli Esposti M, Morselli D, Fava F, et al. (2021) The role of biotechnology in the transition from plastics to bioplastics: An opportunity to reconnect global growth with sustainability. FEBS Open Bio 11:967–983. https://doi.org/10.1002/2211-5463.13119

Devadas VV, Khoo KS, Chia WY, et al. (2021) Algae biopolymer towards sustainable circular economy. Bioresour Technol 325:124702.

Ehrenhauser FS (2015) PAH and IUPAC nomenclature. Polycycl Aromat Compd 35:161–176. https://doi.org/10.1080/10406638.2014.918551

Fabra MJ, Martínez-Sanz M, Gómez-Mascaraque LG, et al. (2017) Development and characterization of hybrid corn starch-microalgae films: Effect of ultrasound pre-treatment on structural, barrier and mechanical performance. Algal Res 28:80–87. https://doi.org/10.1016/j.algal.2017.10.010

Ganapathy K, Ramasamy R, Dhinakarasamy I (2018) Polyhydroxybutyrate production from marine source and its application. Int J Biol Macromol 111:102–108. https://doi.org/10.1016/j.ijbiomac.2017.12.155

Gross M, Jarboe D, Wen Z (2015) Biofilm-based algal cultivation systems. Appl Microbiol Biotechnol 99:5781–5789.

Guidotti G, Soccio M, Lotti N, et al. (2018) Poly(propylene 2,5-thiophenedicarboxylate) vs. poly(propylene 2,5-furandicarboxylate): Two examples of high gas barrier bio-based polyesters. Polymers (Basel) 10. https://doi.org/10.3390/polym10070785

Günerken E, D'Hondt E, Eppink MHM, et al. (2015) Cell disruption for microalgae biorefineries. Biotechnol Adv 33:243–260.

Harmon VL, Wolfrum E, Knoshaug EP, et al. (2021) Reliability metrics and their management implications for open pond algae cultivation. Algal Res 55:102249. https://doi.org/10.1016/j.algal.2021.102249

Jariyasakoolroj P, Leelaphiwat P, Harnkarnsujarit N (2020) Advances in research and development of bioplastic for food packaging. J Sci Food Agric 100:5032–5045. https://doi.org/10.1002/jsfa.9497

Jiang Y, Loos K (2016) Enzymatic synthesis of biobased polyesters and polyamides. Polymers (Basel) 8. https://doi.org/10.3390/polym8070243

Johnsson N, Steuer F (2018) Bioplastic material from microalgae: Extraction of starch and PHA from microalgae to create a bioplastic material. KTH R Inst Technol Sch Ind Engineering Manag 42.

Karan H, Funk C, Grabert M, et al. (2019) Green bioplastics as part of a circular bioeconomy. Trends Plant Sci 24:237–249.

Koller M (2020) "Bioplastics from microalgae"—Polyhydroxyalkanoate production by cyanobacteria. Elsevier Inc.

Koopmans RJ (2014) Polyolefin-based plastics from biomass-derived monomers. Bio-Based Plast Mater Appl 2010:295–310. https://doi.org/10.1002/9781118676646.ch11

Kuddus MR, Roohi (2021) Bioplastics for sustainable development. Springer Nature. https://doi.org/10.1007/978-981-16-1823-9

Lemos Machado Abreu AS, de Moura IG, de Sá AV, Alves Machado AV (2017) Biodegradable polymernano-composites for packaging applications. Elsevier Inc.

Lutzu GA, Ciurli A, Chiellini C, et al. (2021) Latest developments in wastewater treatment and biopolymer production by microalgae. J Environ Chem Eng 9:104926. https://doi.org/10.1016/J.JECE.2020.104926

Mathiot C, Ponge P, Gallard B, et al. (2019) Microalgae starch-based bioplastics: Screening of ten strains and plasticization of unfractionated microalgae by extrusion. Carbohydr Polym 208:142–151. https://doi.org/10.1016/j.carbpol.2018.12.057

Mekonnen T, Mussone P, Khalil H, Bressler D (2013) Progress in bio-based plastics and plasticizing modifications. J Mater Chem A 1:13379–13398. https://doi.org/10.1039/c3ta12555f

Mendieta CM, Vallejos ME, Felissia FE, et al. (2020) Review: Bio-polyethylene from wood wastes. J Polym Environ 28. https://doi.org/10.1007/s10924-019-01582-0

Misra M, NagaraJan V, Reddy J, Mohanty AK (2011) Bioplastics and green composites from renewable resources: Where we are and future directions! ICCM Int Conf Compos Mater 1–5.

Mutanda T, RameSh D, Karthikeyan S, et al. (2011) Bioprospecting for hyper-lipid producing microalgal strains for sustainable biofuel production. Bioresour Technol 102:57–70. https://doi.org/10.1016/j.biortech.2010.06.077

Noreen A, Zia KM, Zuber M, et al. (2016) A critical review of algal biomass: A versatile platform of bio-based polyesters from renewable resources. Int J Biol Macromol 86:937–949.

Otoni CG, Azeredo HMC, Mattos BD, et al. (2021) The food-materials nexus: Next generation bioplastics and advanced materials from agri-food residues. Adv Mater 33:2102520.

Pandit P, Nadathur GT, Maiti S, Regubalan B (2018) Functionality and properties of bio-based materials. Bio-based Mater Food Packag Green Sustain Adv Packag Mater 81–103. https://doi.org/10.1007/978-981-13-1909-9_4

Patil CK, Jirimali HD, Paradeshi JS, et al. (2018) Synthesis of biobased polyols using algae oil for multifunctional polyurethane coatings. Green Mater 6:165–177. https://doi.org/10.1680/jgrma.18.00046

Patil CK, Jirimali HD, Paradeshi JS, et al. (2019) Functional antimicrobial and anticorrosive polyurethane composite coatings from algae oil and silver doped egg shell hydroxyapatite for sustainable development. Prog Org Coatings 128:127–136. https://doi.org/10.1016/j.porgcoat.2018.11.002

Pawar MS, Kadam AS, Dawane BS, Yemul OS (2016) Synthesis and characterization of rigid polyurethane foams from algae oil using biobased chain extenders. Polym Bull 73:727–741. https://doi.org/10.1007/s00289-015-1514-1

Qin S, Lin H, Jiang P (2012) Advances in genetic engineering of marine algae. Biotechnol Adv 30:1602–1613.

Rajpoot AS, Choudhary T, Chelladurai H, et al. (2022) A comprehensive review on bioplastic production from microalgae. Mater Today Proc 56:171–178. https://doi.org/10.1016/j.matpr.2022.01.060

RAND E (2021) Product packaging. Ellis Isl Snow Globe 207–238. https://doi.org/10.2307/j.ctv11cw45p.12

Reddy RL, Reddy VS, Gupta GA (2008) International journal of emerging technology and advanced engineering study of bio-plastics as green & sustainable alternative to plastics. Certif J 9001:82–89.

Richardson JW, Johnson MD, Zhang X, et al. (2014) A financial assessment of two alternative cultivation systems and their contributions to algae biofuel economic viability. Algal Res 4:96–104. https://doi.org/10.1016/j.algal.2013.12.003

Rosenboom JG, Langer R, Traverso G (2022) Bioplastics for a circular economy. Nat Rev Mater 7:117–137.

Roy Chong JW, Tan X, Khoo KS, et al. (2022) Microalgae-based bioplastics: Future solution towards mitigation of plastic wastes. Environ Res 206:112620. https://doi.org/10.1016/j.envres.2021.112620

Sashiwa H, Fukuda R, Okura T, et al. (2018) Microbial degradation behavior in seawater of polyester blends containing poly(3-hydroxybutyrate-co-3-hydroxyhexanoate) (PHBHHx). Mar Drugs 16:1–11. https://doi.org/10.3390/md16010034

Shandilya KK, Pattarkine VM (2019) Using microalgae for treating wastewater. In: Advances in Feedstock Conversion Technologies for Alternative Fuels and Bioproducts: New Technologies, Challenges and Opportunities. Woodhead Publishing, pp. 119–136.

Sreenikethanam A, Bajhaiya A (2022) Algae based bio-plastics: Future of green economy. In: Biorefineries—Selected Processes. IntechOpen.

Tan FHP, Nadir N, Sudesh K (2022) Microalgal biomass as feedstock for bacterial production of pha: Advances and future prospects. Front Bioeng Biotechnol 10. https://doi.org/10.3389/fbioe.2022.879476

Wang K, Mandal A, Ayton E, et al. (2016) Modification of protein rich algal-biomass to form bioplastics and odor removal. In: Protein Byproducts: Transformation from Environmental Burden into Value-Added Products. Academic Press, pp. 107–117.

Zhang C, Show PL, Ho SH (2019) Progress and perspective on algal plastics—a critical review. Bioresour Technol 289:121700. https://doi.org/10.1016/j.biortech.2019.121700

8 Third-Generation Bioplastics from Municipal Waste Material

Krishna Gautam, Pallavi Gupta, Shreya Dwivedi, and Vivek Kumar Gaur

CONTENTS

8.1 INTRODUCTION

Growing demand for plastic goods is a result of population growth. On average, 150 million tonnes of fossil fuel are needed to produce 140 million tonnes of plastic that are used annually. If this pattern persists, plastic is predicted to outweigh all marine life by 2050 (Nanda and Bharadvaja, 2022). Because of their long history of usage and advanced technology, petroleum-based plastics provide comfort, adaptability, durability, and affordability. However, because they are not biodegradable, they amass in the environment and pose harm to the ecosystems. One of the primary obstacles to the economic transformation has been recognized as the difference between the demand and supply for biomass. There is not enough arable land available to fulfill the anticipated demand in the future, given existing usage and harvesting methods. In order to get over the obstacle of a biomass scarcity, there are three issues that need to be resolved. First, while ensuring that resources are not overexploited, sufficient biomass production is required. Second, it is inevitable that greenhouse gas emissions brought on by the production of biomass and the resulting changes in land use must be controlled. Third, biomass-generating processes must be economically viable. There are significant research efforts being made in this regard. Biopolymers such as polyhydroxyalkanoates (PHA) and polylactic acid (PLA) have consequently lately become more well-known. They have the combined benefits of being biodegradable and having mechanical and physical qualities that are on par with those of synthetic polymers. In comparison to petrochemical-based plastics, which take a thousand years to disintegrate, bioplastics break down by at least 60% into CO_2 and water after 3 months of composting (Saratale et al., 2020).

Living in both marine and freshwater habitats, microalgae are microscopic in size that use solar energy to make adenosine triphosphate (ATP). A growing number of people are interested

DOI: 10.1201/9781003344018-8

in microalgae because of its implications for the bioeconomy. In the European bioeconomy, they might be crucial in alleviating the biomass scarcity. Cyanobacteria are one type of microalgae that has the ability to close the biomass supply-demand gap, making them an important feedstock for the switch to a sustainable green economy. Because of their exponential growth rates and potential for year-round production, microalgae can meet the rising need for biomass. Furthermore, microalgae require less area than terrestrial crops, avoiding both land rivalry with food plants and the environmental effects of changing land usage. Microalgae may reduce the use of a variety of different sources for the generation of biofuel, as well as being employed in the food and cosmetics sectors and in medicinal formulations (Cinar et al., 2020). Production of fundamental components like saccharides, proteins, and lipids, together with secondary components like pigments, bioplastics, antioxidants, or vitamins, should be taken into consideration in order to create more environmentally friendly and economically viable processes. Bio-based plastics are one of them and are becoming more and more popular as an alternative to fossil-based plastics in the market (Cinar et al., 2020). However, during the past 10 years, there have been several issues with the large-scale marketing of bio-based substances. Microalgae may be a preferable biomass source for the synthesis of bioplastics because it does not compete with food sources, can grow on waste resources, and can accumulate significant levels of lipids. Usually synthesized from raw feedstock, the first generation of biopolymers competes with food production since they need nutrients, arable land, and fresh water (Madadi et al., 2021). For the synthesis of bioplastic, these resources are constrained and inadequate. Microalgal biomass has recently been identified as a possible source of raw materials for numerous industries, including the production of bioplastics called as third-generation bioplastics. It may be utilized directly as biomass or as a feedstock for secondary processes. Because of this, research in bioplastics derived from quickly reproducing microorganisms like bacteria and microalgae has surged (Madadi et al., 2021). If microalgae are effectively transformed into bioplastics and biopolymers, which are desperately needed on a worldwide scale to replace petroleum-based plastics, their sustainability potential will be further increased. This chapter discusses the different types, significance, and strategies for the cultivation and harvesting with the use of additives for the production of third-generation bioplastics using appropriate and efficient approaches. The life cycle assessment for the production of bioplastic using microalgae and their suitability in terms of removing CO_2 from the environment are also examined and addressed with the relevant challenges and future perspectives.

8.2 MUNICIPAL WASTE MATERIAL AS SUBSTRATE FOR MICROALGAE CULTIVATION

The microalgal biomass has been widely studied as a feedstock for the production of third-generation bioplastics that increases the biomass productivity. Microalgae cultivation is increasing in demand in different industrial sectors. Biorefineries and renewable resources are different strategies that implement the advancement of microalgae cultivation. Despite their many advantages, the production of microalgae biomass possesses various drawbacks or restrictions such as the small size, huge distribution of cultivation medium, and cultivation cost that ranges between USD 20 and USD 200 per kg of microalgal biomass (Gupta et al., 2015) Such huge economy expenses in the microalgae cultivation in photobioreactors (PBRs) are produced mainly due to the following factors: CO_2 supplementation, mixing cultivation, medium preparation, and biomass harvesting where 20%–30% of the total cost of production consists of separation and harvesting of biomass. Hence it is important to select the appropriate method for biomass separation in the financial aspect of cultivation (Garbowski et al., 2020). It has been studied that microalgae utilizes inorganic phosphorus and nitrogen for their development and has the ability to reproduce in various types of wastewaters (Garbowski et al., 2020). Thus municipal waste that serves as a good source of phosphorus and nitrogen can be used as a substrate to reduce the costs of microalgal cultivation. Several studies reported the use of wastewater from different sources, such as domestic, industrial, agricultural, municipal, and effluents from landfills for microalgal cultivation (Figure 8.1; Garbowski et al., 2020). The Department of Environment Malaysia reported that industrial effluent did not hold on to standards and was improperly treated prior to being discharged into aquatic or marine environment. Researchers studied the waste discharge of palm oil mill effluent, a rich source of carbon for utilization in microalgae cultivation produced from the palm oil industry, generated various environmental toxic by-products. Microalgae have gained potential in treating wastewater and producing biomass; it also helps in uptake and removal of toxic chemicals that are discharged from industrial effluents into water bodies. These wastes can serve as substrate to produce high lipid microalgae content as a raw material for the production of bioenergy (Kamyab et al., 2018).

Several prokaryotic microalgae such as cyanobacteria have been considered best for the bioplastics production owing to their restricted nutritional necessities and survival in unsuitable environment such as wastewater. Municipal wastewater treatment plants (WWTPs) have been widely studied as a source of raw material and biomass for the synthesis of PHA as they provide an opportunity to indemnify the drawbacks of PHA production using starchy plants as a substrate. A microalgae consortium cultivated and harvested in a WWTP was used as biomass in addition with glycerol as a plasticizer to produce bioplastics (Coppola et al., 2021). Researchers implemented a two-step process for the extraction of PHAs from microbial consortium utilizing fermented oil mill wastewater. In a two-step process, pure microbial cultures and consortium aerobically converted volatile fatty acids (VFA) from anaerobic fermentation to produce PHA. Additionally, two- and three-step processes have also been studied for the production of PHAs by utilizing cheese whey agro-industrial wastewater and paper mill wastewater, respectively (Mannina et al., 2019). Several researchers also studied the sustainable production of bioplastics from organic fraction of municipal solid wastes (OF-MSW) that were profitable for the production of biofuels and biodegradable PHA with a yield of 257.4 mL/kg and 40g/kg, respectively (Ebrahimian et al., 2020). Other municipal waste sources such as casein-rich dairy wastewater, potato processing industry wastewater, activated sludge, wood mill effluents, and municipal sewage sludge have been investigated to be used as a possible substrate for the production of third-generation bioplastics (Coppola et al., 2021). Thus for bioplastic production, microalgae biomass can serve as a good source that possesses the capability to produce bioplastic on municipal waste resources. Moreover, the production of bioplastics from microalgae is sustainable and contributes toward the circular economy and the bioeconomy. Therefore the produced bioplastics have wide applications in food packaging, cosmetics, and pharmaceuticals.

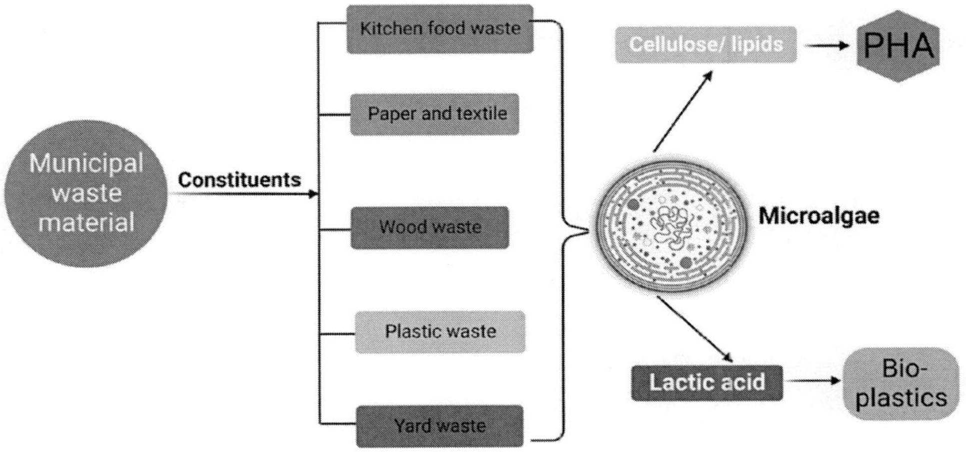

FIGURE 8.1 The utilization of different municipal waste materials for the production of bioplastic.

8.3 MICROALGAE

Microalgae biomass serves as a good substrate for the production of bioplastics as it is biodegradable, cost-effective, and produces a high quantity of biomass. The algae-based bioplastics have the inherent property of easy biodegradability in nature due to the presence of PHB molecules in the algae cell. Microalgae can be either used directly or indirectly as a biomass for the production of bioplastics or extraction of PHBs within algal cells (Coppola et al., 2021). Microalgae have become an interesting topic for study as a source for the production of various beneficial products for industrial applications. This section highlights the microalgae species utilized for bioplastic production (Table 8.1).

TABLE 8.1
Bioplastic Production Utilizing Microalgae under Different Conditions

Algal Biomass	Operational Conditions	Type of Polymer Produced	Polymer (% Dry Cell Weight)	References
Spirulina platensis	6% MAH and 30% glycerol	PVA-blend film	–	(Gozan and Noviasari 2018)
Synechococcus subsalsus	–	PHAs	16%	(Costa et al. 2018)
Spirulina sp.	–	PHAs	12%	(Costa et al. 2018)
Aulosira fertilissima	under phosphorus limited conditions	P(3HB)	77%	(Madadi et al. 2021)
Microalgae consortium	cultivated in wastewater	PHA production	43%	(Chakravarty et al. 2010)
Synechocystis salina	CO_2 as a substrate	P(3HB)	7.5%	(Madadi et al. 2021)
Nostoc muscorum	0.2% acetate as a substrate	P(3HB)	35%	(Madadi et al. 2021)
Spirulina platensis	CO_2/acetate as a substrate	P(3HB)	10%	(Jau et al. 2005)
N. muscorum	under nitrogen deficiencies	P(3HB-CO-3HV)	78%	(Bhati and Mallick 2015)
N. muscorum	under phosphorus deficiencies	P(3HB-CO-3HV)	71%	(Bhati and Mallick 2015)
Chlorella pyrenoidosa	–	PHB	27%	(Das et al. 2018)
Phaeodactylum tricornutum	acetate supply	PHB	10.6%	(Hempel et al. 2011)
Chlorogloea fritschii	–	poly(3-hydroxybutyrate)	17%	(Monshupanee et al. 2016)
Botryococcus braunii	sewage wastewater	PHB	–	(Monshupanee et al. 2016)
Chlorella sp. and *Scenedesmus* sp.	fermentation with wastewater treatment	PHB, lipids, and bioethanol	–	(Kumar et al. 2020)
Chlorella	ultrasonic *Chlorella* treatment	*Chlorella*-PVA–blend films	–	(Sabathini et al. 2018)
Synechococcus elongates	phosphorus and nitrogen deficiency	PHA	17.15%	(Mendhulkar and Shetye 2017)
Synechococcus sp.	nitrogen deficiency and using acetate as carbon source	PHB	26%	(Nikel et al. 2006)
Cupriavidus necator	–	PHAs	75.4%	(Khomlaem et al. 2021)

8.3.1 Chlorella

Chlorella, green algae mainly found in freshwater bodies, have high protein content (~58%) with a cell wall consisting of cellulose and pectin (Khalis, 2018). *Chlorella* is one of the main microalgae involved in the synthesis of bioplastics. Various *Chlorella* species were studied for biomass-polymer blends containing different additive and polymers. *Chlorella vulgaris, Chlorella sorokiniana*, and *Chlorella pyrenoidosa* are some of the species that have been widely studied for the production of bioplastics (Coppola et al., 2021). *Chlorella* possesses several properties such as high thermal stability and exceedingly small cell size that allows them to be suitable for the production of bioplastics without any pre-treatment, thus reducing its cost expenses. The addition of polyvinyl alcohol (PVA, a polymeric material) in the formation of bioplastics enhances the flexibility, durability, and strength (Khalis, 2018). The addition of 6% maleic anhydride as a compatibilizer during production of composites by *C. vulgaris* showed better quality by increasing the tensile strength and other mechanical characteristics (Onen Cinar et al., 2020). The production of P(3HB) from different strains of *Chlorella* microalgae (*Chlamydomonas reinhardtii, Chlorella pyrenoidosa, Chlorella vulgaris, and Chlorella fusca*) were examined for recyclability, biodegradability, biocompatibility, and plasticizing capacity, and it has been reported that the bioplastics produced from *C. pyrenoidosa* showed higher biodegradability. This difference in biodegradability was attributed to the differences in their structure. Das et al. (2018) also studied the production of bioplastics from *C. pyrenoidosa* and reported that after 2 weeks of growth, 27% of PHB content were observed which was eco-friendly and easily biodegradable. *Chlorella* species has been greatly focused for the production of bioplastics due to its fast growth rate and cultivation ease. Khomlaem et al. (2021) utilized defatted *Chlorella* biomass (DCB) as a substrate for different bacterial strains (*Haloferax mediterranei* DSM, *Cupriavidus necator*, and *Bacillus megaterium*) for the accumulation of PHAs and reported that the *C. necator* strain employed 75.4% of DCB and produced the highest amount of PHA (7.51 g/L). Kamyab et al. (2018) studied *Chlorella pyrenoidosa* cultivation utilizing different municipal wastes (palm oil mill effluent [POME], domestic, piggery, and kitchen wastes) as substrates and reported that the optimum growth rate was found in POME, suggesting that it is more suitable as a nutrient substrate for microalgae cultivation. Thus microalgae as a third-generation feedstock has potential for sustainable bioenergy production such as biogas, bioplastics, and biodiesel (Khomlaem et al., 2021).

8.3.2 Spirulina

Spirulina, a blue-green algae, is a good protein source, having 60% of protein content, and is well-known for its potential to adapt to extreme harsh environmental conditions. One of the known species of *Spirulina*, namely *Spirulina platensis*, has high protein content and can be efficiently used for bioplastics production. *S. platensis* is a spiral cyanobacteria that can be cultivated in brackish water, freshwater, and seawater. Like *Chlorella*, *Spirulina* also possess small cell size, which is an attractive feature for bioplastic blend production, but both are dissimilar in bioplastic properties during blending due to the difference in their amino acid components. Thus, due to its small cell size, the high protein content in *S. platensis* does not requires extraction and can be directly utilized, thereby reducing the cost of extraction (Gozan and Noviasari, 2018). *S. platensis* and other salt-rich *Spirulina* spp. have been reported for the production of biodegradable bioplastics using PVA as a plasticizer. It has better blending performance, so the blending of PVA with *S. platensis* reduces the cost of material for biosynthesis of plastics. Furthermore, the addition of a compatibilizer to the *S. platensis* and PVA mixture enhances the tensile strength and elongation capacity of the produced bioplastics (Onen Cinar et al., 2020). Gozan and Noviasari (2018) studied the effect of addition of different concentrations of glycerol as a plasticizer on bioplastics produced from *S. platensis* and found that the addition of 30% glycerol enhances the flexibility and tensile strength of the produced bioplastics that can be effectively used in different industrial sectors. Therefore the addition of glycerol in *S. platensis* showed indistinguishable properties between commercial plastic bags and the produced bioplastics; besides this, the produced bioplastics have better properties than the commercial ones. Researchers used *Spirulina* algal biomass for bioplastics production using compression molding. They also studied the difference between the plasticized or blending *Spirulina* biomass with ultra-high-molecular-weight polyethylene (UHMW-PE); they compatibilized *Spirulina* biomass with polyethylene-graft-maleic anhydride (PE-g-MA) to enhance bioplastics performance and reported that plasticization of *Spirulina* with UHMW-PE obtained high intersurface attachment and consistent phase separation (Onen Cinar et al., 2020). *Spirulina* are good source of PHB-accumulating microorganisms and can grow on different mediums such as modified organic medium, chemical medium and on human excreta that minimizes the cost of cultivation of *Spirulina*-based bioplastics. Therefore it can be concluded that the high cost needed for the cultivation of *Spirulina* can be reduced by growing them in low-cost medium such as modified chemical medium or cheap organic medium that produced 100% biodegradable bioplastics utilizing cyanobacterial derived starch, oil, and PHB.

8.3.3 Other Microalgae Species

Apart from *Chlorella* and *Spirulina*, other microalgae species involved in the biosynthesis of bioplastics are *Scenedesmus almeriensis, Phaeodactylum tricornutum, Neochloris oleoabundans, Nannocloropsis gaditana*, and *Calothrix scytonemicola.* Various cyanobacteria or blue-green algae belonging to class Cyanophyceae (*Synechocystis* sp., *Nostoc* sp., *Oscillatoria* sp.) produced 1%–10% of PHA that acts as a carbon and energy storage complex, whereas other cyanobacterial species (*Gloeothece* spp., *Calothrix scytonemicola, Aphanothece* spp., *Synechococcus* spp.) produced P(3HB) using CO_2 as a carbon source (Madadi et al., 2021). Researchers studied the potential of *Nostoc muscorum* microalgae for P(3HB-co-3HV) copolymer films and P(3HB) homopolymer production and suggested that the PHAs produced from *N. muscorum* have better mechanical and thermal properties as compared to those obtained from bacterial origin (Bhati and Mallick, 2015). *Microcystis aeruginosa* biomass isolated from high-rated algal ponds have been studied for the bioplastic production, which represented it as a promising source for the sustainable production of bioplastics. It was also noticed that the PHB content obtained were highly biodegradable and biocompatible, which are suitable factors for the strength of PHBs, making them applicable in medicinal industrial, agricultural, and other fields. Three microalgae species (*Calothrix scytonemicola*, considered PHA-rich microalgae; *Scenedesmus almeriensis*; and *Neochloris oleoabundans*, starch-rich algae) were focused on for bioplastics production; *C. scytonemicola* showed the best results (Johnsson and Steuer, 2018). The diatom *P. tricornutum*, a marine unicellular microalga, acts as a bioreactor for the synthesis of different biopolymers such as PHB with the relatively high yield of 10.6% of algal dry weight (Hempel et al., 2011). The thermal characterization of PHA polymer synthesized from different microalgae species (including *Chlorella, Leptolyngbya valderiana, Oscillatoria salina*, and *Synechococcus elongatus*) was studied, and PHA obtained from the *L. valderiana* had the highest thermal stability (Roja et al., 2019). All this suggests that various species of microalgae are promising sources for production of bioplastics due to the presence of a large amount of protein, lipid, and carbohydrate polymers.

8.4 IMPORTANCE AND USE OF ADDITIVES

Additives are low-molecular-weight plasticizers with a non-volatile nature that are employed to enhance the mechanical strength, durability, tensile strength, flexibility, and biodegradable potential of bioplastics (Madadi et al., 2021). Microalgae and bio-petroleum products are combined with additives via thermo-mechanical methods to generate biopolymers; this is considered an environmentally sustainable alternative over synthetic polymers (Devadas et al., 2021). It has been reported that bioplastics produced from microalgae maintain better mechanical properties when compared to other petroleum-based synthetic polymers, and in industrial application, additives are mixed with microalgae to improve the quality of bioplastics. Polyvinyl acetate is one of the additives used in conjugation with *Chlorella* species of microalgae, for the production of water-soluble biofilms (Ali et al., 2021). *Chlorella*-polyvinyl acetate mixtures are subjected to pre-treatment by using an ultrasonic homogenization procedure that results in enhancement of surface properties of the bioplastic, hence it can be used in packaging of materials as an alternative of synthetic polyfilms (Ali et al., 2021). However, it has been studied that blending of additives like PVA with microalgae does not form fully homogenized bonds with each other, and therefore compatibilizers like maleic anhydride (MA) and glycerol are introduced to enhance the bonding and stimulate the polymerization reaction by providing an anhydrous group, which also increases the elongation, mechanical strength, thermal stability, interfacial bonding, and miscibility of biopolymer (Devadas et al., 2021). Cellulose acetate (CA), thermoplastic starch (TPS), PHA, polylactic acid (PLA), polypropylene (PP), and polyethylene (PE) are used as additives and mixed with microalgae to produce biodegradable and good-quality bioplastics which are used in the food packaging, cosmetic, and pharmaceutical industries. Moreover, plasticizers used as additives improve plasticization properties of bioplastics produced from microalgae. According to Zeller et al., microalgae like *Spirulina* blend with PE and glycerol to produce bioplastics with maximum tensile strength of approximate modulus of 1.0×10^8 Pa and Tan δ of 0.8, hence it has been proven that addition of glycerol during bioplastic production from microalgae can augment the extensibility and flexibility of bioplastics (Zeller et al., 2013). Therefore it was suggested that glycerol has the capability to enhance the flexibility of bioplastics as it promotes the crosslinking of biopolymers by reducing the intermolecular interaction of hydrogen molecules.

The performance, morphology, and miscibility of algae-based bioplastics can also be enhanced by the incorporation of surface modifiers like dimethyl silicone oils, gluten, and castor. However, one of the limitations of gluten-based thermoplastic is its thermal-induced crosslinking by S-S bonds that increases the material brittleness and limits its possible final applications (Devadas et al., 2021). This phenomenon can be minimized by reducing processing temperature and time. Thus gluten blended with *S. platensis* improves and innovates the thermoplastic compounds from plasticized wheat gluten (Ciapponi et al., 2019). Algae-based bioplastic are biodegradable in nature; however, it must be taken into consideration that these bioplastics will have environmental impact due to the additives added to improve the mechanical properties of the material. As synthetic plastics are non-degradable and accumulate in the environment, blending of additives with microalgae produce biodegradable plastics, which also helps in mitigation of the greenhouse effect.

8.5 MICROALGAL CULTIVATION STRATEGIES FOR BIOPLASTIC PRODUCTION

Production of bioplastics from microalgae is suggested by many studies, as algal cells have unique property to produce immense number of proteins, PHAs, cellulose, lipids, and carbohydrates (predominately starch), which are main components in the biopolymer production (Madadi et al., 2021). Hence the most suitable cultivation strategy of microalgae is the one that restores these major components required for the production of bioplastics. In this context, copious studies have been done and suggested many cultivation processes that differ from one another in terms of productivity, economical sustainability, feasibility, and applicability. Cultivation of microalgae also depends upon various physicochemical and biological parameters like light intensity, temperature, pH of the culture medium, and size and structure of algal cells. The most optimal temperature and pH for the cultivation of microalgae is 20–30°C and 6–9, respectively; however, some species can tolerate high temperatures of up to 40°C (Suparmaniam et al., 2019). These conditions can greatly influence the growth and characteristics of microalgae, as production and concentration of major components (proteins, lipids, carbohydrates, cellulose, and PHAs) can significantly affect the bioplastic production. It has been studied that a semi-continuous strategy for cultivation of *Scenedesmus* and *Cyanobacterium aponinum* microalgae can produce twice the amount of protein as compared to other batches of microalgae (Veerabadhran et al., 2021). Apart from this, cultivation of microalgae for bioplastic production is sustainable as it requires lesser amount of water than other food crops and as it is not a primary food source, its exploitation for biopolymer production did not affect the food chain. However, cultivation of microalgae is simple and can be done in freshwater, seawater, wastewater, or brackish water and does not require addition of any destructive chemicals (Veerabadhran et al., 2021). As attributed, despite the great number of strategies available for cultivation of microalgae, their industrial application is still limited. For sustainable production of algal biomass, there is a requirement of certain nutrients like iron, potassium, nitrogen, sulfur, carbon, and phosphorous in good quantity (Veerabadhran et al., 2021). Interestingly, wastewater treatment plants have excessive nutrients which prove to be an appropriate medium for microalgal cultivation.

For large-scale cultivation of microalgae, there are two main systems classified: open pond cultivation and closed cultivation (photo bioreactors). However, the better of the two systems is not known. Some researchers reported that open pond cultivation systems are better than a closed system as the former requires lesser investment and gives high output, whereas the latter has minor chances of contamination but is cost-effective (Mata et al., 2010). As per study, cultivation of *Chlorella* species of microalgae is subjected to both open and closed system cultivation, and it has been observed that in open pond system the biomass production is about 35,000 L while in closed system (pulsed magnetic field), the productivity is found to be just doubled (Baldev et al., 2021). Hence it is clear from the study that closed system cultivation is more efficient in producing more quantity of biomass. But as already mentioned, a closed system is not economically sound; most industries prefer open pond system over closed system so as to decrease the overall expenditure. On the other hand, photo-bioreactors (PBR) play an important role in cultivation of microalgae in eco-friendly manner, as this system prevents contamination and excess loss of water by evaporation; as a result it increases the efficiency and productivity of algal cells. A tubular bioreactor is an example of a closed system for cultivation of microalgae, which is considered as one of the favorable techniques for cultivation of *Synechocystis* species as it provides more surface area for illumination, which tends to increase the productivity of microalgae (Veerabadhran et al., 2021). However, this system is not applicable for filamentous algae as there will be chances of culture crash due to high stress, and additionally heating sterilization is also not possible.

Overall demand of plastic is increasing at a global scale, and bioplastics contributes to this only around 1% (European Bioplastics, 2020). To overcome this, microalgae proves to be a good alternative for the production of bioplastics which need to be explored in the near future.

8.5.1 Microalgae Harvesting

Harvesting of microalgae involves a plethora of processes and techniques, but the selection of most feasible and cost-effective techniques is still unknown. Reportedly, in the total cost for the production of microalgae, about 20%–30% is accountable by harvesting alone (Salim et al., 2011). Hence development of an integrated approach for harvesting of microalgae is the need of the hour. It has been studied that morphological properties of microalgae, such as size, density and surface charges of the algal cells, play an important role in the selection of harvesting process. On the basis of morphological characteristics, harvesting of microalgae is divided into four major processes that include biological, mechanical, chemical, and electrical methods; they have their own pros and cons (Morais Junior et al., 2020).

First, the mechanical method is the most established and commonly used method, including centrifugation, filtration, separation, flocculation, sieving, flotation, and sedimentation (Chen et al., 2011). Out of these, centrifugation is considered that most practicable, proficient, profligate, and common process for harvesting of microalgae at commercial scale with 98% proficiency (Morais Junior et al., 2020). But this technique has some limitations, as during the process the cells may get damaged due to high centrifugal and gravitational force, and it is costly because it requires high energy (Harun et al., 2010). Filtration is another commonly used harvesting technique, and in this process membranes or filters are used to separate liquid culture medium and solid algal biomass (Morais Junior et al., 2020). Filtration technique can be subdivided into tangential flow filtration, vacuum filtration, ultrafiltration, pressure filtration, microfiltration, and dead-end filtration (Dragone et al., 2010). Cells of *Arthrospira* and *Dunaliella* have been conventionally harvested by this technique (Cuellar-Bermudez et al., 2020). Mechanical separation of microalgae has been considered as one of the most efficient methods, but Knuckey et al. (2006) reported that cells may be damaged by the separation method due to gravitational forces.

Second, harvesting of microalgae from wastewater treatment plants can be achieved by biological flocculation, which involves flocculation of microalgae in the presence of bacteria. However, this approach of harvesting is costly as it requires additional substrate, which increases the cost of the process, and bacterial cultures may infect the algal colony (Salim et al., 2011). Bio-flocculation is categorized into four sub-categories on the basis of biological material used: bio-flocculation by microalgal-fungal association, microbial flocculation, plant-based bio-flocculation, and auto-flocculation. In the microalgal-fungal association bioflocculation process, it has been reported that *Aspergillus fumigatus*, a fungus associated with *Botryococcus braunii*, a microalgae, recovered about 98% biomass (Al-Hothaly et al., 2015).

Third, microalgae can also be harvested by the chemical method by combining organic and inorganic solvents. This technique is known as chemical flocculation (Zheng et al., 2012).

The fourth method is the electrical method. As already mentioned, algal cells have negative charges on their surface; hence they can be separated by exposing the cells to an electrical field. The process is known as electrophoresis, and the combination of the electrical method with flocculation is known as electro-flocculation (Vandamme et al., 2011). It has been reported that harvesting of microalgae *Dunaliella salina* by the electro-flocculation technique has a 97.44% proficiency rate (Zenouzi et al., 2013). Despite its high efficiency, this technique is not economically approachable for large-scale application as it requires more energy and has high maintenance and operational costs. However, combining two or more techniques proves to be more affordable and also increases the quality of the harvested microalgae. In support of this, in some papers it has been reported that by combining sedimentation or floatation technique with flocculation followed by centrifugation or filtration proves to be most efficient technique and requires less energy. So far there is not a single method for harvesting of microalgae which can be used both from small- and large-scale harvesting of microalgae, which preserves the integrity of the cells and does not hinder the property of the desired product. Hence further technological interventions are required for cost-effective and sustainable harvesting of microalgae.

8.6 LIFE CYCLE ASSESSMENT FOR MICROALGAL BIOPLASTIC PRODUCTION

Life cycle assessment (LCA) evaluates the full life cycle of bioplastics based upon several impact categories, giving a comprehensive picture of their socioeconomic and environmental efficacy. Human toxicity, ecotoxicity, water and land usage, acidification potential, eutrophication, ozone depletion potential, global warming potential, and other impact categories are indicators for LCA quantification. Due to their significant ecological impact, plastics are without a doubt a major subject in international forums. The search for environmentally friendly alternatives led to the development of bioplastics from renewable resources. Bio-based polymers replace synthetic ones and cut CO_2 emissions by 1.8 tonnes (Nanda and Bharadvaja, 2022). Researchers frequently point out the potential for enhancement of microalgal production systems; however, the environmental advantages of microalgal-based production are not proven. LCA research that is especially focused on microalgae-based bioplastics is not very common. For instance, researchers evaluated the synthesis of PLA from microalgae and plant-based sources and discovered considerable reduction potential for adverse ecological effect in terms of land utilization and soil ecotoxicity as a result of using microalgae. Additionally, by using and absorbing the inorganic and organic nitrogen and carbon for production, microalgae have exceptional ability to lower the chemical oxygen demand (COD) of wastewater. It also aids in economically and sustainably recovering phosphorus (Stávková and

Maroušek, 2021). This offers microalgae a competitive advantage over other bioplastic manufacturing sources. It may now adhere to a zero-waste policy and utilize biorefineries to transform leftover microalgal biomass and current garbage into useful goods. Compared to their petrochemical equivalents, the manufacturing of PHB from microalgae will provide the greatest social advantages at a fair cost and with no adverse environmental effects (Nanda and Bharadvaja, 2022). Oceans are the biggest CO_2 sink, in part because of microalgae. Algae directly absorb atmospheric carbon and sequesters it in biopolymers, making them a perfect resource for the production of bioplastics. For instance, *Chlorella vulgaris* is an extensively researched plant for the manufacture of bioplastics and can absorb the most CO_2, and up to 960 kg of CO_2 may be captured by algae (Rahman and Miller, 2017). Based on a study, the emissions of CO_2 from PLA are 1600 kg/tonne, compared to 7150, 4140, and 2740 kg/tonne for PA, PET, and PP, respectively. This shows that the production of PLA uses a closed-loop method that emits less greenhouse gases (Nanda and Bharadvaja, 2022). Therefore using bioplastics for this purpose will turn leftover biomass into a common purpose while also reducing climate change.

8.7 CHALLENGES AND FUTURE PERSPECTIVE

In comparison to its traditional equivalents, the production of biopolymers from waste strategies might be more difficult. The application of suitable and effective methods necessitates cooperation among several parties involved in waste generation, collection, conversion, manufacturing, and distribution, and the interaction of multiple sectors may result in a number of challenges. As an example, waste is produced in a primary process, but its quantity and qualities are different, necessitating its conversion in a two-step procedure (secondary processes). So it is necessary to analyze territorial flow of nutrients, materials, energy, cross-sector connections, and exploiting knowledge and information. Furthermore, it is also required to examine how microalgae affect the characteristics of bioplastics and to improve the algal biomass growing conditions. Hydrocarbon extracted from waste plastics can now only be used on a small scale in the lab, but further research should be done to increase its productivity and profitability. The use of waste by-products from refineries, including glycerol and other cellulosic sugars, provides a reliable route for the production of PHAs in a sustainable manner. However, comprehensive economic strategy has not yet been developed to fully use the fermentable sugars generated from lignocellulose (Yadav et al., 2020). Microalgae may produce more bioproducts if contemporary genetic engineering is combined with novel growth methods. Microalgae-derived bioplastics will have a role as long as there is a market for plastic goods. There are several ways to develop an economically viable bioproduct, including utilizing microalgae directly, employing derivatives of microalgae, or genetically altering microalgae strains. Future research may involve a hybrid of blending petrochemical plastics or other bio-based plastics with genetically modified microalgae strains to produce PHAs. The least expensive microalgae feedstocks for a biorefinery system will also be those produced on waste streams. However, in order to attain the productivity and resilience required to make the technology cost-competitive, algal biorefineries must overcome technological challenges related to both upstream processing and microalgae culture. In light of this, environmental applications and microalgal biomaterials serve as a key factor in the commercialization of microalgal biofuels.

Despite the extremely favorable results that algae-based bioplastics showed on a laboratory scale, there are still several challenges that prevent the commercialization and mass manufacture of algae-based plastics. The best algae must be chosen in the first place in order to produce polymers for different types of bioplastics with different properties. Choosing the best polymers made from algae requires careful study to design sustainable bioplastics that outperform conventional plastics in terms of durability and strength. When choosing the best polymers, it is important to take a number of factors into account, including the ability of the feedstock to be recycled, the rate at which the polymer degrades or biodegrades, its size, molecular weight, brittleness, moisture content, and most importantly, the acceptance or satisfaction of the end user. Managing bioplastic waste appropriately is also crucial. Composting, digesters, landfilling, and incineration are other techniques for processing bioplastic waste in addition to mechanical and chemical recycling. Because biodegradable bioplastics break down more quickly than conventional plastics do, composting is the approach that works best for managing bioplastic waste and helps to reduce the waste burden from environment.

8.8 CONCLUSION

Due to the longevity and widespread manufacturing, chemical-based plastics have remained a dominant force in the market throughout time, which drives up the cost of first-generation bio-based plastics manufacture. Though they are mostly made from renewable resources and may compete with human or animal food while using less land, the next generation of bio-based materials, known as the third generation (3G), are potential alternatives. Although the market price prospection for bio-based goods is now low, it has been made clear that their inexpensive basic substrates have significant potential to be incorporated into the green circular economy. A microalgal biorefinery that combines various value-added product streams is the most appealing strategy in this regard. The most popular species for the production of biopolymers and plastic blends are *Chlorella* and *Spirulina*. Additives including compatibilizers, plasticizers, and other compounds are used as blending components to enhance the quality of the end product. Synergies, for instance, may be attained through biorefineries that produce a variety of products and the improvement of agricultural methods. The utilization of microalgal waste for the manufacturing of bioplastics has the potential to raise overall LCA ratings. Currently, the key area where microalgae production methods excel is in terms of decreased land use.

8.8.1 Acknowledgment

The authors received no financial support for the publication of this article.

8.8.2 Conflict of Interest Statement

The authors declare no competing financial or personal interests.

REFERENCES

Al-Hothaly, K.A., Adetutu, E.M., Taha, M., Fabbri, D., Lorenzetti, C., Conti, R., May, B.H., Shar, S.S., Bayoumi, R.A., Ball, A.S., 2015. Bio-harvesting and pyrolysis of the microalgae *Botryococcus braunii*. Bioresour Technol 191, 117–123. https://doi.org/10.1016/j.biortech.2015.04.113

Ali, S., Paul Peter, A., Chew, K.W., Munawaroh, H.S.H., Show, P.L., 2021. Resource recovery from industrial effluents through the cultivation of microalgae: A review. Bioresour Technol 337, a125461. https://doi.org/10.1016/j.biortech.2021.125461

Baldev, E., MubarakAli, D., Sivasubramanian, V., Pugazhendhi, A., Thajuddin, N., 2021. Unveiling the induced lipid production in *Chlorella vulgaris* under pulsed magnetic field treatment. Chemosphere 279, 130673, https://doi.org/10.1016/j.chemosphere.2021.130673

Bhati, R., Mallick, N., 2015. Poly (3-hydroxybutyrate-co-3-hydroxyvalerate) copolymer production by the diazotrophic cyanobacterium *Nostoc muscorum* Agardh: Process optimization and polymer characterization. Algal Res 7, 78–85.

Chakravarty, P., Mhaisalkar, V., Chakrabarti, T., 2010. Study on poly-hydroxyalkanoate (PHA) production in pilot scale continuous mode wastewater treatment system. Bioresour Technol 101, 2896–2899.

Chen, C.Y., Yeh, K.L., Aisyah, R., Lee, D.J., Chang, J.S., 2011. Cultivation, photobioreactor design and harvesting of microalgae for biodiesel production: A critical review. Bioresour Technol 102(1), 71–81. https://doi.org/10.1016/j.biortech.2010.06.159

Ciapponi, R., Turri, S., Levi, M., 2019. Mechanical reinforcement by microalgal biofiller in novel thermoplastic biocompounds from plasticized gluten. Materials 12(9), 1476. https://doi.org/10.3390/ma12091476

Cinar, S.O., Chong, Z.K., Kucuker, M.A., Wieczorek, N., Cengiz, U., Kuchta, K., 2020. Bioplastic production from microalgae: A review. Int J Environ Res Public Health 17(11), 3842. https://doi.org/10.3390/ijerph17113842

Coppola, G., Gaudio, M.T., Lopresto, C.G., Calabro, V., Curcio, S., Chakraborty, S., 2021. Bioplastic from renewable biomass: A facile solution for a greener environment. Earth Syst Environ 5, 231–251.

Costa, S.S., Miranda, A.L., Andrade, B.B., Assis, D. de J., Souza, C.O., de Morais, M.G., Costa, J.A.V., Druzian, J.I., 2018. Influence of nitrogen on growth, biomass composition, production, and properties of polyhydroxyalkanoates (PHAs) by microalgae. Int J Biol Macromol 116, 552–562. https://doi.org/10.1016/J.IJBIOMAC.2018.05.064

Cuellar-Bermudez, S.P., Kilimtzidi, E., Devaere, J., Goiris, K., Gonzalez-Fernandez, C., Wattiez, R., Muylaert, K., 2020. Harvesting of *Arthrospira platensis* with helicoidal and straight trichomes using filtration and centrifugation. Sep Sci Technol (Philadelphia) 55(13), 2381–2390. https://doi.org/10.1080/01496395.2019.1624573

Das, S.K., Sathish, A., Stanley, J., 2018. Production of biofuel and bioplastic from *Chlorella pyrenoidosa*. Mater Today Proc 5, 16774–16781.

Devadas, V.V., Khoo, K.S., Chia, W.Y., Chew, K.W., Munawaroh, H.S.H., Lam, M.K., Lim, J.W., Ho, Y.C., Lee, K.T., Show, P.L., 2021. Algae biopolymer towards sustainable circular economy. Bioresour Technol 325, 124702. https://doi.org/10.1016/j.biortech.2021.124702

Dragone, G., Fernandes, B., Vicente, A., Teixeira, J., 2010. Third generation biofuels from microalgae. Current Research, Technology and Education Topics in Applied Microbiology and Microbial Biotechnology 2, 1355–1366.

Ebrahimian, F., Karimi, K., Kumar, R., 2020. Sustainable biofuels and bioplastic production from the organic fraction of municipal solid waste. Waste Management 116, 40–48.

European Bioplastics, 2020. Report—bioplastics market data 2019, in: European Bioplastics. Berlin.

Garbowski, T., Pietryka, M., Pulikowski, K., Richter, D., 2020. The use of a natural substrate for immobilization of microalgae cultivated in wastewater. Sci Rep 10, 1–9.

Gozan, M., Noviasari, C., 2018. The effect of glycerol addition as plasticizer in *Spirulina platensis* based bioplastic, in: E3S Web of Conferences. EDP Sciences, p. 03048. https://doi.org/10.1051/e3sconf/20186703048

Gupta, P.L., Lee, S.-M., Choi, H.-J., 2015. A mini review: Photobioreactors for large scale algal cultivation. World J Microbiol Biotechnol 31, 1409–1417.

Harun, R., Singh, M., Forde, G.M., Danquah, M.K., 2010. Bioprocess engineering of microalgae to produce a variety of consumer products. Renew Sustain Energy Rev 14(3), 1037–1047. https://doi.org/10.1016/j.rser.2009.11.004

Hempel, F., Bozarth, A.S., Lindenkamp, N., Klingl, A., Zauner, S., Linne, U., Steinbüchel, A., Maier, U.G., 2011. Microalgae as bioreactors for bioplastic production. Microb Cell Fact 10, 1–6.

Jau, M.-H., Yew, S.-P., Toh, P.S.Y., Chong, A.S.C., Chu, W.-L., Phang, S.-M., Najimudin, N., Sudesh, K., 2005. Biosynthesis and mobilization of poly (3-hydroxybutyrate)[P (3HB)] by *Spirulina platensis*. Int J Biol Macromol 36, 144–151.

Johnsson, N., Steuer, F., 2018. Bioplastic material from microalgae: Extraction of starch and PHA from microalgae to create a bioplastic material. KTH Royal Institute of Technology.

Kamyab, H., Chelliapan, S., Lee, C.T., Rezania, S., Talaiekhozani, A., Khademi, T., Kumar, A., 2018. Microalgae cultivation using various sources of organic substrate for high lipid content, in: International Conference on Urban Drainage Modelling. Springer International Publishing, pp. 893–898.

Khalis, S.A., 2018. The Effect of compatibilizer addition on *Chlorella vulgaris* microalgae utilization as a mixture for bioplastic, in: E3S Web of Conferences. EDP Sciences, p. 03047.

Khomlaem, C., Aloui, H., Kim, B.S., 2021. Biosynthesis of polyhydroxyalkanoates from defatted *Chlorella* biomass as an inexpensive substrate. Appl Sci 11, 1094.

Knuckey, R.M., Brown, M.R., Robert, R., Frampton, D.M.F., 2006. Production of microalgal concentrates by flocculation and their assessment as aquaculture feeds. Aquac Eng 35(3), 300–313. https://doi.org/10.1016/j.aquaeng.2006.04.001

Kumar, A.N., Chatterjee, S., Hemalatha, M., Althuri, A., Min, B., Kim, S.-H., Mohan, S.V., 2020. Deoiled algal biomass derived renewable sugars for bioethanol and biopolymer production in biorefinery framework. Bioresour Technol 296, 122315.

Madadi, R., Maljaee, H., Serafim, L.S., Ventura, S.P.M., 2021. Microalgae as contributors to produce biopolymers. Mar Drugs 19, 466. https://doi.org/10.3390/MD19080466

Mannina, G., Presti, D., Montiel-Jarillo, G., Suárez-Ojeda, M.E., 2019. Bioplastic recovery from wastewater: A new protocol for polyhydroxyalkanoates (PHA) extraction from mixed microbial cultures. Bioresour Technol 282, 361–369. https://doi.org/10.1016/j.biortech.2019.03.037

Mata, T.M., Martins, A.A., Caetano, N.S., 2010. Microalgae for biodiesel production and other applications: A review. Renew Sustain Energy Rev 14(1), 217–232. https://doi.org/10.1016/j.rser.2009.07.020

Mendhulkar, V.D., Shetye, L.A., 2017. Synthesis of biodegradable polymer polyhydroxyalkanoate (PHA) in *Cyanobacteria synechococcus* elongates under mixotrophic nitrogen-and phosphate-mediated stress conditions. Ind Biotechnol 13, 85–93.

Monshupanee, T., Nimdach, P., Incharoensakdi, A., 2016. Two-stage (photoautotrophy and heterotrophy) cultivation enables efficient production of bioplastic poly-3-hydroxybutyrate in auto-sedimenting cyanobacterium. Sci Rep 6, 1–9.

Morais Junior, W.G., Gorgich, M., Corrêa, P.S., Martins, A.A., Mata, T.M., Caetano, N.S., 2020. Microalgae for biotechnological applications: Cultivation, harvesting and biomass processing. Aquaculture 528, 735562. https://doi.org/10.1016/j.aquaculture.2020.735562

Nanda, N., Bharadvaja, N., 2022. Algal bioplastics: Current market trends and technical aspects. Clean Technol Environ Policy 24(9), 2659–2679

Nikel, P.I., Pettinari, M.J., Galvagno, M.A., Méndez, B.S., 2006. Poly (3-hydroxybutyrate) synthesis by recombinant *Escherichia coli* arcA mutants in microaerobiosis. Appl Environ Microbiol 72, 2614–2620.

Onen Cinar, S., Chong, Z.K., Kucuker, M.A., Wieczorek, N., Cengiz, U., Kuchta, K., 2020. Bioplastic production from microalgae: review. Int J Environ Res Public Health 17(11), 3842.

Rahman, A., Miller, C.D., 2017. Microalgae as a source of bioplastics, in: Algal Green Chemistry: Recent Progress in Biotechnology. Elsevier, pp. 121–138. https://doi.org/10.1016/B978-0-444-63784-0.00006-0

Roja, K., Sudhakar, D.R., Anto, S., Mathimani, T., 2019. Extraction and characterization of polyhydroxyalkanoates from marine green alga and cyanobacteria. Biocatal Agric Biotechnol 22, 101358.

Sabathini, H.A., Windiani, L., Gozan, M., 2018. Mechanical physical properties of *Chlorella*-PVA based bioplastic with ultrasonic homogenizer, in: E3S Web of Conferences. EDP Sciences, Russia, p. 03046.

Salim, S., Bosma, R., Vermuë, M.H., Wijffels, R.H., 2011. Harvesting of microalgae by bio-flocculation. J Appl Phycol 23, 849–855. https://doi.org/10.1007/s10811-010-9591-x

Saratale, R.G., Cho, S.K., Ghodake, G.S., Shin, H.S., Saratale, G.D., Park, Y., Lee, H.S., Bharagava, R.N., Kim, D.S., 2020. Utilization of noxious weed water hyacinth biomass as a potential feedstock for biopolymers production: A novel approach. Polymers (Basel) 12(8), 1704. https://doi.org/10.3390/POLYM12081704

Stávková, J., Maroušek, J., 2021. Novel sorbent shows promising financial results on P recovery from sludge water. Chemosphere 276, 130097. https://doi.org/10.1016/j.chemosphere.2021.130097

Suparmaniam, U., Lam, M.K., Uemura, Y., Lim, J.W., Lee, K.T., Shuit, S.H., 2019. Insights into the microalgae cultivation technology and harvesting process for biofuel production: A review. Renew Sustain Energy Rev 115, 109361. https://doi.org/10.1016/j.rser.2019.109361

Vandamme, D., Pontes, S.C.V., Goiris, K., Foubert, I., Pinoy, L.J.J., Muylaert, K., 2011. Evaluation of electro-coagulation-flocculation for harvesting marine and freshwater microalgae. Biotechnol Bioeng 108(10), 2320–2329. https://doi.org/10.1002/bit.23199

Veerabadhran, M., Natesan, S., MubarakAli, D., Xu, S., Yang, F., 2021. Using different cultivation strategies and methods for the production of microalgal biomass as a raw material for the generation of bioproducts. Chemosphere 285, 131436. https://doi.org/10.1016/j.chemosphere.2021.131436

Yadav, B., Pandey, A., Kumar, L.R., Tyagi, R.D., 2020. Bioconversion of waste (water)/residues to bioplastics—a circular bioeconomy approach. Bioresour Technol 298, 122584. https://doi.org/10.1016/j.biortech.2019.122584

Zeller, M.A., Hunt, R., Jones, A., Sharma, S., 2013. Bioplastics and their thermoplastic blends from *Spirulina* and *Chlorella* microalgae. J Appl Polym Sci 130(5), 3263–3275. https://doi.org/10.1002/app.39559

Zenouzi, A., Ghobadian, B., Hejazi, M.A., Rahnemoon, P., 2013. Harvesting of microalgae *Dunaliella salina* using electroflocculation. J Agric Sci Technol 15(5), 879–887.

Zheng, H., Gao, Z., Yin, J., Tang, X., Ji, X., Huang, H., 2012. Harvesting of microalgae by flocculation with poly (γ-glutamic acid). Bioresour Technol 112, 212–220. https://doi.org/10.1016/j.biortech.2012.02.086

9 Biodegradation of Second- and Third-Generation Bioplastics

Gustavo Amaro Bittencourt, Walter José Martínez-Burgos, Lucas Lourenço Castiglioni Guidoni, Érico Kunde Corrêa, and Carlos Ricardo Soccol

CONTENTS

9.1 INTRODUCTION

Although it is said that most conventional polymers and plastics ever produced are non-biodegradable, conceptually speaking, all organic compounds are biodegradable (Innocenti, 2003). Biodegradation is a process that microbial respiration oxidizes organic carbon of any substance into carbon dioxide and/or methane in anaerobic conditions; together with physical and chemical factors which compose different degradation pathways, such as temperature, irradiation, oxidation, and depolymerization, biodegradation leads to carbon mineralization and the closed loop of its bio-geo-chemical cycle (Danso et al., 2019; Emadian et al., 2017). The problems related to the degradability of conventional plastics lie in its degradation time, that is much longer than would be necessary to the productive systems to correctly manage their disposal. For instance, in PET polymers, only 0.1% of carbon per year can be transformed into CO_2 via biodegradation, and only under ideal laboratory conditions (Hahladakis et al., 2018).

As a matter of fact, it was exactly with that property in mind, the long-term stability, that conventional engineered polymers were manufactured. Due to the difficulty to break it down and its easy handling, plastic production had an extraordinary rapid growth during 20th century, surpassing all other industrial materials except cement and steel from the construction sector (Geyer et al., 2017). Nowadays, a world without organic polymers is very difficult to conceive. In 2015, the global annual production of plastics was 380 million metric tons, and it was estimated that the demand for plastic polymers can reach 2000 million metric tons by 2050 (Zheng and Suh, 2019).

DOI: 10.1201/9781003344018-9

As a result of the high demand for polymers use and the difficulty to decompose them properly, they accumulate in landfills and natural environments. It was estimated that 80% of all plastics ever made were accumulated in different environment compartments, while 4.8–12.7 million metric tons entered the ocean in 2010 (Geyer et al., 2017). Recent works reported plastic pollution consequences in ocean, freshwater, and terrestrial environments, highlighting their ubiquitous presence, such as in edible crops, seafloor sediments, post-wastewater treatment plants, among others, and their ecological and health risks (Kurniawan et al., 2021). Due to their limited degradability, outputs of plastics fragmentation in environment results in small particles as microplastics (<5 mm) and nanoplastics (<1 μm), which involve concerns related to emerging contaminants (Hahladakis et al., 2018; Kurniawan et al., 2021). Moreover, during the production of plastic polymers, there are certain chemical compounds that are added to enhance performance, functionality, appearance, and aging properties of the final product, which can potentially migrate to the environment and lead to human exposure (Bishop et al., 2021; Hahladakis et al., 2018). Some materials contain additives to undergo oxidative degradation by an accelerated process by light and heat. The presence of these additives leads to a more challenging recycling and composting processes, as well as to environmental impacts over time, by the spread of easier leachable microparticles into soil, land, and surrounding water resources.

To fulfill the problems related to plastic pollution, dumping wastes in landfills was the most common way to dispose plastics before the modernization of public waste management; however, this practice still happens across the globe, mainly in developing countries (Nandakumar et al., 2021). The key factor for plastic end-of-life options is the relation between the rate of waste production and the rate of system waste treatment. Thus the objective of such system must be to recover materials and energy and dispose of wastes at the same rate that plastics are produced. However, a global evidence-based strategy, with practical and measurable interventions, does not yet exist (Lau et al., 2020). In that way, plastics were meant to be recyclable and reusable, but global recycling strategies are insufficient, besides it delays rather than avoids final disposal. The only known way to permanently eliminate conventional plastic wastes is through destructive thermal treatments, which presents a very fast mineralization process (Geyer et al., 2017). However, each technology can own significative impacts on the environment and on human health, in terms of public and occupational concerns. In relation with incineration, N-containing, S-containing and Cl-containing polymers produce toxic NO_x, SO_x, and HCl, respectively, which require high-cost capture and interventions.

Plastic recycling processes involve four different steps: primary, secondary, chemical depolymerization, and thermal recovery. The first steps face technical and economic difficulties related to poor separation, which causes contamination and mixing of different polymer types. Also, mechanical properties of recycled material are often inferior from those of the virgin polymers, which limits the economical return of recycling. Thus only 9% of plastics ever made, since the beginning of 20th century, were recycled; and only 10% of them more than once (Geyer et al., 2017). In addition, limited regulatory frameworks and uncontrolled processes can result in significant contamination of ambient areas where plastic is recycled, and in the transfer of potentially harmful substances from certain groups of additives into recycled plastics for sensitive uses (Hahladakis et al., 2018). For example, brominated and phosphorous flame retardants and phthalates have been found in children's toys, food contact materials, and household products after recycling (Chen et al., 2009; Hahladakis et al., 2018). Furthermore, according to the work of He et al. (2015), which evaluated different recycling workshops and their health risks in China, a total of 64 volatile organic compounds (VOCs) are emitted during the melting extrusion procedure. Workers from the facilities suffered acute and chronic health risks in polystyrene recycling workshops, while lifetime cancer risk assessment indicated that local residents suffered from definite cancer risk.

Even if all of those risks were mitigated in the near future and incineration methods were rigidly controlled via multiple filters to prevent massive air pollution, the extensive dependency on fossil fuels for polymers materials will continuously increment greenhouse gas (GHG) emissions and carbon footprints, which will raise the problems related to global warming. Thus the market shift to bio-based plastics can mitigate major concerns of plastic industry, as its carbon neutrality provides a renewable resource to the closed loop of the carbon bio-geo-chemical cycle.

9.2 THE REPLACEMENT OF PLASTICS BY BIOPLASTICS

The production of plastics consumes about 5%–7% of the annual global oil supply, representing 2% of the global atmospheric CO_2 output, with >850 million tons (CIEL, 2019). If the growing tendency presented by the last four decades continue, the GHG emissions from plastics would reach 15% of the currently global carbon budget by 2050, which means 6.5 Gt of CO_2-equivalent (CO_2-eq; Zheng and Suh, 2019). The raw material extraction accounts for 61% of CO_2 emitted by the global plastic industrial chain, while the production of the polymer itself accounts for 31%, and only 9% is associated with different end-of-life options (Rosenboom et al., 2022). Due to the largest carbon footprint associated with raw material extraction and polymer production, there are lot of potentialities to considerably improve the global circular plastic economy only substituting the inputs of existing processes by renewable feedstocks and low-carbon energy. Thus the improved circularity and potential carbon neutrality are among the main advantages to replace fossil fuel–based plastics with bio-based plastics.

The biodegradability capacity of bio-based plastics is the most important feature to its utilization. While it encompass a series of advantages to achieve the UN Sustainable Development Goals, introducing new recycling and degradation pathways and using less harmful substances in production processes (Karan et al., 2019), their use can also provide solutions to the trends of widespread contamination of microplastics, through safe disposal in natural environments. Biodegradation introduces alternative routes to the end-of-life options of plastics, acting as a result of the action of enzymes and chemical deterioration associated with living microorganisms, often leading to the assimilation of the polymer fragments and its subsequent mineralization to form CO_2 and/or methane, water, biomass, and mineral salts (Danso et al., 2019).

However, these benefits depend on several factors, such as the manufacturing process and the chemical structure of the resulted polymer, as well as the end-of-life pathway applied. The environmental efficiency of these factors must be accurately evaluated via life cycle assessments (LCA), in terms of climate impact, ecotoxicity, and recyclability, to fairly compare the benefits of such bioplastics over petrochemicals and other alternatives (Bishop et al., 2021). Similar to the already discussed problems related to conventional plastics, bioplastics also raise concerns related to the leaching of monomers, oligomers, and their additives. Zimmermann et al. (2020) reported that nine different types of bioplastics characterized using in vitro bioassays showed the potential to be similarly toxic as conventional plastics, exclusively related to their monomers, oligomers, additives, lubricants, and non–intentionally added substances. This shows that bioplastic evaluation deserves the same scrutiny in product design and formulation as petrochemical plastics. Otherwise, Razza et al. (2015) reported that a starch-based bioplastic for food packaging use resulted in a lower consumption of non-renewable energy resources (–50%) and less GHG emissions (–60%) throughout all the production stages when compared to a polystyrene product for the same purpose. Thus, the evaluation of biodegradability extent and impacts, and life cycle performance of each product, are essential to provide clear evidence on the comparative sustainability of different plastics.

One important aspect of the environmental impacts arising from bio-based plastics is the carbon footprint associated with land use changes from crop feedstocks. Land-use change can have considerable effects on product life cycle due to the significant GHG emissions via disturbance of carbon stocks in soil and vegetation (Bishop et al., 2021). Despite renewable resources of bioplastic crop feedstocks, it requires land that could serve other functions, and increased use of fertilizers that can raise nitrogen and phosphorus leachates; the soil organic carbon stocks under cropland are normally lower than those under natural forests and pastures (Bishop et al., 2021). The "carbon cost" is the difference between the accounting of carbon storage from existing uses and the sequestration sacrificed by changing the land use. In 2017, land use for bioplastics was reported to be 0.82 million hectares, which means 0.016% of global land area (Zheng and Suh, 2019). Assuming a shift in 250 million tonnes of plastic to crop-based bioplastics, it would be required such as 5% of all arable land in the world, which provokes ethical concerns about the potential competition with food resources, as well as it could undermine carbon benefits from bioplastics utilization.

In a biorefinery context, these above-mentioned resources are defined as first-generation feedstocks, as they are mainly constituted by readily fermentable sugars from edible polysaccharide sources, such as sugarcane sucrose and corn starch. Second-generation feedstocks comprise various non-edible biomass, such as lignocelluloses and other wastes, that offer widely available resources to mitigate arable land pressure and associated GHG emissions related to land-use change. Also, third-generation microalgae-based bioplastics can address many issues, due to their capacity to be located in nonarable lands and their increased ability to absorb CO_2, for a global photosynthetic capacity expansion (Karan et al., 2019). Third-generation bioplastics have the potential to be designed as a complete biodegradable material, in natural and industrial composting conditions. As can be seen, due to high variations in production processes, in an integrated approach point of view, each bioplastic has different life cycle performances and disadvantages that can vary substantially (Bishop et al., 2021).

But, first of all, which is exactly the conception of a "bioplastic"? The generic definition used for bioplastics often leads to misleading understandings on what exactly it means. The term has been used for plastics made from a biological resource, including the biodegradable ones and those which are durable and non-degradable (neat or partial blends), but also for petrochemical plastics that have, for some reason, increased ability to biodegradation (Nandakumar et al., 2021). This misconception implies positive connotations as a "natural material" for all bioplastics, compared to their petrochemical counterparts, even to certain non-bio–based plastics, such as poly(butylene adipate-co-terephthalate) (PBAT) and polycaprolactone (PCL), and also to bio-based polymers that are non-biodegradable, such as bio-PET. Another representative of the misunderstood terms is the oxo-biodegradable plastics, which are conventional polymers physically degradable due to the addition of metal compounds, which are not broken down or taken up by microorganisms (Nandakumar et al., 2021). Similar concerns are related to some so-called enzyme-mediated degradable bioplastic, which are conventional polymers (normally polyethylene or polypropylene) enriched with small amounts of an organic additive. While its organic additive fraction passes through an enzyme-mediated degradation, their "conventional fraction" is fragmented to smaller particles of microplastic (European Bioplastics, 2014).

Actually, even bio-based biodegradable plastics need further research to completely clarify the extent of their biodegradability under managed and unmanaged environments. According to Napper and Thompson (2019), some plastic bags labeled as "biodegradable" have presented similar deterioration rates in marine and soil environments to conventional plastics, even after 3 years. Also, biodegradability essays of bioplastic food packaging in simulated home composting systems suggest that it is important to distinguish bio-based plastics that are biodegradable under home composting conditions from those that only biodegrade under more severe conditions of industrial composting facilities (Nandakumar et al., 2021). Some weaknesses of wider public opinion and regulatory frameworks related to specific characteristics of bioplastics may be explored in a marketing context of greenwashing. Thus these distinctions about the biodegradability of bioplastics must be communicated and regulated by proper standards.

9.3 BIODEGRADABILITY PROPERTIES AND ISSUES

The most common bioplastics evaluated in terms of their biodegradability are polylactic acid (PLA), polybutylene succinate (PBS), polyhydroxyalkanoates (PHA), polyhydroxybutyrate (PHB), thermoplastic starches (TPS), bio-polyethylene (bio-PE), bio-polyethylene terephthalate (bio-PET), bio-polyurethane (bio-PU), bio-polyvinyl chloride (bio-PVC), cellulose-based polymers, polyamide-based polymers (PA-based), and polycaprolactone-based polymers (PCL-based). According to Danso et al. (2019), the biodegradation process is carried out in three consecutive steps (Figure 9.1): (1) biodeterioration, (2) biofragmentation, and (3) assimilation. The first step of biodegradation is characterized by the decomposition or surface embrittlement of the polymer; this action is carried out by different types of enzymes or microorganisms. The most notable effects in this first step are

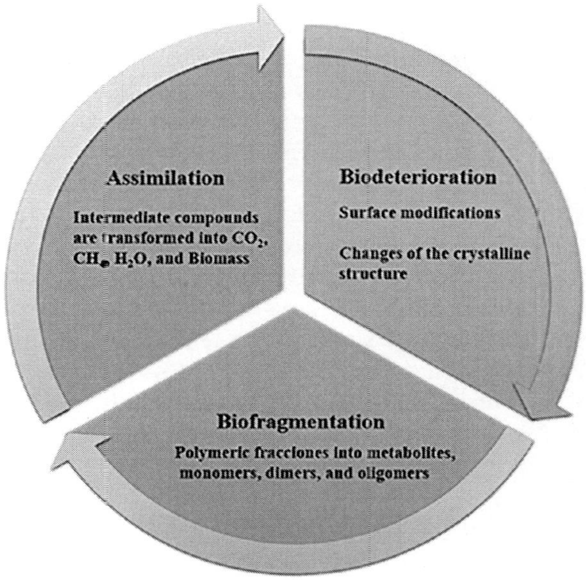

FIGURE 9.1 The consecutive steps of the biodegradation process.

the surface and crystal structure modifications of the biopolymer. In the second step, the enzymes secreted by the microorganisms can cause depolymerization, which causes the fragmentation of the polymer into oligomers, dimers, and monomers (Danso et al., 2019). Finally, in step three, the intermediate metabolites produced in the biofragmentation are used by the microorganisms to produce the final compounds, such as carbon dioxide, methane, water, and biomass.

Biodegradation processes are affected by abiotic factors, such as chemical and physical processes (UV, temperature, moisture, pH), and by biotic factors related to microbial activity (Karamanlioglu et al., 2017; Nandakumar et al., 2021). The characteristics and properties of the polymer, such as molecular weight, optical purity, and crystallinity are other variables that must be considered in the biodegradation of bioplastics (Karamanlioglu et al., 2017). In relation to that, aromatic polyesters are the most sensitive to microbial degradation, while aliphatic polyesters are degraded through hydrolysis of ester linkages. For example, when the molecular weight of PLA is low (<100,000 Da), it presents brittle and opaque characteristics, whereas at higher molecular weights, PLA is more transparent and less susceptible to biodegradation. Also, biodegradation reaches different efficiencies depending on which structures of biopolymer are related; for instance, the crystalline regions of PLA hydrolyze slower than the amorphous regions, since in the latter water can diffuse more easily (Kale et al., 2007b; Karamanlioglu et al., 2017).

9.3.1 Abiotic Factors

Abiotic factors are key in the degradation of conventional plastics and bioplastics, as these factors are the first to induce changes in the polymer. Abiotic factors facilitate the internal modifications of polymers and directly affect its mechanical, chemical, and thermal properties (Danso et al., 2019). Among the predominant abiotic factors in the biodegradation process, chemical processes (hydrolysis and oxidative degradation) and photodegradation stand out. Chemical degradation processes are favored by the presence of hydrolysable covalent bonds such as ester, amide, and ether chemical groups (Quecholac-Piña et al., 2020). In addition, the attack of oxidizing agents such as atmospheric oxygen is also favored by the molecular structure of the polymer.

9.3.2 Biotic Factors

It is necessary to understand how the differences between ecosystems and different microorganisms affect the biodegradability of polymers. Different microbial families participate interactively in the biodegradation of polymers to enable subsequent steps (Gan and Zhang, 2019). Therefore it is necessary to know in detail the microorganisms or enzymes that act in each stage of the biodegradation of plastics (Gan and Zhang, 2019). Bioplastics biodegradation processes are generally carried out in three environments: (1) compost, (2) soil, and (3) aquatic systems. In composting, plastics are degraded by groups of microorganisms to H_2O, CO_2, and humus. However, not all bioplastics can be biodegraded via compost (Kale et al., 2007b). Among the bioplastics that have been biodegraded using composting systems are PLA, PHB, and those that are PA-based or PCL-based. In the case of aquatic systems, the biodegradation of bioplastic is an extremely complex process, since this ecosystem is composed by different micro habitats with different ambient conditions (pelagic and buried in the sediments, deep sea benthic, sublittoral benthic, supralittoral, eulittoral). Thus to understand the process of biodegradation in this system, it is necessary to know the phenomena that happen in each micro habitat. According to Emadian et al. (2017), polymer biodegradation is more efficient in the pelagic habitat than in the eutrophic one. Some of the bioplastics that have been biodegraded in these ecosystems are starch-based and PHB. Finally, bioplastics can also be biodegraded in soil, mainly because this is an environmental matrix with a wide range of microorganisms, so these processes can become more viable. The bioplastics that have been biodegraded in this ecosystem are PHA, PLA, and PCL-based.

More than 90 microorganisms have been identified among prokaryotes (aerobes, anaerobes, photosynthetic bacteria, archaebacteria) and eukaryotes (filamentous fungi and yeasts), responsible for the biodegradation and catabolism of bioplastics (Emadian et al., 2017). In the case of PLA, microorganisms can only biodegrade after the PLA molecule undergoes hydrolysis and the weight of the molecule is less than 10,000 Da (Karamanlioglu et al., 2017). PLA has been degraded with different bacterial strains, among which are strains of the genera *Amycolatopsis, Saccharothrix, Kibdelosporangium, Bacillus*, and *Thermomonospora* (Jarerat et al., 2002). Fungal strains have also been identified as potential PLA degraders, such as *Fusarium moniliforme, Penicillium roqueforti, Tritirachium album, Cryptococcus sp., Aspergillus fumigatus*, and *Trichoderma viride* (Karamanlioglu et al., 2017). Other microorganisms capable of biodegrading other bioplastics are shown in Figure 9.2.

In the biodegradation processes, the intra- or extracellular enzymes of the microorganisms play an important role as they are the catalysts of the degradation reactions. Thus, for example, the enzyme intercellular depolymerase from *Rhodospirillum rubrum* participates in the biodegradation process of polyhydroxybutyrate (PHB). Other enzymes, such as lipase, esterase, and serine from *Alcaligenes faecalis, Comomonas acidivorans*, and *Pestalotiopsis microspora*, respectively, are also involved in biodegradation processes of bioplastics (Emadian et al., 2017).

9.3.3 International Standards to Assess the Biodegradation of Plastics

The main method used to evaluate the degree of degradation of plastic materials is respirometry, which is based on the cumulative measurement of the biogas generated in bioreactors that contain the polymer to be evaluated, in comparison with controls without the presence of tested material (Quecholac-Piña et al., 2020). International organizations such as the International Organization for Standardization (ISO) and the American Society for Testing and Materials (ASTM) have defined some standard methods to evaluate the biodegradation of polymers under anaerobic conditions (Table 9.1). These methods have been tested in the biodegradation of some plastic materials. For example, Yagi et al. (2013) achieved degradation rates of 75%, 80%, and 90% for PLA, polycaprolactone, and PHB plastics, respectively, using the ISO 13975 standard method. Gómez and Michel (2013) used the ASTM D5511 standard method to evaluate the rate and extent of mineralization of a wide range of commercially available materials, such as PLA, PHA, and starch-based materials, during anaerobic digestion and soil incubation. The results showed that after 660 days of soil

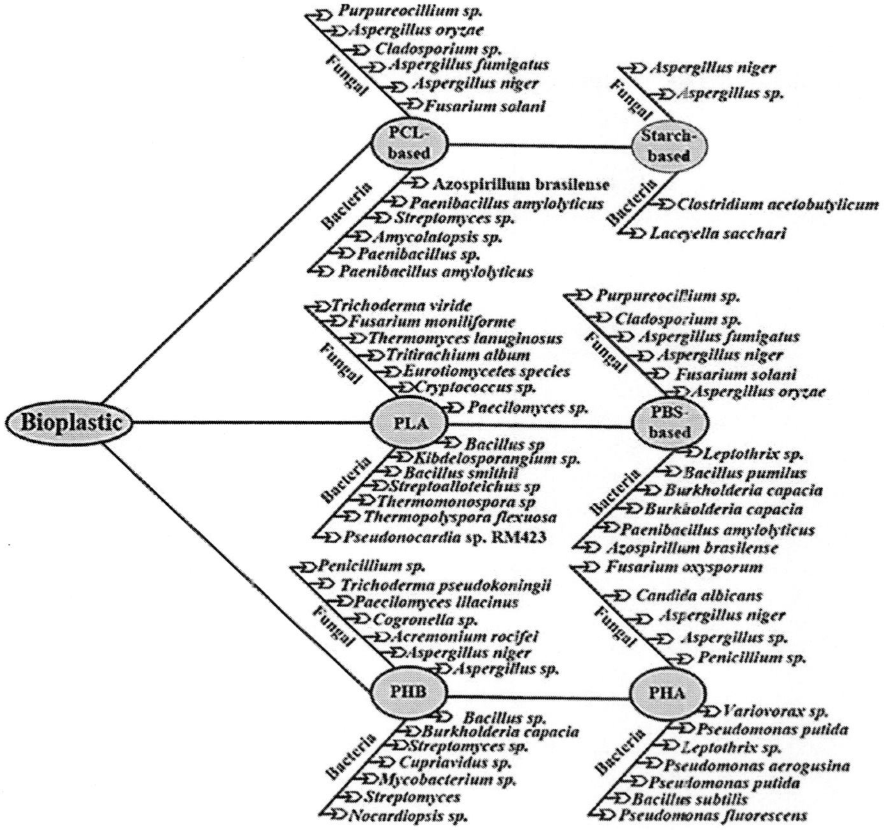

FIGURE 9.2 Degrading microorganisms of different types of bioplastics.

Sources: Danso et al. (2019); Emadian et al. (2017); Karamanlioglu et al. (2017).

TABLE 9.1
International Standards for Evaluating the Biodegradation of Bioplastics

Method	Description	Environmental Conditions	Observations
ASTM D5526	Plastic materials are pre-treated and exposed to a methanogenic consortium. Anaerobic decomposition takes place under static conditions and the process is carried out with solids concentrations greater than 30%.	35 ± 2°C (minimum 7 days)	There is no known ISO equivalent to this method.
ASTM D7475	The method determines aerobic or anaerobic biodegradation of plastic materials under accelerated bioreactor landfill conditions.	35 ± 2°C (minimum 7 days)	This is a modification of Test Method D5526. There is no known ISO equivalent to this method.
ASTM D5511	The method determines the degree or rate of biodegradation under anaerobic conditions. Plastic materials are exposed to a methanogenic inoculum. The process is carried out under static conditions and under conditions of high solids content.	52 ± 2°C (minimum 10 days)	This test method is equivalent to ISO 15985.

(*Continued*)

TABLE 9.1 (CONTINUED)
International Standards for Evaluating the Biodegradation of Bioplastics

Method	Description	Environmental Conditions	Observations
ISO 15985	Specifies a method for evaluating the anaerobic biodegradability of plastics based on organic compounds under high solids conditions by measuring biogas. The method is designed to give the percentage of carbon in the test material and its rate of conversion to carbon dioxide and methane.	52±2°C (minimum 15 days)	
ISO 14853	Specifies a method for determining the biodegradability of plastic material using anaerobic microorganisms. The method suggests exposing the material to be evaluated for a period of up to 90 days.	35°C (maximum of 90 days)	The method applies to the following materials: natural or synthetic polymers.
ISO 13975	Specifies a method for evaluating the anaerobic biodegradability of plastic materials in a controlled anaerobic digestion system. The system operates with solids concentrations less than 15%.	55±5°C (for thermophilic digestion) and 35±3°C (for mesophilic digestion) for a maximum of 90 days.	The method can be applied to the following materials, provided they have a known carbon content: natural or synthetic polymers, copolymers, or mixtures.

incubation, substantial mineralization was observed for PHA and starch-based plastics. In the case of commercial conventional plastics with additives to improve biodegradation, no significant degradation was observed for polyethylene and polypropylene. During anaerobic digestion, after 50 days, only 20%–25% of the bio-based materials were converted to biogas, compared to less than 2% of the additive-containing conventional plastics. This shows that some other strategies must to be evaluated together with soil incubation and anaerobic digestion to fulfill the biodegradation of bio-based plastics.

9.4 COMPOSTABILITY OF DIFFERENT BIOPLASTICS

Composting is a biological treatment that aims for the degradation and mineralization of heterogeneous solid materials, under controlled moisture content, in a self-heating aerobic process. Composting recycles organic wastes to be used as fertilizers for agricultural production, reducing the waste volume to be disposed, and consequently extending landfill lifetime. However, regarding plastic wastes, large amounts are still disposed in landfills, which can potentially generate toxic compounds and leachates (Emadian et al., 2017). An alternative to reduce the disposal volume from plastics on landfills and the environment is the evaluation of the biodegradability of bio-based polymers (García-Depraect et al., 2021).

It should be noted that all compostable plastic is biodegradable, whereas not all biodegradable plastic is compostable (Kale et al., 2007b). Although, the degradation process of bioplastics can take several months or years. On the other hand, when compostable plastics are treated under suitable operational conditions, they can be degraded within few months.

9.4.1 International Standards

Currently, there are several organizations that establish standards and regulations for the plastics composting issues, such as the North American ASTM D 6400, the European EN 13432, the international ISO 14855, and the Australian AS 5810.

9.4.1.1 North American Standard (ASTM D 6400)

One of the standards that set criteria to label a plastic as "compostable" is ASTM D 6400 (Standard Specification for Compostable Plastics); the first version was published in 1999 by the ASTM. The updated regulation establishes standards to identify products and materials as able to be composted satisfactorily in industrial and municipal composting facilities. It determines mandatory characteristics for compostable plastics including disintegration during composting, inherent biodegradation, absence of phytotoxicity effects, and impacts of the final compost on plant growth. According to this standard, compostable plastics must allow degradation by biological processes during composting to produce CO_2, water, inorganic compounds, and biomass at a rate consistent with other known compostable materials and leave no visible, distinguishable, or toxic residues at the end of process.

In order to confirm if plastics and plastic packaging are truly compostable, the materials must undergo specific laboratory tests with controlled conditions, similar to those found in composting facilities. In addition, the quantity evaluated must be representative and products must be tested in the same condition as they are used. For example, for products that have different thicknesses or densities, such as foams, films, and containers, as long as the chemical composition and structure remain the same, only the thickest or densest material needs to go through testing.

The parameters that the plastic must reach, after controlled composting on a laboratory scale, are:

- No more than 10% of its original dry weight must remain after a 2.0 mm sieving.
- If the test sample consists of a single polymer, it must show a satisfactory biodegradation rate and convert 60% of the organic carbon into carbon dioxide, during the standard time of 45 days, and not exceeding 6 months.
- When the evaluated sample consists of more than one polymer, it must show a satisfactory biodegradation rate and convert 90% of the organic carbon into carbon dioxide, during the standard time of 45 days, and not exceeding 6 months.
- The test period must not exceed 180 days for non-radiolabeled material.
- To be considered safe in aquatic and terrestrial environments, the plastic material must have concentrations lower than 50% of heavy metals, of those prescribed in CFR Part 503.13 (Standards for the Use or Disposal of Sewage Sludge), and present no phytotoxicity for seed germination tests.

9.4.1.2 European Standard (EN 13432)

Another standard that defines the requirements and procedures to determine the plastic compostability is the European standard EN 13432 (Requirements for packaging recoverable through composting and biodegradation). First, this standard defines limits for the presence of potentially harmful constituents in products and packages that will be destined for organic recovery (Table 9.2).

TABLE 9.2
Maximum Allowed Content of Chemical Elements in Plastics and Compostable Plastic Packaging, according to European Standard EN 13432

Element	Dry Weight (mg/kg)
Zn	150
Cu	50
Ni	25
Cd	0.5
Pb	50
Hg	0.5
Cr	50
Mo	1
Se	0.75
As	5
F	100

That said, the criteria to determine if plastics and plastic packaging can be considered compostable are:

- *Biodegradability*: It is determined through a controlled composting test on a laboratory scale, during the standard time of 45 days, not exceeding 6 months, according to ISO 14855.
- *Disintegration*: After going through a composting test for a maximum of 12 weeks, the material must be sieved with a 2 mm mesh, leaving no more than 10% of the original dry weight. The composting test should be conducted simulating as closely as the real conditions obtained in a composting facility.
- *Quality*: It can be determined by evaluating the ecotoxicological effects of the plastic after degradation by composting, considering a control treatment with no plastic. The compost obtained with plastic must not change the physical-chemical parameters: volumetric weight (density), total dry solids, volatile solids, salt content, pH, total nitrogen, ammonia nitrogen, phosphorus, magnesium, and potassium, and must not interfere with the growth of two higher plants.

9.4.1.3 Standard ISO 14855–2

The standard ISO 14855–2 specifies a method to determining the biodegradation of compostable plastics. It recommends to conduct the test with a small-scale reactor and controlled moisture content, aeration rate, and temperature. The method aims the gravimetric measurement of the amount of carbon dioxide released during composting.

The ISO 14855–2 is applied to the following materials: natural and/or synthetic polymers and copolymers and their mixtures; plastic materials that contain additives such as plasticizers or colorants; water-soluble polymers; and materials that, under the test conditions, do not inhibit the activity of the microorganisms present in the inoculum.

9.4.1.4 Australian Standard AS 5810:2010

The AS 5810 standard specifies requirements and procedures to determine the biodegradability parameters of plastic under home composting conditions. In this way, it allows plastic materials to be labeled as "home compostable." This standard also determines criteria for biodegradability, disintegration during biological treatment, effect on the biological treatment process, and effect on the quality of the resulting homemade compost.

The ISO 14855 may be used as the standard test method unless it is inappropriate to the type and properties of the plastic tested. However, the test environment must be kept below 30°C during the test.

9.4.2 Different Types of Composting Systems and Their Potential as Bioplastic Waste Treatment

Bioplastic composting systems can be divided into two groups: industrial/municipal composting systems and home composting systems. These systems present differences related to the volume of material in degradation. For this reason, larger volumes, such as large-scale composting systems, provide different temperature conditions, carbon dioxide, and water (steam) emissions to the environment. In addition, large-scale composting also presents different microbial activity from that observed in home composting systems.

Generally, temperatures observed in home composting systems are lower (<35°C) than those observed on a large scale (50–60°C; Nandakumar et al., 2021). A study comparing home and industrial-scale composting on the biodegradation of PLA bioplastic, showed that a slower biodegradation was reached by a home composting system, even after 11 months of treatment. The authors attributed the temperature as the main factor for the speed of biodegradation of the plastic (Rudnik and Briassoulis, 2011). Solano et al. (2022) evaluated a home composting system with PLA and other oxo-biodegradable plastics, and even after 18 months, which is the maximum time suggested in the AS 5810 standard, the authors still detected the presence of unchanged plastic materials. However, both home composting and large-scale composting may achieve degradation conditions for bioplastics such as PHA, PBAT, and those consisting of starch or chemical pulp.

The important factors that affect plastic biodegradation in the environment are the chemical structure, polymer chain, crystallinity, and the complexity of the polymer formula. Polymers with a shorter chain, more amorphous parts, and less complex formula are more susceptible to microorganism biodegradation (Emadian et al., 2017). In addition, the environmental conditions in which bioplastics are treated is fundamental for efficient biodegradation. The pH, temperature, moisture content, and oxygen content are among the most significant environmental factors that must be considered in polymer composting (Kale et al., 2007b).

The biodegradation of polymers in compost is caused by microorganisms belonging to the three domains of life (Bacteria, Archaea and Eukarya), but fungal and bacterial species are the most important (Gan and Zhang, 2019). The bioplastics biodegradation process can be split in three stages. In the first stage, there are chemical, mechanical, and physical changes caused by the biological activity of microorganisms on the surface of the material and by abiotic factors, such as thermal degradation and chemical oxidation (García-Depraect et al., 2021). In the second, microbial attack by extracellular enzymes lead to the breakdown of the polymer into oligomers and monomers. And last, the biofragmented compost is used as a source of carbon and energy by the microbiota and converted into CO_2, H_2O, and microbial biomass. The biological mechanisms of biodegradation normally occur due to the ability of microorganisms to attach and eventually form biofilms on the surface of the solid polymer, through mechanisms of depolymerization and hydrolysis by intra- and extracellular enzymes.

Both fossil-based and bio-based bioplastics, including PLA, PHA, starch-based, PBS, PES, and PCL, are compostable under suitable conditions, such as temperature, pH, and moisture content (Emadian et al., 2017). Table 9.3 shows composting systems for different types of bioplastics, in processes that reached a biodegradability rate equal or greater than 80% in a period up to 120 days.

TABLE 9.3
Composting Tests for Different Bioplastics (Processes That Showed ≥80% Biodegradability in Up to 120 Days)

Bioplastic/Product	Experimental Unit	Conditions	Time (days)	Reference
PLA bottles (500 mL)	432 m^3 composting pile (6 m × 24 m × 3 m)	Peak temperature: 65°C Initial pH: 8.5 Initial humidity: 63%	58	(Kale et al. 2007a)
PLA film; PLA/PHB film (30 mm × 30 mm × 0.2 mm)	Laboratory-scale reactor	Controlled temperature: 58°C Controlled humidity	35	(Arrieta et al. 2014)
PLA/corn pellets (90/10)	8 L reactor	Controlled temperature: 58°C Controlled humidity	90	(Sarasa et al. 2009)
PHA film	2.5 L compost pot	Controlled temperature: 58°C Controlled humidity	35	(Weng et al. 2011)
PLA/PHB film (5 cm × 5 cm)	Laboratory scale	Controlled temperature: 58°C Controlled humidity	28	(Tabasi and Ajji 2015)
Starch-based (potato)	Laboratory scale	Controlled temperature: 58°C Controlled humidity	90	(Javierre et al. 2015)
Sponge cloth (cellulose/cotton)	Industrial composting facility	Peak temperature: 66°C	84	(Adamcová et al. 2017)
PBS/soybean	1 L hermetic glass bottle	Controlled temperature: 58°C pH: 7–8 Humidity: 50%–55%	100	(Anstey et al. 2014)
PBS/canola	0.5 L bottle	Controlled temperature: 25°C	44	(Cho et al. 2011)

Composting is highly recommended to bioplastic treatment. However, the use of proper compostable materials is necessary, according to current legislation, as well as suitable conditions for the process to reach optimal efficiency. For different types of bioplastics, natural resource–based such as PLA, PHA, starch and cellulose-based, or petroleum-based, including PBS and PCL, enough rates of biodegradability can be technically achieved. In addition to guaranteed biodegradable materials, it is essential to reach the appropriate conditions in the composting process, such as temperature, humidity, and oxygen. At the end, final compost (fertilizer) must have homogeneous characteristics, with indistinguishable particles and no phytotoxicity.

9.5 OUTLOOKS FOR BIOPLASTIC WASTE MANAGEMENT

Traditional end-of-life pathways of conventional plastics, when applied, mainly include recycling, incineration, and landfilling, while the use of biodegradable bioplastics adds the possibility of biological treatments, such as composting and anaerobic digestion technologies. The use of bio-based and biodegradable plastics also improves the circularity of plastic chain, due to their potential carbon neutrality and product recovery, as fertilizers, energy resources, and raw materials. However, these advantages must be explored properly by waste management systems due to existing non-designed facilities to explore the full potential of biodegradable materials.

The disposal of bioplastics in landfills could cause methane emissions due to their biodegradability, which means that depending on the landfill gas capture technology available, this could lead to harmful ramifications when transitioning to biodegradable bioplastics. In the case of recycling, drop-in bioplastics, such as bio-PET, bio-PA, bio-PE, and bio-PP could be treated together with conventional polymers, exploring current recycling infrastructures, due to their similar chemical structure. However, current global systems caused a mismanagement of almost all conventional polymers ever made. In 2015, roughly 58% of global plastic disposals were in landfills, and 18% were recycled (Zheng and Suh, 2019). Also, traditional recycling methods present huge barriers for the treatment of second- and third-generation bioplastics due to stream contamination. After mixing 10% of PLA or TPS into petrochemical plastics, the resulted recycled material showed greatly lower mechanical properties, which could increase the already existing problems of economical suitability of recycling plants (Van Roijen and Miller, 2022). Thus the energy recovery through anaerobic digestion could be a feasible option for bio-based polymers, highlighting the potential carbon neutrality of these materials. Composting is another great option for single-use bioplastics, for example, since they can be decomposed and assimilated to provide a humus-rich material for agriculture use.

Regarding to the LCA of bioplastics end-of-life options, the overall GHG emissions for composting bioplastics are found to be higher than anaerobic digestion and incineration due to CO_2 emissions that are not captured. For instance, PLA and PHB composting resulted in a 7%–50% and 40%–85% increase in GHGs emissions, respectively, compared to the other alternatives (Van Roijen and Miller, 2022). In comparison, various end-of-life options were analyzed for PLA bioplastic, and the use of anaerobic digestion reduced GHG emissions by 55%, assuming that the biogas would be burned to support energy demand and the digestate produced would be used in soils to displace traditional conditioners and chemical fertilizers (Van Roijen and Miller, 2022). Although composting is not an effective treatment to be applied alone, the combination of this technology with an energy recovery treatment in an integrated approach could provide safer fertilizers with no phytotoxicity to be used in agriculture. Gadaleta et al. (2022) evaluated a bio-based single-use plastic film of cellulose acetate during a sequential treatment of anaerobic digestion and composting. The authors reported a 66.8% disintegration of the material through anaerobic digestion, with 2.2 L CH_4 produced after 21 days, and a 74.6% of theoretical CH_4 production. After composting, bioassays on seeds germination confirmed the absence of toxic effects of the final compost, plus a 7.1% disintegration, reaching a 73.8% complete biodegradation of cellulose acetate plastic film.

Despite the great potentialities presented by the biodegradation pathways to reduce GHG emissions and solve the problems related to plastic pollution, these technologies still are in the early stages of comprehension and applicability, necessitating further research on their microbial activities and bioplastic formulation from a "beginning-of-life" approach. Many studies have showed inefficient biodegradation of many bioplastics. Shrestha et al. (2020) reported a biodegradability less than 50% of cellulose bioplastic film and PLA coffee capsules after 35 days of anaerobic digestion, while a complete biodegradation was not achieved after final treatment. Meanwhile, Bandini et al. (2022) reported a disintegration of ~65% for PLA and TPS under industrial conditions of anaerobic digestion, without meeting the requirements imposed by the EN 13432. Although physicochemical characterizations showed that biopolymers were partially biodegraded and composting process had proper effects on phytotoxicity tests, the biological treatment was not efficient to ensure full waste biodegradation, and bioplastic residues were found in the final compost. This shows that pretreatment strategies could be applied in some cases; the application of anaerobic co-digestion could be another alternative to reach higher efficiency bioenergy recoveries, combining different characteristics of co-substrates, such as nitrogen-rich materials including food wastes, animal manure, and silage. Also, product design has to be formulated in conformance with available end-of-life options and its efficiency.

In relation with that, second-generation bioplastic production approach can develop strong bioplastics with great biodegradability capacity and also improve product life cycles. Xia et al. (2021) reported an in situ lignin regeneration strategy to synthesize a high-performance bioplastic from poplar wood. The proposed process used a deep eutectic solvent (1:1 molar ratio of chloromethylene and oxalic acid) at a mass ratio of 15:1 at 110°C to deconstruct the porous matrix of natural wood to form a homogeneous slurry between cellulose nanofibrils and regenerated lignin. The resulting bioplastic not only showed high mechanical strength, excellent water stability, UV resistance, and improved thermal stability but also exhibited great biodegradability in the natural environment. The bioplastic was buried in soil at a depth of 5 cm, and after 2 months it underwent a swelling process and started to get fractured, becoming completely biodegraded after 3 months.

Algae are also a promising alternative to synthetize biopolymers for third-generation bioplastic production, through different routes in a biorefinery context. Natural biopolymers can be produced inside microalgal biomass by cellular factories, or the microalgae biomass can be fermented by microorganisms or directly melted with additives to form composites (Nandakumar et al., 202). Torres et al. (2015) showed that the use of residual microalgae biomass from the biodiesel production process with PBAT as an additive improved the mechanical properties, tensile modulus, and elongation of the biocomposites, while the end product easily decomposed in the soil, suggesting that it can be used as sustainable agricultural films. In another study, Kalita et al. (2021) evaluated a PLA blending with algae biomass as a biodegradation accelerator, using ASTM international D5338–15 as comparative assessment. The authors found that a blend with 5% of deoiled algae biomass in PLA showed an incremental effect on biodegradation of PLA biocomposites due to a higher content of nitrogen that fueled the microbial growth.

9.6 SUMMARY AND CONCLUSIONS

The production of biodegradable bio-based plastics has tremendous potentialities to reduce the environmental impacts of plastic industrial chain and solve plastic pollution, reduce GHG emissions, and enable more efficient biological treatments. However, further research is necessary for each different bioplastic product entering the market related to the degradation of bioplastics in the natural environments. In some cases, bioplastics were found to behave similarly to petrochemical plastics in environmental compartments, which suggests a slowly and limited degradation over time and the possibility to generate microplastics pollution.

The biodegradability of different bioplastics is strongly influenced by their properties and by environmental conditions. Among the different end-of-life options available for bioplastic treatment, the most prominent are composting and anaerobic digestion due to the possibility of product and energy recovery exploring their biodegradability capacities. For this, waste management practices that incorporate bioplastics in municipal organic streams must be applied, with efficient separate collection management, and treatment in a proper separated biowaste treatment system, which may be composed of anaerobic digestion followed by aerobic composting of the digestates. In some cases, the efficiency of composting and anaerobic digestion of bioplastics are limited, bringing concerns related to their end-of-life options, also for eco-friendly formulated products. Thus the formulation of novel bioplastic products has to be designed taking into account international standards in a "beginning-of-life" approach, highlighting their efficiencies in biological treatment plants and comparing them with available commercial products.

Further research in the biodegradability of novel bioplastics must be applied within the scope of international standards discussed in this chapter. Among these novel bioplastics, second- and third-generation products, as well as blends of them, can be developed and compared with already existing commercial bioplastic products. To prevent misuse of bioplastics and misleading consumers related to that, more informative labeling must be required for bioplastic materials based on international standards of biodegradability. Various commercial bioplastics contain additives and plasticizers to improve their characteristics and performance. These additives have to be analyzed by their origins and impacts, on the rate and extent of their biodegradability, as well as on human health and ecological risks, and novel compounds must be evaluated to improve functionalities of bioplastics in a sustainable way.

REFERENCES

Adamcová, D., Elbl, J., Zloch, J., Vaverková, M.D., Kintl, A., Juřička, D., Hladký, J., Brtnický, M., 2017. Study on the (bio)degradation process of bioplastic materials under industrial composting conditions. Acta Univ. Agric. Silvic. Mendelianae Brun. 65, 791–798. https://doi.org/10.11118/actaun201765030791

Anstey, A., Muniyasamy, S., Reddy, M.M., Misra, M., Mohanty, A., 2014. Processability and biodegradability evaluation of composites from poly(butylene succinate) (PBS) bioplastic and biofuel co-products from Ontario. J. Polym. Environ. 22, 209–218. https://doi.org/10.1007/s10924-013-0633-8

Arrieta, M.P., López, J., Rayón, E., Jiménez, A., 2014. Disintegrability under composting conditions of plasticized PLA-PHB blends. Polym. Degrad. Stab. 108, 307–318. https://doi.org/10.1016/j.polymdegradstab.2014.01.034

AS 5810–2010. Biodegradable plastics—biodegradable plastics suitable for home composting. Standards Australia by Committee EV-017, Degradability of Plastics. Sydney, July, 2010.

ASTM D6400–04, 2004. Standard specification for compostable plastics. American Society for Testing and Materials International. West Conshohocken, PA, September, 2004.

Bandini, F., Taskin, E., Vaccari, F, Soldano, M., Piccinini, S., Frache, A., Remelli, S., Menta, C., Sandro Cocconcelli, P., Puglisi, E., 2022. Anaerobic digestion and aerobic composting of rigid biopolymers in bio-waste treatment: Fate and effects on the final compost. Bioresour. Technol. 351. https://doi.org/10.1016/j.biortech.2022.126934

Bishop, G., Styles, D., Lens, P.N.L., 2021. Environmental performance comparison of bioplastics and petrochemical plastics: A review of life cycle assessment (LCA) methodological decisions. Resour. Conserv. Recycl. 168. https://doi.org/10.1016/j.resconrec.2021.105451

Chen, S.J., Ma, Y.J., Wang, J., Chen, D., Luo, X.J., Mai, B.X., 2009. Brominated flame retardants in children's toys: Concentration, composition, and children's exposure and risk assessment. Environ. Sci. Technol. 43, 4200–4206. https://doi.org/10.1021/es9004834

Cho, H.S., Moon, H.S., Kim, M., Nam, K., Kim, J.Y., 2011. Biodegradability and biodegradation rate of poly(caprolactone)-starch blend and poly(butylene succinate) biodegradable polymer under aerobic and anaerobic environment. Waste Manag. 31, 475–480. https://doi.org/10.1016/j.wasman.2010.10.029

CIEL, Center for International Environmental Law, 2019. Plastic & climate. The hidden costs of a plastic planet. Creative Commons Attribution 4.0 International License, May, 2019. https://www.ciel.org/plasticandclimate/

Danso, D., Chow, J., Streita, W.R., 2019. Plastics: Environmental and biotechnological perspectives on microbial degradation. Appl. Environ. Microbiol. 85. https://doi.org/10.1128/AEM.01095-19

Emadian, S.M., Onay, T.T., Demirel, B., 2017. Biodegradation of bioplastics in natural environments. Waste Manag. 59, 526–536. https://doi.org/10.1016/j.wasman.2016.10.006

EN 13432, 2000. EN 13432:2000—Packaging—Requirements for packaging recoverable through composting and biodegradation—test scheme and evaluation criteria for the final acceptance of packaging. European Committee for Standardization. Brussels, September, 2000.

European Bioplastics, 2014. Enzyme-mediated degradable plastics. Berlin, December, 2014. https://docs.european-bioplastics.org/2016/publications/EUBP_QA_enzyme-mediated_degradable_plastics.pdf

Gadaleta, G., De Gisi, S., Picuno, C., Heerenklage, J., Cafiero, L., Oliviero, M., Notarnicola, M., Kuchta, K., Sorrentino, A., 2022. The influence of bio-plastics for food packaging on combined anaerobic digestion and composting treatment of organic municipal waste. Waste Manag. 144, 87–97. https://doi.org/10.1016/j.wasman.2022.03.014

Gan, Z., Zhang, H., 2019. PMBD: A comprehensive plastics microbial biodegradation database. Database 2019, 1–11. https://doi.org/10.1093/database/baz119

García-Depraect, O., Bordel, S., Lebrero, R., Santos-Beneit, F., Börner, R.A., Börner, T., Muñoz, R., 2021. Inspired by nature: Microbial production, degradation and valorization of biodegradable bioplastics for life-cycle-engineered products. Biotechnol. Adv. 107772. https://doi.org/10.1016/j.biotechadv.2021.107772

Geyer, R., Jambeck, J.R., Law, K.L., 2017. Production, use, and fate of all plastics ever made. Sci. Adv. 3, 3–8. https://doi.org/10.1126/sciadv.1700782

Gómez, E.F., Michel, F.C., 2013. Biodegradability of conventional and bio-based plastics and natural fiber composites during composting, anaerobic digestion and long-term soil incubation. Polym. Degrad. Stab. 98, 2583–2591. https://doi.org/10.1016/j.polymdegradstab.2013.09.018

Hahladakis, J.N., Velis, C.A., Weber, R., Iacovidou, E., Purnell, P., 2018. An overview of chemical additives present in plastics: Migration, release, fate and environmental impact during their use, disposal and recycling. J. Hazard. Mater. 344, 179–199. https://doi.org/10.1016/j.jhazmat.2017.10.014

He, Z., Li, G., Chen, J., Huang, Y., An, T., Zhang, C., 2015. Pollution characteristics and health risk assessment of volatile organic compounds emitted from different plastic solid waste recycling workshops. Environ. Int. 77, 85–94. https://doi.org/10.1016/j.envint.2015.01.004

Innocenti, F.D., 2003. Biodegradability and Compostability—The International Norms, in: Chiellini, E., Solaro, R. (Eds.), Biodegradable Polymers and Plastics. Springer Science+Business Media, New York, pp. 33–45.

ISO 14855-2:2018—Determination of the ultimate aerobic biodegradability of plastic materials under controlled composting conditions—Method by analysis of evolved carbon dioxide—Part 2: Gravimetric measurement of carbon dioxide evolved in a laboratory-scale test. The International Organization for Standardization, Technical Committee ISO/TC 61, Plastics, Subcommittee SC 14, Plastics and environment, July, 2018. https://www.iso.org/obp/ui/#iso:std:iso:14855:-2:ed-2:v1:en

Jarerat, A., Pranamuda, H., Tokiwa, Y., 2002. Poly (L-lactide)-degrading activity in various actinomycetes. Macromol. Biosci. 2, 420–428.

Javierre, C., Sarasa, J., Claveria, I., Fernández, A., 2015. Study of the biodisintegration on a painted bioplastic material waste. Mater. Plast. 52, 116–121.

Kale, G., Auras, R., Singh, S.P., Narayan, R., 2007a. Biodegradability of polylactide bottles in real and simulated composting conditions. Polym. Test. 26, 1049–1061. https://doi.org/10.1016/j.polymertesting.2007.07.006

Kale, G., Kijchavengkul, T., Auras, R., Rubino, M., Selke, S.E., Singh, S.P., 2007b. Compostability of bioplastic packaging materials: An overview. Macromol. Biosci. 7, 255–277. https://doi.org/10.1002/mabi.200600168

Kalita, N.K., Damare, N.A., Hazarika, D., Bhagabati, P., Kalamdhad, A., Katiyar, V., 2021. Biodegradation and characterization study of compostable PLA bioplastic containing algae biomass as potential degradation accelerator. Environ. Challenges. 3. https://doi.org/10.1016/j.envc.2021.100067

Karamanlioglu, M., Preziosi, R., Robson, G.D., 2017. Abiotic and biotic environmental degradation of the bioplastic polymer poly (lactic acid): A review. Polym. Degrad. Stab. 137, 122–130. https://doi.org/10.1016/j.polymdegradstab.2017.01.009

Karan, H., Funk, C., Grabert, M., Oey, M., Hankamer, B., 2019. Green bioplastics as part of a circular bioeconomy. Trends Plant Sci. 24, 237–249. https://doi.org/10.1016/j.tplants.2018.11.010

Kurniawan, S.B., Said, N.S.M., Imron, M.F., Abdullah, S.R.S., 2021. Microplastic pollution in the environment: Insights into emerging sources and potential threats. Environ. Technol. Innov. 23, 1–14. https://doi.org/10.1016/j.eti.2021.101790

Lau, W.W.Y., Shiran, Y., Bailey, R.M., Cook, E., Stuchtey, M.R., Koskella, J., Velis, C.A., Godfrey, L., Boucher, J., Murphy, M.B., Thompson, R.C., Jankowska, E., Castillo, A.C., Pilditch, T.D., Dixon, B., Koerselman, L., Kosior, E., Favoino, E., Gutberlet, J., Baulch, S., Atreya, M.E., Fischer, D., He, K.K., Petit, M.M., Sumaila, U.R., Neil, E., Bernhofen, M.V., Lawrence, K., Palardy, J.E., 2020. Evaluating scenarios toward zero plastic pollution. Science (80-.). 369, 1455–1461. https://doi.org/10.1126/SCIENCE.ABA9475

Nandakumar, A., Chuah, J.A., Sudesh, K., 2021. Bioplastics: A boon or bane? Renew. Sustain. Energy Rev. 147, 111237. https://doi.org/10.1016/j.rser.2021.111237

Napper, I.E., Thompson, R.C., 2019. Environmental deterioration of biodegradable, oxo-biodegradable, compostable, and conventional plastic carrier bags in the sea, soil, and open-air over a 3-year period. Environ. Sci. Technol. 53, 4775–4783. https://doi.org/10.1021/acs.est.8b06984

Quecholac-Piña, X., Hernández-Berriel, M.D.C., Mañón-Salas, M.D.C., Espinosa-Valdemar, R.M., Vázquez-Morillas, A., 2020. Degradation of plastics under anaerobic conditions: A short review. Polymers (Basel). 12, 1–18. https://doi.org/10.3390/polym12010109

Razza, F., Innocenti, F.D., Dobon, A., Aliaga, C., Sanchez, C., Hortal, M., 2015. Environmental profile of a bio-based and biodegradable foamed packaging prototype in comparison with the current benchmark. J. Clean. Prod. 102, 493–500. https://doi.org/10.1016/j.jclepro.2015.04.033

Rosenboom, J.G., Langer, R., Traverso, G., 2022. Bioplastics for a circular economy. Nat. Rev. Mater 7, 117–137. https://doi.org/10.1038/s41578-021-00407-8

Rudnik, E., Briassoulis, D., 2011. Degradation behaviour of poly(lactic acid) films and fibres in soil under Mediterranean field conditions and laboratory simulations testing. Ind. Crops Prod. 33, 648–658. https://doi.org/10.1016/j.indcrop.2010.12.031

Sarasa, J., Gracia, J.M., Javierre, C., 2009. Study of the biodisintegration of a bioplastic material waste. Bioresour. Technol. 100, 3764–3768. https://doi.org/10.1016/j.biortech.2008.11.049

Shrestha, A., van-Eerten Jansen, M.C.A.A., Acharya, B., 2020. Biodegradation of bioplastic using anaerobic digestion at retention time as per industrial biogas plant and international norms. Sustain. 12. https://doi.org/10.3390/su12104231

Solano, G., Rojas-Gätjens, D., Rojas-Jimenez, K., Chavarría, M., Romero, R.M., 2022. Biodegradation of plastics at home composting conditions. Environ. Challenges 7, 100500. https://doi.org/10.1016/j.envc.2022.100500

Tabasi, R.Y., Ajji, A., 2015. Selective degradation of biodegradable blends in simulated laboratory composting. Polym. Degrad. Stab. 120, 435–442. https://doi.org/10.1016/j.polymdegradstab.2015.07.020

Torres, S., Navia, R., Campbell Murdy, R., Cooke, P., Misra, M., Mohanty, A.K., 2015. Green composites from residual microalgae biomass and poly(butylene adipate-co-terephthalate): Processing and plasticization. ACS Sustain. Chem. Eng. 3, 614–624. https://doi.org/10.1021/sc500753h

Van Roijen, E.C., Miller, S.A., 2022. A review of bioplastics at end-of-life: Linking experimental biodegradation studies and life cycle impact assessments. Resour. Conserv. Recycl. 181. https://doi.org/10.1016/j.resconrec.2022.106236

Weng, Y.X., Wang, X.L., Wang, Y.Z., 2011. Biodegradation behavior of PHAs with different chemical structures under controlled composting conditions. Polym. Test. 30, 372–380. https://doi.org/10.1016/j.polymertesting.2011.02.001

Xia, Q., Chen, C., Yao, Y., Li, J., He, S., Zhou, Y., Li, T., Pan, X., Yao, Y., Hu, L., 2021. A strong, biodegradable and recyclable lignocellulosic bioplastic. Nat. Sustain. 4, 627–635. https://doi.org/10.1038/s41893-021-00702-w

Yagi, H., Ninomiya, F., Funabashi, M., Kunioka, M., 2013. Thermophilic anaerobic biodegradation test and analysis of eubacteria involved in anaerobic biodegradation of four specified biodegradable polyesters. Polym. Degrad. Stab. 98, 1182–1187. https://doi.org/10.1016/j.polymdegradstab.2013.03.010

Zheng, J., Suh, S., 2019. Strategies to reduce the global carbon footprint of plastics. Nat. Clim. Chang. 9, 374–378. https://doi.org/10.1038/s41558-019-0459-z

Zimmermann, L., Dombrowski, A., Völker, C., Wagner, M., 2020. Are bioplastics and plant-based materials safer than conventional plastics? In vitro toxicity and chemical composition. Environ. Int. 145, 106066. https://doi.org/10.1016/j.envint.2020.106066

10 Downstream Processing and Formulation of Bioplastics

Júlio Cesar de Carvalho, Luciana Porto de Souza Vandenberghe, and Carlos Ricardo Soccol

CONTENTS

10.1 INTRODUCTION

Synthetic, oil-based plastics tend to be replaced by bioplastics. Synthetic plastics are ubiquitous in our society because of their versatility and low cost. However, most of these plastics degrade very slowly, taking decades to decompose fully. The global manufacturing of plastics was about 370 million tons in 2019 (Plastics Europe, 2020) and could reach 500 million tons by 2030 if the average annual growth of about 4% per year (Vigneswari et al., 2021) is maintained. The inadequate disposal of plastics creates the enormous problem of recalcitrant waste scattered in all biomes and the increasing transformation in microplastics. While recycling properly discarded plastics is an important way to tackle the problem, it is not enough to prevent inadequate disposal; also, the very raw materials currently used to produce plastics are non-renewable and require a sustainable source. The volume of bioplastics produced worldwide is growing in response to these environmental concerns.

In a broader sense, bioplastics are bio-based plastics such as sugarcane ethanol converted into polyethylene, biodegradable synthetic plastics such as PBAT (polybutylene adipate terephthalate), or both bio-based and biodegradable, such as PLA (polylactic acid) or PHA (polyhydroxyalkanoates). Figure 10.1 illustrates these classifications and bioplastic production capacity, about 2.42 million tonnes—less than 1% of the global production of fossil-based plastics. The market share of bioplastics may seem small, but the CAGR projected for bioplastics is about 16%–18% for 2022–2031 (Research Nester, 2022; Vandenberghe et al., 2021), which will mean an increment of four to five times in a decade, reaching 10–12 million tonnes by 2032.

DOI: 10.1201/9781003344018-10

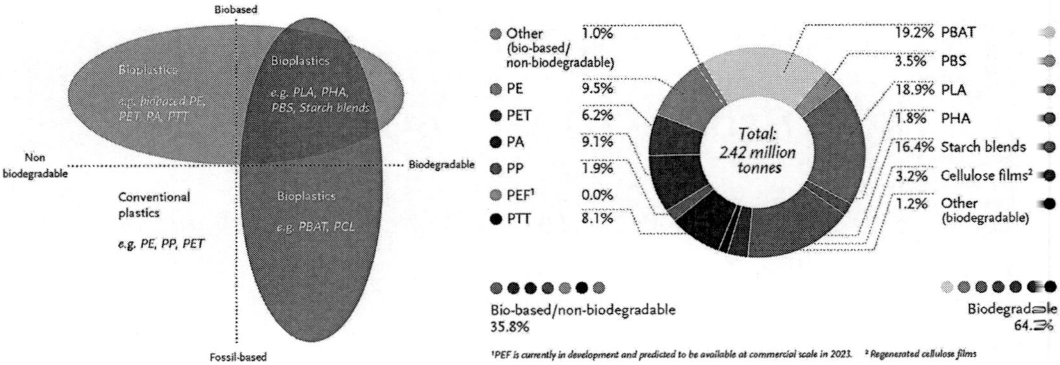

FIGURE 10.1 Bioplastics as categorized by European Bioplastics (www.european-bioplastics.org), an association of bio-based plastics industries, and bioplastics production capacity in 2021.

Source: Reproduced with permission.

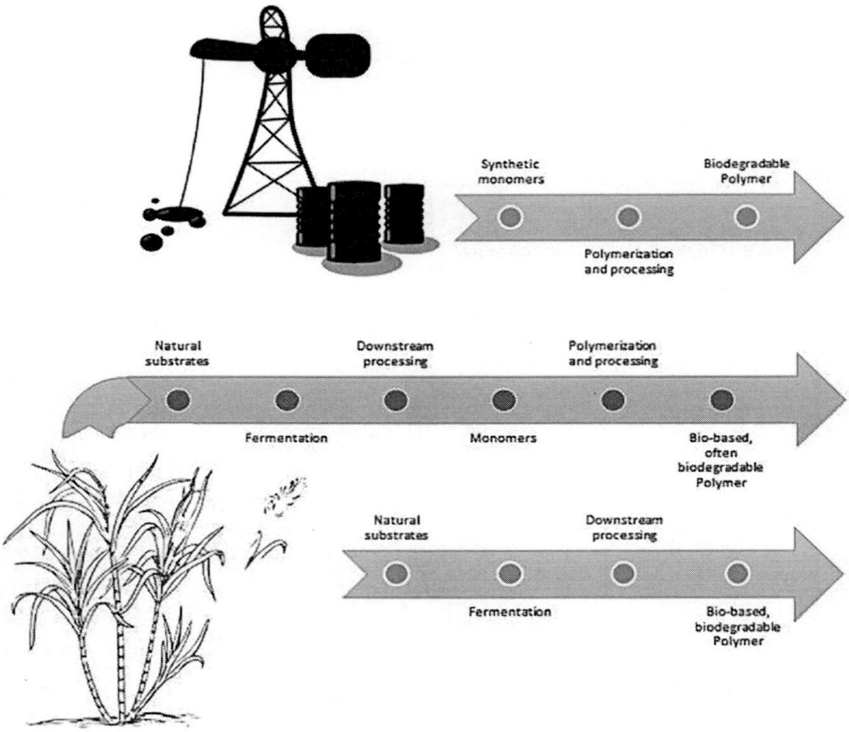

FIGURE 10.2 Broad processing steps for production of bioplastics.

With market growth, it is expected that part of the substrates currently used for biofuel production may be used for biopolymer production. To support the anticipated market growth, it is important to develop applications—bioplastics have specific properties and must be properly formulated to substitute synthetic plastics. This development requires both research for novel or improved materials, and market development—packaging, fabrics, and building materials manufacturers need time to know novel alternatives and adapt current practices.

Figure 10.2 shows broad routes for bioplastic production. Bioplastics produced from petrochemical feedstocks are synthesized using existing chemical technology. In contrast, bioplastics from renewable monomers, such as bio-polyethylene and polylactic acid, have a preliminary step of bulk monomer production and the second step of polymer synthesis using chemical routes.

Some biopolymers can be synthesized directly, as macromolecules, by microorganisms, in a process somewhat similar to the formation of starch granules in plants—in fact, modified starches and regenerated cellulose are important biopolymers, constituting about 20% of the bioplastic market share. They are not, however, usually produced by fermentation. The prototypical bioplastic directly produced by fermentation is PHA (polyhydroxyalkanoate), and this important bioproduct will be the focus of the following sections.

10.1.1 PHA Structure and Biosynthesis

PHA, or polyhydroxyalkanoate, is a class of polyesters derived from the condensation of hydroxy acids, such as 3-hydroxybutyric acid (giving PHB, polyhydroxybutyrate), 3-hydroxypentanoic (or hydroxyvaleric) acid (giving PHV, polyhydroxyvaleric acid), and combinations (copolymers) of these and other similar monomers (Figure 10.3). Because the monomers are bifunctional, the derived polymers are usually linear. Even so, the variability is enormous, with different side chain lengths, depending on precursor molecules and even PHAs with non-alkyl side chains (Kosseva & Rusbandi, 2018).

There are many options if biopolymers are to be produced through partially synthetic routes. Although the production of PHAs can be done enzymatically using esterases such as a bacteria-derived PHA synthase (Kosseva & Rusbandi, 2018), this is neither economic nor common. All established processes for PHA production use microorganisms—and more than 300 species can produce PHAs (Khatami et al., 2021), perhaps because of its efficiency as a dense energy reserve. Usually, PHAs are made using pure selected cultures, sometimes even modified organisms. However, using mixed cultures directly from wastewater-coupled processes is gaining interest. It might lower the cost of biomass production, but it has a more complicated downstream processing and is not yet used on pilot or industrial scales (Mannina et al., 2020). There is extensive research on the use of waste streams for both pure cultures and mixed microbial, community-based PHA production (de Donno Novelli et al., 2021; Estévez-Alonso et al., 2021). Pure cultures have the advantage of better

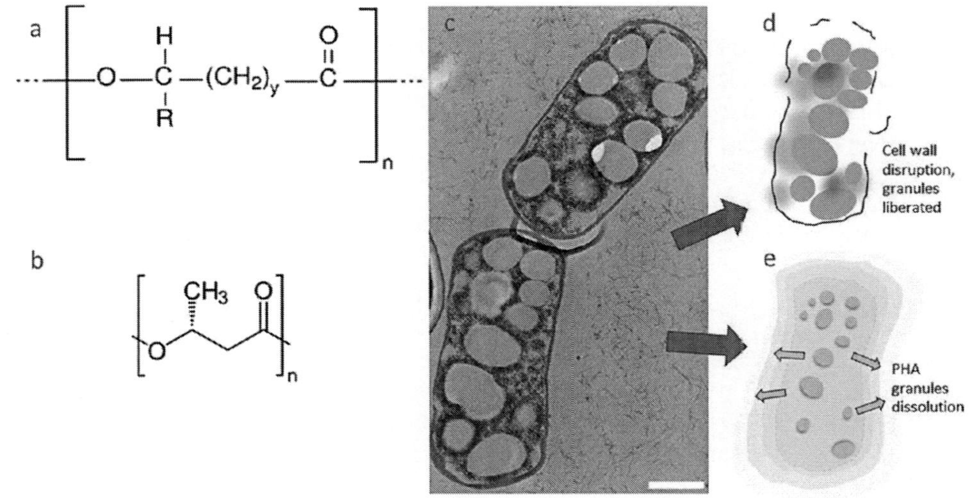

FIGURE 10.3 PHA structure and granules in cells. (a) general structure of PHAs; (b) structure of PHB, the most common polyhydroxyalkanoate; (c) SEM of *Cupriavidus* sp. cells, after (Venkateswar Reddy et al., 2015)—the scale bar represents 0.5 μm; (d) concept of mechanical disruption of cells, liberating PHA granules; and (e) concept of chemical digestion and solvent dissolution of PHA and other cell components.

control in laboratory environments, but the recovery of bioresources from waste streams may be better done with mixed cultures.

Industrially, the most common organism used for PHA production is *Cupriavidus necator* (Koller et al., 2017), a gram-negative, facultatively anaerobic, mesophilic bacterium with a size of 1.8–2.6 × 0.7 μm, and several synonym names (Soˇ et al., 2022) (Figure 10.3). As with other reserve substances in microorganisms, the biosynthesis and accumulation of PHA are improved in nutrient imbalance conditions, such as a high C:N ratio (Ahn et al., 2015). Other strategies, such as phosphate concentration control (Grousseau et al., 2014), can modulate production and composition. After fermentation, PHA concentrations intracellularly are typically in the range of 60%–70% (from synthetic media; Fernández-Dacosta et al., 2015), 50% (from lignocellulosic sources; Vigneswari et al., 2021) and 25% (from algae; Afreen et al., 2021). However, the cell concentration obtainable with heterotrophs is much higher than with microalgae, especially in fed-batch processes.

10.1.2 Downstream Processing as a Bottleneck in PHA Production

The fermentative step is the most important for conferring the desirable properties to PHAs and producing it with competitive concentration and productivity. However, the separation and purification steps (downstream processing) add a high cost to the whole process because PHA is produced intracellularly and requires separation from the cell components to be used as a polymer. It is estimated that PHA downstream constitutes between 30% and 50% of the total production cost—proportionally much more than the downstream of traditional polymers (Khatami et al., 2021; Pala-Ozkok et al., 2022). Upstream operations also significantly impact the production of PHAs from lignocellulosic residues.

10.1.3 PHA Properties

To discuss the downstream processing, it is essential to understand the properties of the fermentation broth, the microorganism, in which form it accumulates PHAs, the processing properties of the polymer, and the properties of undesired contaminants.

Because of the linear chain structures and the presence of ester bonds, PHAs are thermoplastics (i.e., they melt when heated) and are insoluble in water. They are, however, soluble in several organic solvents, most commonly halogenated, such as chloroform (Aramvash et al., 2018), and somewhat resistant to acids and bases.

As already stated, PHA is accumulated in granules with relatively high purity. The material density is about 1.15–1.25 g/mL, with granule sizes of 100–500 nm (Goswami et al., 2021; Mas et al., 1985), rendering the settling velocity of the material low.

Being gram-negative, PHA-producing bacteria have cell wall lipopolysaccharides known as endotoxins that are liberated upon cell death (Dutt Tripathi et al., 2021), eventually co-purifying with the polymer, contaminating it (Gahlawat & Kumar Soni, 2019).

10.2 DOWNSTREAM PROCESSING

As an intracellular product, PHA production requires cell disruption and separation of the polymer from the other cell components. There are several possible routes, each with its advantages for specific fermentation processes, microorganisms, and product uses. However, it is usually best to work with concentrated streams; therefore, the first step in most processes is the separation of solids using filtration or centrifugation. After this, the biomass can be disrupted or digested, or the polymer can be dissolved; and after polymer liberation as granules or solution, it is necessary to recover the polymer by removing the solvent (de Donno Novelli et al., 2021; Estévez-Alonso et al., 2021; Mannina et al., 2020; Pala-Ozkok et al., 2022). The options are further complicated by eventual steps of decolorization or endotoxin degradation are used. Importantly, not all methods used on a laboratory scale can be economically translated to large-scale production (de Carvalho, Medeiros, Vandenberghe, et al., 2017). Table 10.1 shows the main routes, possible processing steps, and scalability for PHA downstream from the fermented broth.

Different processing routes lead to different product characteristics and require quite different OPEX and CAPEX. Ultimately, the product destination (e.g., polymers for packaging with or without contact with food, polymers for biomedical devices, or polymers for fibers and tissues) will affect the economic evaluation and the process chosen. Energy usage and solvent recovery are also crucial in cost analysis: in solvent-based processes, drying and solvent recovery consumes much energy; however, water-based systems have increased costs of chemicals and generate effluents with a higher organic load, which must be treated (Estévez-Alonso et al., 2021). The following sections briefly discuss possible downstream steps (also termed *unit operations*) used in the different routes for PHA production.

10.2.1 Cell Concentration

Separation of cells from the fermented broth is typically the first unit operation in a downstream sequence because of volume reduction and the possibility of biomass washing. The biomass titers in fermentation broths depend highly on the substrate, microorganism, and fermentation strategy. Therefore a wide range of concentrations in PHA production is reported in the literature, from less than 10 g/L up to 250 g/L (Afreen et al., 2021; Mudenur et al., 2019; Vigneswari et al., 2021). The higher titers are obtained with selected microorganisms under fed-batch fermentation. While very concentrated suspensions can proceed to the subsequent separation step, the much more common, relatively diluted suspensions can benefit from volume reduction by centrifugation, filtration, or tangential flow filtration.

Three critical parameters in bacterial cell concentration are viscosity, density, and particle size. Since the polymer is produced intracellularly, the viscosity of the fermentation broths is low. The biomass density is relatively average for bacteria, at 1.14 g/L (Pedrós-Alió et al., 1985), despite the accumulation of up to 75% of the dry biomass as PHA granules, whose density ranges from 1.15 to 1.25 g/mL (Mas et al., 1985; Oshiki et al., 2010). With this size and density, the natural sedimentation velocity of the cells can be as low as 0.03 cm/h (de Carvalho et al., 2013), considering a broth viscosity of 1 cP and an average cell diameter of 1 μm, impractical for natural sedimentation. Flocculation can help increase the apparent particle size (Cho et al., 2000) but adds a flocculant whose fate must be evaluated during the subsequent operations.

Centrifugation is one of the most common operations on a laboratory scale. It is also routinely used for biomass harvesting in industry. However, for bacteria separation, the residence times must be high, or the equipment must operate at high speed, requiring a relatively high capital cost. For example, a centrifugation step performed at 15000 ×g and 10 minutes on a lab scale with an average sedimentation radius of 10.5 cm (Arikawa et al., 2017) can be replicated with a tubular centrifuge with a similar speed, which limits bowl volume; a continuous disk centrifuge, which limits the final biomass concentration; or a decanter, operating at a lower speed, with a higher residence time.

Direct filtration is comparatively low-cost and requires high pressures because of the high resistance of the filter cake. Even with filtering aids, direct (dead-end) filtration of bacteria is difficult because of the small particle size comparable to that of the filter aid. However, filtration can be helpful for MMCs (mixed microbial cultures) produced in new, alternative processes.

One way of avoiding cake formation is using a tangential flow filter, a well-known filtration system adequate for the concentration of suspensions as long as the fluid viscosity is moderate. Tangential flow filtration has the advantage of no moving parts in the filter itself (but requires pumps) and high modularity. The energy input in the process is lower than that for centrifugation. Still, the method suffers from membrane fouling (Liu et al., 2020), and the periodic replacement of membranes add to the operational costs.

TABLE 10.1
Most Common Strategies and Operations for PHA Downstream Processing

	Approach			Unit Operations	
	Solvent Based	**Water Based**	**Water Based, with Digestion**	**Process**	**Scalability**
	Cell concentration	Cell concentration	Cell concentration	Centrifugation	Scalable, energy-intensive
				Microfiltration	Scalable
				Dead-end filtration	Scalable, may require filter aid
	Cell drying	↓	↓	Spray or fluid bed drying	Scalable
				Freeze drying	Scalable, more complex
				Tray drying	Scalable, may require post-milling
	↓	Cell disruption	↓	High pressure homogenization	Scalable, energy intensive
				Bead mill	Scalable, energy intensive
				Sonication	Not economically scalable
	↓	↓	Cell digestion	Enzymatic digestion	Scalable, costly
				Alkaline digestion	Scalable, cheap, affects PHA, reduces contamination
				Acid digestion	Scalable, cheap
				Hypochlorite digestion	Scalable, cheap, affects PHA, reduces contamination
	↓	Granules washing and recovery	Granules washing and recovery	Centrifugation	Scalable
				Microfiltration	Scalable
				Dead-end filtration	Difficult because of particle size
	PHA solubilization	↓	↓	Solid-liquid extraction	Scalable, requires distillation for solvent recovery
				Supercritical fluid extraction	Scalable, more expensive
	PHA precipitation	↓	↓	Precipitation and filtration	Scalable
	Solvent recovery	↓	↓	Distillation	Scalable
	Product drying	Product drying	Product drying	Spray drying or fluid bed drying	Scalable; solvent drying requires safety measures
Recovery	High	Lower	Lower		
Purity	High	Lower	High		
Degradation	Low	Lower	Higher		

10.2.2 Drying

Biomass drying is usually necessary for processes that will use solvent extraction. Water in the biomass impedes proper contact of PHA granules with the solvent and complicates the polymer solution separation, forming emulsions with the complex solution and cell debris. The ideal dry solids have large surface areas to enhance contact with the solvent and, thus, the mass transfer from the cell granules to the bulk of the solution. On a small scale, processes such as lyophilization and thermal drying are standard (Pala-Ozkok et al., 2022).

Lyophilization, the slow drying of frozen solids at low pressures, gives porous, solvent-accessible biomass that is adequate for research but whose cost can be prohibitive on a large scale. Other traditional methods, such as tray or tumble drying, are cheaper to establish but may give granulated biomass that will be harder to extract. This problem can be solved by milling the dry solids or using high-shear stirring during the subsequent extraction. However, it is possible to dry the solids directly as powders using spray drying. The process requires a cell slurry instead of a cell paste and therefore consumes more energy than other drying methods. Still, it has a simple setup, robust equipment, and gives high-quality biomass.

Another advantage of drying biomass is that it stabilizes the material for further processing, allowing storage and later pooling or standardization of the batches. Interestingly, when the solvent is similarly or less volatile than water, this moisture can be removed during the extraction through evaporation.

10.2.3 Cell Disruption

Cell disruption or lysis involves exposing the intracellular content, with the cell membrane disruption and eventually the cell wall disruption (de Carvalho, Medeiros, Letti, et al., 2017). PHA granules are much larger than cell wall pores; thus effective cell wall disruption is necessary for polymer recovery. This can be done by numerous methods, from sonication to the chemical digestion of biomass. However, some methods used on a laboratory scale are difficult to establish on a large scale: sonication, for example, is highly energy intensive (de Carvalho et al., 2020), requiring refrigeration during processing.

Mechanical disruption by bead milling and high-pressure homogenization is the most common industrial cell disruption method in PHA production (Kosseva & Rusbandi, 2018), with several equipment vendors and a well-established process. These mechanical methods are both capital and energy intensive, compared to chemical methods, but do not necessarily require the use of chemicals.

High-pressure homogenization involves pumping the cell slurry at high speeds through narrow passages in a disruption valve. High velocity requires high pressure (typically 500–1500 atm) for pumping, and the high shear and sharp pressure changes in the disruption valve cause enough mechanical deformation to the cell to tear cell walls. It was shown that high-pressure homogenization done at 620 atm and two passes was enough to disrupt *R. eutropha* cells in the stationary, N-depleted phase, and that disruption was enhanced by using SDS for membrane degradation, NaCl to modulate the ionic strength and solubility of cell components, EDTA as chelator, urea as denaturant, and high pH (Pérez-Rivero et al., 2019).

Bead milling is a process where the cell slurry is agitated at high speeds in the presence of a solid abrasive, typically constituting about 50% of the chamber volume. This intense mixing causes enough shear through the suspension to disrupt cells. The process, however, generates heat that must be dissipated (Jacquel et al., 2008). Industrial bead mills are jacketed for refrigeration, and the feed—which can be continuous—can also be pre-chilled to reduce the temperature in the chamber. The process can also be combined with chemical agents (Kosseva & Rusbandi, 2018).

Since cell disruption exposes and mixes otherwise compartmentalized intracellular components, it allows for the degradation of PHA granules by enzymes and throws in the mix the endotoxins from the bacterial outer membrane. Using detergents such as SDS may help inactivate enzymes such as PHA depolymerase and nucleic acids that would otherwise increase lysate viscosity and solubilize NPCM (non-PHA cellular materials). Heat treatment can also help enzyme inactivation, although specific conditions must be tested to reduce damage to the polymer (Jacquel et al., 2008).

Autolysis and cell fragilization are strategies in development that can lower disruption and digestion costs. Autolysis, programmable cell lysis, has been attempted in modified organisms, using holin-endolysin systems (bacteriophage T4 enzymes), heterologous lysozyme expression after cultivation, and eliciting osmotic membrane disruption (Acuña & Poblete-Castro, 2022). Cell fragilization can be achieved through modification of medium composition, such as modifying the inorganic salts content (Kosseva & Rusbandi, 2018).

10.2.4 Cell Digestion

Cell digestion is the aqueous solubilization of part or all of the cell components for PHA granule exposure or NPCM dissolution. A wide range of substances can do it, often combined, such as acids or alkalis, surfactants as dispersing agents, and membrane-dissolving aids. The PHA concentration inside cells can be as high as 90%, and therefore digesting the cells to free PHA granules is effective and does not require prior drying (de Donno Novelli et al., 2021; Mannina et al., 2020), although care must be taken to avoid polymer degradation.

Acid digestion is the nonspecific hydrolysis of cell material using strong mineral acids, such as HCl or H_2SO_4. The process is simple, scalable—since conditions of temperature, concentration and agitation are simple to scale up—and low-cost, considering that an acid such as sulfuric acid is a commodity sold at about USD 150/ton. However, the acid can catalyze ester hydrolysis and thus lowers the average molecular weight of PHA and, therefore, its properties such as strength. The process was found to be competitive with NaOH digestion (EUR 1.11/kg compared to EUR 1.02/kg for the alkaline process, and with 98% recovery and lower polymer degradation; López-Abelairas et al., 2015). However, this process was developed using freeze-dried biomass.

Alkaline digestion is another low-cost process for cell digestion; NaOH, the most common strong base used in alkali treatment, costs about USD 400/ton. Again, the treatment impacts PHA quality; Mohammadi et al. (2012) successfully recovered PHAs with 96% purity and a molecular weight reduction of 13% using 0.05 M NaOH at 4°C for 3 hours. However, the process used previously lyophilized cells on a small scale. Although their initial PHA content was modest (38.4%), the results had higher yield and purity than alkaline treatments with NaOH at higher concentration and shorter time (0.1–1 M at 30–60°C) but a lower purity compared to chloroform extraction of dry material. A conceptual LCA (life cycle analysis) of the process showed that alkali-surfactant treatment might be better than either acidic or solvent-based processes (Fernández-Dacosta et al., 2015).

One advantage of alkaline digestion is that it lowers endotoxin levels: long alkali digestion (6 h, 2.5 N NaOH) successfully reduced endotoxins to less than EUR 5/g for medical applications (Gahlawat & Kumar Soni, 2019).

Hypochlorite digestion can be very effective in releasing PHA from cells. Not only it is a mildly alkaline condition (bleach at 5% concentration has a pH of around 11, the same as 0.001 N NaOH), but most importantly, it is a potent oxidizing agent that attacks several reactive groups in organic matter, inactivating enzymes, destroying chromophores, and shortening macromolecules. The process must be done with care: treatment of lyophilized *C. necator* (*Alcaligenes eutrophus* in the original) with 10.5% NaOCl for 1 h at 25°C, reaching 95% PHA purity and reduction of the molecular weight from 1200 kDa and PI (polydispersity index) of 3–600 kDa and PI 4 (Berger et al., 1989). Other authors made similar observations over the years (López-Abelairas et al., 2015; Pérez-Rivero et al., 2019), which indicates that lyophilization and hypochlorite digestion may be adequate for analysis of PHA content and composition and maybe when short-chain, highly pure polymers are sought. However, hypochlorite treatment is not adequate for producing structural material for fibers or films unless crosslinkers or modifiers are used in the final polymer formulation. Hypochlorite can, however, be an essential and low-cost aid in purification for endotoxin removal, or color reduction, as part of multiple polymer treatments.

Enzymatic digestion was initially developed by ICI (later Metabolix) in the 1990s and consisted of using a cocktail of proteases to digest NPCM, with modest efficiency (Jiang et al., 2018). More than 30 years later, many hydrolases are commercially available at a more competitive cost. The process can be quite effective but is still expensive, usually involving cocktails combined with heat and surfactants (de Donno Novelli et al., 2021; Koller, 2020). Besides proteolytic enzymes that can degrade PHA depolymerases, the use of phospholipases enhances the membrane degradation, to a synergic outcome. Because of the protection by the outer lipid membrane, gram-negative bacteria are less susceptible to cell wall–degrading enzymes. Still, it is conceivable that a cocktail containing phospholipases and muraminidases could enhance the effect of proteases in PHA granules liberation.

Digestion of MMCs seems to be even more complicated, perhaps because of the complex matrix that holds the biomass flocs, produced from wastewaters (Mannina et al., 2020), besides the diversity in microorganisms and thus in polymeric structures that must be solubilized.

Digestion using the mealworm *Tenebrio molitor* is perhaps the most curious process used for PHA recovery from *C. necator* cells, as tested by Murugan et al. (2016). At first, it seems that the modest purity of 89% (but 100% after washing with 1% SDS), the need to separate the PHA granules from fecal matter, and the use of freeze-dried biomass makes the process unpractical. However, it can be argued that this process converts NPCMs directly into valuable (nutritional) biomass, besides the polymer, in a potentially circular process.

10.2.5 Granules Washing and Recovery

After granules exposure, the resulting suspension from mechanical disruption consists of cell debris, cytoplasm contents, and PHA granules. The suspension resulting from digestion consists of low-molecular-weight NPCM and PHA granules. In both cases, with the granules ranging from 100 to 500 nm and density of 1.15–1.25, the settling velocity of the granules would be in the range of 0.0003–0.0125 cm/h, which is 3–100 times lower than the settling velocity of bacteria (de Carvalho, Medeiros, Vandenberghe, et al., 2017). This makes recovery by centrifugation challenging, requiring long residence times or rotational speeds. Tangential flow filtration is still a good option for granules recovery, requiring only a lower membrane cutoff than cell separation. Dead-end filtration is further complicated because of the small particle size. Air flotation is a less common option that can be useful for separating the low-polarity granules (Pala-Ozkok et al., 2022). Metabolix suggests intensive use of washing and digestion steps in diafiltration (an operational mode of TFF) for obtaining a PHA latex (patent WO 1999/051760 A1).

10.2.6 Polymer Extraction (Solubilization) and Recovery

In the extractive route of PHA recovery, dry biomass can be treated with a suitable solvent that will dissolve the polymeric granules. Solvent extraction is the most common method (Dutt Tripathi et al., 2021) and a significant processing step requiring adequate solvents. Many reports describe the extraction from granules using halogenated solvents such as chloroform, 1,2-dichloromethane and methylene chloride (Mannina et al., 2020). However, solvents employed to determine PHA content on a small scale are not always economical on a large scale. Patents from Kaneka Co., Japan (US 2005/0222373 A1, US 2005/0287654) suggest using several solvents, but extensive use of 3-hydroxyhexanoate in the examples. The solvent 1,2-propylene carbonate seems promising (Pala-Ozkok et al., 2022), eventually with NaOH digestion and aiding surfactants (Koller, 2020).

Solvents are under intense scrutiny because of health and safety concerns and recovery economics, especially when thinking of the mass production of PHAs (Kosseva & Rusbandi, 2018). Therefore greener (alcohols, esters, and ketones) and novel (ionic liquids, supercritical processes, pressurized solvents) are being actively investigated and have been reported in the literature (Gahlawat & Kumar Soni, 2019; Rosengart et al., 2015).

Solvent selectivity is also an important factor for choosing a solvent system: it is reported that chloroform has the advantage of reducing the endotoxin level in the extraction (de Donno Novelli et al., 2021), presumably because of comparatively low solubility. Considering that there are dozens of common, well-known industrial solvents (Cheremisinoff, 2003) and that properties can be tailored to an extent by using solvent mixtures, it is conceivable that better solvent solutions can be developed in the future.

Solid-liquid extraction is done by simply contacting the solvents and the dry biomass, using high solvent proportions (40–150 g of the dehydrated biomass for each liter of solvent, in patents WO 2005/052175 A2 and WO 2013/016558 A1). Some agitation can help in the process, but since the solubilization of granules takes place at a microscopic scale, the process is mostly diffusion controlled—it needs time for the polymer to dissolve and diffuse to the bulk of the solution. Mild heating can help. The resulting polymer solution, containing cell debris and suspended NPCM, can then be filtrated (filter aids are not a problem in this step), decanted, or centrifuged if low-density solvents are used. The solution viscosity depends on polymer molecule size and concentration but was observed to be low for dichloromethane solutions (Patent WO 2013/016558 A1). The PHA solution can pass through purification steps such as decolorization or endotoxin adsorption, but careful precipitation can give PHA pure enough to be used as a bioplastic. Hypochlorite coextraction with chloroform was described as an effective route for PHA recovery with 97% purity and 91% yield (Koller, 2020).

Recovery of PHA from the solution could be done by evaporation and spray drying, but lipids and other components co-extracted with the polymer would remain in the final product. Therefore the typical recovery route is precipitation with a miscible nonsolvent such as methanol or ethanol (for chlorinated solvents), eventually with solvent evaporation. Again, the nonsolvent use is intensive: the proportion is approximately 1:1 (volume of nonsolvent:volume of polymer solution). If the solubility of PHA is highly affected by temperature, as with some alcohol or ketones, precipitation can be achieved by cooling the solution. The precipitation rate can be controlled to increase crystallinity, and the process can leave impurities in the solvent phase, selectively precipitating PHA (Patents WO 2005/052175 A2 and WO 2013/016558 A).

Supercritical fluid extraction is another option that can prove helpful in solvent extraction of PHA (Dutt Tripathi et al., 2021). The process is similar to a solid-liquid extraction, except that instead of a liquid, it is used a fluid above its critical point (a supercritical fluid), usually CO_2, at temperatures above 31.1°C and 73 atm ($SCCO_2$). In this condition, CO_2 has a high density and, thus, a high solubilization capacity while keeping the high mobility of a gas. It is common practice to use co-solvents to modulate the $SCCO_2$ properties. After contacting the solvent mixture and the solid matrix, the extract containing PHA proceeds to an expansion chamber where CO_2 vaporizes, giving a solid material or liquid suspension of the polymer in the co-solvent. Despite the relatively high purity obtained, from 85% to 99%, supercritical fluid technology has significant capital and maintenance costs (Kosseva & Rusbandi, 2018; Mannina et al., 2020). The technology must not be confused with cell disruption with $SCCO_2$, which is akin to "pressure bomb" disruption and consists of the incubation of cells under highly pressurized gas, followed by the sudden release of pressure, causing a rapid expansion of the intracellular content and consequent cell bursting, as done by Hejazi et al. (2003).

Non-halogenated and green solvents are being actively researched (Aramvash et al., 2018) and have high recoveries (95%) and purities (98%)—ethylene carbonate seems to be an especially efficient solvent at 100–150°C and, being a high-boiling point solvent, has low emissions and can be recycled many times. Cyclohexanone was found to be efficient at 120°C (the boiling point of the pure solvent is 147°C), reaching a comparable recovery of 95% and even higher purity at 99.5% (Jiang et al., 2018). These are preliminary laboratory tests that must be replicated and scaled up.

10.2.7 Solvent Recovery and Waste Streams

The large volumes of solvents used in PHA extraction must be recovered to reduce the use of makeup solvent, which is costly, and also to avoid solvent emissions in the air and in liquid effluents. Solvent recovery is typically done by distillation, a well-known technology. Recovery adds energy costs to the process (Fernández-Dacosta et al., 2015). However, the technology is well-known, being used in many other industries, and the thermal energy it requires is cheaper than electricity. Even so, reducing solvent volumes and easier recovery can improve the competitiveness of the process. It is conceivable that switchable solvents and surfactants that could exclude the polymer from the solution by refrigeration, for example, could reduce costs (Lawley et al., 2022).

Waste streams from the process must be appropriately disposed of. Even if solvent emissions are limited, the process generates at least another stream of non-polymeric organic matter, usually a liquid phase with a high biological oxygen demand (BOD), requiring proper treatment. The NPCMs generated in cell digestion by chemical routes can be used for methane production by biodigestion or completely hydrolyzed for use in alternative culture media or biofertilizers (Kosseva & Rusbandi, 2018).

10.2.8 Final Drying

The PHAs obtained as wet granules or wet precipitates usually have a satisfactory purity—if not, proper treatment, such as extra washing steps or treatment with oxidizing agents, can be done before solids separation. The wet solids can be quickly dried by spray or fluid-bed drying for solid particles, suspensions, or solutions or by belt, cone, or other high-solids drier for fibrous precipitate masses. Drying is a well-established technology, but it adds costs to the process because of the thermal energy required. Drying from mixtures containing solvents may require negative pressures to reduce emissions, and using inert gases instead of air, in the case of flammable solvents. The vapors must be condensed, maybe with a previous compression step, to recover solvents and reduce emissions.

10.3 FORMULATION AND PROCESSING FOR APPLICATION

Polymer application is a world apart from downstream processing: the ways that polymers can be applied in extrusion coatings, films, injection molding resins, fibers, and additives is staggering. PHAs can have the properties modulated by the fermentation ingredients but also by mixing with compatible polymers such as PLA (polylactic acid), another biodegradable polymer, to form blends with tailored properties (Bugnicourt et al., 2016); however, like most thermoplastics, PHAs are bulk products, sold business to business, typically as granules or masterbatches that can be processed in regular machinery.

The bulk PHA granules produced at the end of water-based processes can, in principle, be directly marketed for dissolution and mixing with other plastics or extruded into pellets. Because of the large surface area, dissolution is facilitated, but bulk density is low. Plastic pellets of 1–5 mm can be used for mold injection or blended with other components, such as plasticizers, to enhance properties and film-forming capacity. The final step of precipitation or crystallization from solutions may lead to microparticles that are larger than the natural PHA granules but still small enough to be used as exfoliating or UV-blocking agents, such as Bio-On Minerv (Koller & Mukherjee, 2022).

PHA also has excellent miscibility with PVC, working as a plasticizer; it can be transformed into aqueous latex without other additives for coating cups and cardboards, and into micro powders with tailored crystallinity for personal care products—if there is no microbial contamination, it will not degrade even in aqueous media, according to information from Metabolix (Metabolix, 2022).

Table 10.2 presents the properties of polypropylene compared to those of PHB, PHV, and three commercial PHA-based products: Metabolix Mirel, a family of PHA-doped PLAs with better performance in extrusion for woven or nonwoven fabrics; ColorFabb PHA-PLA filaments for extrusion 3D printers; and Kaneka's rigid PHBH, a polyhydroxybutyrate-co-hydroxyhexanoate.

Besides tailoring PHAs through fermentation with co-substrates and controlled crystallization and blending PHA with PLA, as already commercially done (Table 10.2), it is possible to blend PHAs with other natural polymers such as starch, cellulose, and lignin, and its derivatives, further modulating the polymer properties and the blend biodegradability (Dutt Tripathi et al., 2021). Also, fibrous biodegradable materials can be added as micro- or nanoparticles, forming biodegradable nanocomposites. Because of the abundance and low price of modified starches and celluloses, blends with these polymers may help reduce the cost of PHA-based packaging materials.

Novel technologies are unraveling formulation and biomaterial possibilities. PHAs can be impregnated with many substances using supercritical CO_2: active pharmaceutical ingredients, functional finishing agents, colorants, and other substances can be diffused into the particle, producing modified polymers, masterbatches, and biocompatible carriers for drug release (Abate et al., 2019; Herek et al., 2006). Another exciting application of the bioproduct is its direct use as amalgamated biomass. Since the polymer constitutes up to 90% of the bacterial biomass, researchers attempted mixing raw biomass with additives (melt mixing at 130–160°C), followed by compression molding at the same temperatures with pressures ranging from 0.2 to 100 atm. Mixtures with solvents can be used to prepare plastic films (Mal et al., 2022).

TABLE 10.2
PHAs and Polypropylene Selected Properties

Polymer Properties	PP*	PHB*	Metabolix Mirel (PHA-Doped PLA)	ColorFabb PLA:PHA	Kaneka Rigid PHBH
Aspect			Off-white pellets	1.75 and 2.85 filaments, many colors	Off -white granules
Odor			Mild		
Crystalline melting point, °C	176	175	100–190	>155	142
Crystallinity, %	70	80			
Molecular weight × 10^5, Da	2	5	>1		
Glass transition temperature, °C	−10	4			2
Density, g/cm³	0.905	1.25	1.4	1.24	1.2
Tensile strength, MPa	38	40		61.5	28
Extension to break, %	400	6			26
Hydrophobicity	Hydrophobic	Hydrophobic	Hydrophobic	Hydrophobic	Hydrophobic

Note: * Data compiled by (Verhoogt et al., 1994).

https://colorfabb.com/media/datasheets/tds/colorfabb/TDS_E_ColorFabb_PLA_PHA.pdf, accessed 10/15/2022.

www.lookpolymers.com/pdf/Kaneka-PHBH-Rigid-Bio-based-Polymer.pdf, accessed 10/15/2022.

10.4 SUMMARY AND CONCLUSIONS

The research and production of PHAs are already a few decades old. The technology has space for improvement, the cost is still high, and the portfolio of products is limited. However, these bioplastics are versatile, and humankind reached a nexus where the need for biodegradable polymers makes the production of PHA commercially viable, even if initially for higher-end products. Market development, that is, creating demand, allied to improved materials, can increase scale and consequent price reduction.

Specifically, downstream processing has many promising routes ahead for more economical, greener, and integrated processes. Promising approaches in this sense are (1) the use of selected or modified organisms, which may be easier to disrupt, or that would be devoid of endotoxins (e.g., using a probiotic bacterium or a eukaryote chassis), thus simplifying the downstream; (2) fermentation strategies for enhancing PHB accumulation while fragilizing the cells, which also facilitates the cell disruption or digestion in the downstream processing; (3) the development of enzymatic cocktails for cell wall disruption, NPCM solubilization, and even in solvent-based processing, the degradation of endotoxins; (4) the use of novel pure or mixed green solvents; (5) the integration of processes toward complete recycle or biodegradation of waste streams; and (6) the development of thermoplastic blends and composites with other natural polymers, producing PHA-based bioplastics at lower costs.

REFERENCES

Abate, M. T., Ferri, A., Guan, J., Chen, G., Nierstrasz, V., Abate, M. T., Ferri, A., Guan, J., Chen, G., & Nierstrasz, V. (2019). Impregnation of materials in supercritical CO_2 to impart various functionalities. *Advanced Supercritical Fluids Technologies*. https://doi.org/10.5772/INTECHOPEN.89223

Afreen, R., Tyagi, S., Singh, G. P., & Singh, M. (2021). Challenges and perspectives of polyhydroxyalkanoate production from microalgae/cyanobacteria and bacteria as microbial factories: An assessment of hybrid biological system. *Frontiers in Bioengineering and Biotechnology*, *9*. https://doi.org/10.3389/FBIOE.2021.624885

Ahn, J., Jho, E. H., & Nam, K. (2015). Effect of C/N ratio on polyhydroxyalkanoates (PHA) accumulation by *Cupriavidus necator* and its implication on the use of rice straw hydrolysates. *Environmental Engineering Research*, *20*(3), 246–253. https://doi.org/10.4491/EER.2015.055

Aramvash, A., Moazzeni, F., Narges, Z., & Banadkuki, G. (2018). Comparison of different solvents for extraction of polyhydroxybutyrate from *Cupriavidus necator*. *Life Sciences*, *18*, 20–28. https://doi.org/10.1002/elsc.201700102

Arikawa, H., Sato, S., Fujiki, T., & Matsumoto, K. (2017). Simple and rapid method for isolation and quantitation of polyhydroxyalkanoate by SDS-sonication treatment. *Journal of Bioscience and Bioengineering*, *124*(2), 250–254. https://doi.org/10.1016/J.JBIOSC.2017.03.003

Berger, E., Ramsay, B. A., Ramsay, J. A., Chavarie, C., & Braunegg, G. (1989). PHB recovery by hypochlorite digestion of non-PHB biomass. *Biotechnology Techniques*, *3*(4), 227–232. https://doi.org/10.1007/BF01876053

Bugnicourt, E., Cinelli, P., Lazzeri, A., & Alvarez, V. (2016). Polyhydroxyalkanoate (PHA): Review of synthesis, characteristics, processing and potential applications in packaging. *Express Polymer Letters*, *8*(11), 791–808. https://doi.org/10.3144/EXPRESSPOLYMLETT.2014.82

Cheremisinoff, N. P. (2003). Industrial solvents handbook, revised and expanded. In *Industrial Solvents Handbook, Revised and Expanded* (2nd ed., 344 pp.). Boca Raton: CRC Press. https://doi.org/10.1201/9780203911334.

Cho, K. S., Ryu, H. W., Lee, E. G., & Chang, Y. K. (2000). Separation of *Alcaligenes eutrophus* cells containing poly(3-hydroxybutyrate) from fermentation broth with pretreatment using Al- and Fe-based coagulants. *Biotechnology Progress*, *16*(2), 238–243. https://doi.org/10.1021/bp990151+

de Acuña, J. M. B., & Poblete-Castro, I. (2022). Rational engineering of natural polyhydroxyalkanoates producing microorganisms for improved synthesis and recovery. *Microbial Biotechnology*, *16*, 262–285. https://doi.org/10.1111/1751-7915.14109

de Carvalho, J. C., Magalhães, A. I., de Melo Pereira, G. V., Medeiros, A. B. P., Sydney, E. B., Rodrigues, C., Aulestia, D. T. M., de Souza Vandenberghe, L. P., Soccol, V. T., & Soccol, C. R. (2020). Microalgal biomass pretreatment for integrated processing into biofuels, food, and feed. *Bioresource Technology*, *300*, 122719. https://doi.org/10.1016/J.BIORTECH.2019.122719

de Carvalho, J. C., Medeiros, A. B. P., Letti, L. A. J., Kirnev, P. C. S., & Soccol, C. R. (2017). Cell disruption and isolation of intracellular products. In *Current Developments in Biotechnology and Bioengineering: Production, Isolation and Purification of Industrial Products* (1st ed., pp. 807–822). Amsterdam: Elsevier. https://doi.org/10.1016/B978-0-444-63662-1.00035-X

de Carvalho, J. C., Medeiros, A. B. P., Rodríguez-Fernández, D. E., Letti, L. A. J., de Souza Vandenberghe, L. P., Woiciechowski, A. L., & Soccol, C. R. (2013). 9 - Downstream operations of fermented products. In *Fermentation Processes Engineering in the Food Industry* (1st ed., pp. 201–235). Boca Raton, FL: CRC Press.

de Carvalho, J. C., Medeiros, A. B. P., Vandenberghe, L. P. S., Magalhães, A. I., & Soccol, C. R. (2017). Approaches for the isolation and purification of fermentation products. In *Current Developments in Biotechnology and Bioengineering: Production, Isolation and Purification of Industrial Products* (1st ed., pp. 783–805). Amsterdam: Elsevier. https://doi.org/10.1016/B978-0-444-63662-1.00034-8

de Donno Novelli, L., Moreno Sayavedra, S., & Rene, E. R. (2021). Polyhydroxyalkanoate (PHA) production via resource recovery from industrial waste streams: A review of techniques and perspectives. *Bioresource Technology*, *331*, 124985. https://doi.org/10.1016/J.BIORTECH.2021.124985

Dutt Tripathi, A., Paul, V., Agarwal, A., Sharma, R., Hashempour-Baltork, F., Rashidi, L., & Khosravi Darani, K. (2021). Production of polyhydroxyalkanoates using dairy processing waste—a review. *Bioresource Technology*, *326*, 124735. https://doi.org/10.1016/J.BIORTECH.2021.124735

Estévez-Alonso, Á., Pei, R., van Loosdrecht, M. C. M., Kleerebezem, R., & Werker, A. (2021). Scaling-up microbial community-based polyhydroxyalkanoate production: status and challenges. *Bioresource Technology*, *327*. https://doi.org/10.1016/J.BIORTECH.2021.124790

Fernández-Dacosta, C., Posada, J. A., Kleerebezem, R., Cuellar, M. C., & Ramirez, A. (2015). Microbial community-based polyhydroxyalkanoates (PHAs) production from wastewater: Techno-economic analysis and ex-ante environmental assessment. *Bioresource Technology*, *185*, 368–377. https://doi.org/10.1016/J.BIORTECH.2015.03.025

Gahlawat, G., & Kumar Soni, S. (2019). Study on sustainable recovery and extraction of polyhydroxyalkanoates (PHAs) produced by *Cupriavidus necator* using waste glycerol for medical applications. *Chemical and Biochemical Engineering Quarterly*, *33*(1), 99–110. https://doi.org/10.15255/CABEQ.2018.1471

Goswami, M., Rekhi, P., Debnath, M., & Ramakrishna, S. (2021). Microbial polyhydroxyalkanoates granules: An approach targeting biopolymer for medical applications and developing bone scaffolds. *Molecules*, *26*(4), 860. https://doi.org/10.3390/MOLECULES26040860

Grousseau, E., Blanchet, E., Déléris, S., Albuquerque, M. G. E., Paul, E., & Uribelarrea, J. L. (2014). Phosphorus limitation strategy to increase propionic acid flux towards 3-hydroxyvaleric acid monomers in *Cupriavidus necator*. *Bioresource Technology*, *153*, 206–215. https://doi.org/10.1016/J.BIORTECH.2013.11.072

Hejazi, P., Vasheghani-Farahani, E., & Yamini, Y. (2003). Supercritical fluid disruption of *Ralstonia eutropha* for poly(β-hydroxybutyrate) recovery. *Biotechnology Progress*, *19*(5), 1519–1523. https://doi.org/10.1021/BP034010Q

Herek, L. C. S., Oliveira, R. C., Rubira, A. F., & Pinheiro, N. (2006). Impregnation of PET films and PHB granules with curcumin in supercritical CO2. *Brazilian Journal of Chemical Engineering*, *23*(2), 227–234. https://doi.org/10.1590/S0104-66322006000200010

Jacquel, N., Lo, C. W., Wei, Y. H., Wu, H. S., & Wang, S. S. (2008). Isolation and purification of bacterial poly(3-hydroxyalkanoates). *Biochemical Engineering Journal*, *39*(1), 15–27. https://doi.org/10.1016/J.BEJ.2007.11.029

Jiang, G., Johnston, B., Townrow, D. E., Radecka, I., Koller, M., Chaber, P., Adamus, G., & Kowalczuk, M. (2018). Biomass extraction using non-chlorinated solvents for biocompatibility improvement of polyhydroxyalkanoates. *Polymers*, *10*(7), 731. https://doi.org/10.3390/POLYM10070731

Khatami, K., Perez-Zabaleta, M., Owusu-Agyeman, I., & Cetecioglu, Z. (2021). Waste to bioplastics: How close are we to sustainable polyhydroxyalkanoates production? *Waste Management*, *119*, 374–388. https://doi.org/10.1016/J.WASMAN.2020.10.008

Koller, M. (2020). Established and advanced approaches for recovery of microbial polyhydroxyalkanoate (PHA) biopolyesters from surrounding microbial biomass. *The EuroBiotech Journal*, *4*(3), 113–126. https://doi.org/10.2478/EBTJ-2020-0013

Koller, M., Maršálek, L., de Sousa Dias, M. M., & Braunegg, G. (2017). Producing microbial polyhydroxyalkanoate (PHA) biopolyesters in a sustainable manner. *New Biotechnology*, *37*, 24–38. https://doi.org/10.1016/J.NBT.2016.05.001

Koller, M., & Mukherjee, A. (2022). A new wave of industrialization of PHA biopolyesters. *Bioengineering*, *9*(2), 74. https://doi.org/10.3390/BIOENGINEERING9020074

Kosseva, M. R., & Rusbandi, E. (2018). Trends in the biomanufacture of polyhydroxyalkanoates with focus on downstream processing. *International Journal of Biological Macromolecules*, *107*(Part A), 762–778. https://doi.org/10.1016/J.IJBIOMAC.2017.09.054

Lawley, M. D., Boon, D., Stein, L. Y., & Sauvageau, D. (2022). Switchable solvents for the reversible dissolution of poly(3-hydroxybutyrate). *ACS Sustainable Chemistry and Engineering*, *10*(8), 2602–2608. https://doi.org/10.1021/ACSSUSCHEMENG.1C06377/ASSET/IMAGES/LARGE/SC1C06377_0003.JPEG

Liu, R., Wang, L., Yang, L., Liu, Q., Gao, Y., Ye, J., Xiao, J., Hu, Q., & Zhang, X. (2020). Ultrafiltration and microfiltration membrane performance, cleaning, and flux recovery for microalgal harvesting. *Journal of Applied Phycology*, *32*(5), 3101–3112. https://doi.org/10.1007/S10811-020-02204-2/FIGURES/8

López-Abelairas, M., García-Torreiro, M., Lú-Chau, T., Lema, J. M., & Steinbüchel, A. (2015). Comparison of several methods for the separation of poly(3-hydroxybutyrate) from *Cupriavidus necator* H16 cultures. *Biochemical Engineering Journal*, *93*, 250–259. https://doi.org/10.1016/J.BEJ.2014.10.018

Mal, N., Satpati, G. G., Raghunathan, S., & Davoodbasha, M. A. (2022). Current strategies on algae-based biopolymer production and scale-up. *Chemosphere*, *289*, 133178. https://doi.org/10.1016/J.CHEMOSPHERE.2021.133178

Mannina, G., Presti, D., Montiel-Jarillo, G., Carrera, J., & Suárez-Ojeda, M. E. (2020). Recovery of polyhydroxyalkanoates (PHAs) from wastewater: A review. *Bioresource Technology*, *297*, 122478. https://doi.org/10.1016/J.BIORTECH.2019.122478

Mas, J., Pedrós-Alió, C., & Guerrero, R. (1985). Mathematical model for determining the effects of intracytoplasmic inclusions on volume and density of microorganisms. *Journal of Bacteriology*, *164*(2), 749. https://doi.org/10.1128/JB.164.2.749-756.1985

Metabolix—Mirel® PHA Polymeric Modifiers and Additives. (n.d.). Retrieved October 7, 2022, from www.slideshare.net/MetabolixInc/metabolix-mirel-pha-polymeric-modifiers-and-additives

Mohammadi, M., Hassan, M. A., Phang, L. Y., Shirai, Y., Che Man, H., Ariffin, H., Amirul, A. A., & Syairah, S. N. (2012). Efficient polyhydroxyalkanoate recovery from recombinant *Cupriavidus necator* by using low concentration of NaOH. *Environmental Engineering Science*, *29*(8), 783–789. https://doi.org/10.1089/EES.2011.0255

Mudenur, C., Mondal, K., Singh, U., & Katiyar, V. (2019). Production of polyhydroxyalkanoates and its potential applications. In *Advances in Sustainable Polymers* (1st ed., 131–164). Amsterdam: Elsevier https://doi.org/10.1007/978-981-32-9804-0_7

Murugan, P., Han, L., Gan, C. Y., Maurer, F. H. J., & Sudesh, K. (2016). A new biological recovery approach for PHA using mealworm, *Tenebrio molitor*. *Journal of Biotechnology*, *239*, 98–105. https://doi.org/10.1016/J.JBIOTEC.2016.10.012

Oshiki, M., Onuki, M., Satoh, H., & Mino, T. (2010). Separation of PHA-accumulating cells in activated sludge based on differences in buoyant density. *The Journal of General and Applied Microbiology*, *56*(2), 163–167. https://doi.org/10.2323/JGAM.56.163

Pala-Ozkok, I., Zengin, G. E., Taş, D. O., Yağcı, N., Güven, D., Insel, H. G., & Çokgör, E. (2022). Polyhydroxyalkanoate production from food industry residual streams using mixed microbial cultures. In *Clean Energy and Resource Recovery: Wastewater Treatment Plants as Biorefineries* (Volume 2, pp. 265–284). https://doi.org/10.1016/B978-0-323-90178-9.00010-X

Pedrós-Alió, C., Mas, J., & Guerrero, R. (1985). The influence of poly-β-hydroxybutyrate accumulation on cell volume and buoyant density in *Alcaligenes eutrophus*. *Archives of Microbiology*, *143*(2), 178–184. https://doi.org/10.1007/BF00411044

Pérez-Rivero, C., López-Gómez, J. P., & Roy, I. (2019). A sustainable approach for the downstream processing of bacterial polyhydroxyalkanoates: State-of-the-art and latest developments. *Biochemical Engineering Journal*, *150*, 107283. https://doi.org/10.1016/J.BEJ.2019.107283

Plastics Europe. (2020). *Plastics – the Facts 2020*. Brussels: Plastics Europe: Association of Plastics Manufacturers. https://plasticseurope.org/knowledge-hub/plastics-the-facts-2020/

Research Nester. (2022). *Bioplastics market insights, size, trends & growth forecast to 2031. Market Report Preview*. www.researchnester.com/reports/bioplastics-market/3923

Rond'ošová, S., Legerská, B., Chmelová, D., Ondrejovič, M., & Miertuš, S. (2022). Optimization of growth conditions to enhance PHA production by *Cupriavidus necator*. *Fermentation*, *8* https://doi.org/10.3390/fermentation8090451

Rosengart, A., Cesário, M. T., de Almeida, M. C. M. D., Raposo, R. S., Espert, A., de Apodaca, E. D., & da Fonseca, M. M. R. (2015). Efficient P(3HB) extraction from *Burkholderia sacchari* cells using non-chlorinated solvents. *Biochemical Engineering Journal*, *103*, 39–46. https://doi.org/10.1016/J.BEJ.2015.06.013

Vandenberghe, L. P., de Oliveira, P. Z., Bittencourt, G. A., de Mello, A. F. M., Vásquez, Z. S., Karp, S. G., & Soccol, C. R. (2021). The 2G and 3G bioplastics: An overview. *Biotechnology Research and Innovation*, 5(1). https://doi.org/10.4322/BIORI.202104

Venkateswar Reddy, M., Yajima, Y., Mawatari, Y., Hoshino, T., & Chang, Y. C. (2015). Degradation and conversion of toxic compounds into useful bioplastics by *Cupriavidus* sp. CY-1: Relative expression of the PhaC gene under phenol and nitrogen stress. *Green Chemistry*, *17*(9), 4560–4569. https://doi.org/10.1039/C5GC01156F

Verhoogt, H., Ramsay, B. A., & Favis, B. D. (1994). Polymer blends containing poly(3-hydroxyalkanoate)s. *Polymer*, *35*(24), 5155–5169. https://doi.org/10.1016/0032-3861(94)90465-0

Vigneswari, S., Noor, M. S. M., Amelia, T. S. M., Balakrishnan, K., Adnan, A., Bhubalan, K., Amirul, A. A. A., & Ramakrishna, S. (2021). Recent advances in the biosynthesis of polyhydroxyalkanoates from lignocellulosic feedstocks. *Life*, *11*(8). https://doi.org/10.3390/LIFE11080807

11 Biorefineries and Circular Economy in the Production of Second- and Third-Generation Bioplastics

Yoong Kit Leong and Pau Loke Show

CONTENTS

11.1 INTRODUCTION

European Bioplastics defined that "a plastic materials is defined as a bioplastic if it is either bio-based, biodegradable, or feature both properties" (Bioplastics, 2018). Bio-based plastics are plastics made from a wide range of renewable bio-based feedstocks, though not necessarily biodegradable, where the major non-biodegradable bioplastics are bio-polyamide (bio-PA), bio-polyethylene-terephthalate (bio-PET), bio-polytrimethylene-terephthalate (bio-PTT), bio-polyethylene (bio-PE) and bio-polypropylene (bio-PP) (Rahman & Bhoi, 2021). Biodegradable plastics are those that can be biodegraded, though not necessarily derived from bio-based feedstocks, such as polybutylene adipate terephthalate (PBAT) and polycaprolactone (PCL). Polyhydroxyalkanoates (PHAs) and polylactic acid (PLA) are the two representative bioplastics that are both bio-based and biodegradable. Bioplastics considered as part of future circular economies to assist in accomplishing some of the UN Sustainable Development Goals, including introduction of alternative recycling or degradation pathways, diversion from fossil resources and others. Bioplastics offer improved circularity by having a lower carbon footprint, utilizing renewable and sustainable resources, and biodegradation being the alternative end-of-life (EOL) option.

Driven and inspired by the development of biofuel sector, the bioplastics are also categorized into first, second and third generation (Barrett, 2018). Currently, bioplastics are mainly derived from starch-rich, sugar-rich or oil-rich food crops, including sugarcane, wheat, rice, sugar beet, cassava, corn, vegetable oils and others, where these bioplastics are termed "first-generation bioplastics." First-generation feedstocks are currently the most efficient for bioplastics production as they demand less cultivation area and possess high yield. However, first-generation feedstocks remain controversial as they facing the same "food vs. product" debate as the biofuel sector, which

DOI: 10.1201/9781003344018-11

directly competes with the production of food supply. Moreover, the substrate cost contributes to approximately 50% of the overall production cost (USD 4000–15,000/Mt compared with USD 1000–1500/Mt of fossil fuel–derived polymers), thus cheaper substrate is in demand to lower the production cost (Ganesh Saratale et al., 2021).

The feedstock of second-generation bioplastics comes from non-food crops that are not suitable for animal consumption (feed) or human consumption (food), for example waste materials from first-generation feedstocks or lignocellulosic feedstocks such as switchgrass, wood, bagasse, corncobs, wheat straw, short-rotation crops and others. Nevertheless, the use of certain second-generation feedstock still raises the similar concern of "food vs. product" where non-food crops are grown on arable lands that can be used to grow food crops.

On the other hand, third-generation bioplastics is derived from biomass of micro- and macroalgae. The main advantages of algae as feedstocks are they have much higher growth rate and photosynthetic efficiency than terrestrial plants, good adaptability in extreme environments and excellent carbon-fixing ability. However, third-generation bioplastics is currently still in its infant stage as more studies and research is needed to realize commercial-scale applications.

The concept of "circular economy" is currently widely embraced by researchers and industries for the production of wide range of products such as bioplastics, biofuels and other commercially important products from natural resources in a green and sustainable way. In an ideal circular economy, bioplastic would be derived from renewable or recycled resources. Producing bioplastic from valorization of waste streams, such as food waste, agricultural waste, various types of wastewaters, flue gas and others promotes the circular economy of waste recycling, tackling waste management issues as well as mitigating environmental pollutions.

Nevertheless, for effective and sustainable commercial-scale production of bioplastic, several factors needed to be taken into consideration, which are cost of substrate, uniform feedstock composition and characteristic, availability, ease of collection, transportation and storage (Ganesh Saratale et al., 2021). Furthermore, the main challenges in utilizing waste streams as feedstock for microbial bioplastic production are their unsatisfactory carbon content, more time required to adapt to complex composition of waste streams and presence of fermentation inhibitors.

11.2 LOW-COST BY-PRODUCTS AND WASTES AS SUSTAINABLE FEEDSTOCK FOR BIOPLASTIC PRODUCTION

The employment of low-cost and non-food substrates for microbial fermentation of bioplastics has been strongly encouraged. Researchers have put in serious effort to valorize diverse types of wastes such as food waste, agro-industrial waste, wastewater, lignocellulosic biomass and others as substrate for microbial synthesis of PHAs and PLA, which are more environmentally friendly and cost-effective. Usually, these waste streams have high content of chemical oxygen demand (COD), thus resource recovery is encouraged to promote the overall circular economy of bioplastics. Researchers have carried out extensive studies on the valorization of waste stream via biorefinery for production of bioplastics and other co-products. Wild microbial strains have been isolated, screened and employed for effective utilization of waste streams and subsequent production of bioplastics. To further enhance the bioplastic productivity, biotechnological tools such as metabolic and genetic engineering, forced evolution and adaptive mutation can be applied to create a more superior bioplastic-producing strain.

The biotechnological route for bioplastic production from waste usually consisted of two steps where the first step is pretreatment of waste feedstocks to convert the feedstock with complex composition into simple and fermentable carbon source, while the second step is the microbial fermentation for bioplastic production. The choice of pretreatment strategy depends on characteristic and compositions of the waste streams. Nevertheless, Bomrungnok and colleagues have isolated *Bacillus aryabhattai* T34-N4 for direct production of polyhydroxybutyrate (PHB) from waste starch such as oil palm trunk and cassava pulp starch (Bomrungnok et al., 2020). On the other hand, several amylolytic lactic acid bacteria (LAB) strains have been isolated by researchers that are capable for direct conversion of waste into lactic acid without an additional hydrolysis step. On the other hand, fungi, particularly *Rhizopus* sp., have garnered wide interest in production of lactic acid owing to better growth under nitrogen-limited environments, production of L-lactic acid as the sole isomer, and simpler downstream processing (pellet or filamentous forms; Panesar & Kaur, 2015).

11.2.1 Lignocellulosic Biomass

With an estimated annual production of approximately 200 billion tonnes, lignocellulosic biomass, such as woody biomass, forest residues, and agricultural wastes, is a low-cost, abundant and renewable resource that does not compete with food supply (Raj et al., 2022). Lignocellulosic biomass generally consists of lignin (10%–40%), hemicellulose (11%–50%), cellulose (25%–55%), phenolic components and extractives. Nevertheless, the proportions of each component and structure vary depending on the type of biomass, thus, selection of suitable sources is very crucial.

Physical, chemical, physicochemical and biological pretreatment are usually needed to enhance the digestibility of cellulose and obtain high yield of reducing sugars by reducing particle size, disrupting the rigid carbohydrate-lignin matrix, removing the lignin and hemicellulose barrier and enhancing the porosity and specific surface area. The lignocellulosic biomass hydrolysate then can be utilized as feedstock for biopolymer production. The main limitations in utilizing lignocellulosic biomass as feedstock for bioplastic production are the high cost of saccharification, the production of fermentation inhibitors produced during pretreatment or hydrolysis and the relatively low content of sugars (Andler et al., 2021).

11.2.1.1 Lactic Acid Production from Lignocellulosic Biomass

Lactic acid (LA), a monomer of PLAs, has been produced from a diverse range of lignocellulosic feedstock. There are several strategies for biotransformation of lignocellulosic biomass to LA, including separate hydrolysis and fermentation (SHF), simultaneous saccharification and fermentation (SSF) and consolidated bioprocessing (CBP). SHF enable the hydrolysis and fermentation process to be carried out independently, thus each process can be performed at the optimal conditions, giving high efficiency. However, SHF suffers from product inhibition by reducing sugars formed, causing the need for higher input of hydrolytic enzymes or lengthening of total processing time. Furthermore, there might be loss in reducing sugars during the transferring process between equipment. The limitations faced by SHF can be overcome by SSF, where hydrolysis and fermentation are carried out simultaneously, which the reducing sugars formed can be immediately utilized by LA-producing bacteria. This helped in minimizing product inhibition and shortening the total processing time. However, there is a trade-off between optimal operating conditions as the hydrolytic enzymes and fermenting microorganisms usually have different optimal ranges. Saito and colleagues showed that production of lactic acid by *Rhizopus oryzae* NBRC 5378 utilizing wheat straw as substrate achieved yield of 2 g/L and 6 g/L employing SHF and SSF strategy, respectively (Saito et al., 2012).

Compared between SHF and SSF in LA production by *Bacillus coagulans* CC17 from bagasse sulfite pulp, Zhou and coworkers found that SHF demanded for approximately one-third more fungal cellulase than SSF (Zhou et al., 2016). In addition, *B. coagulans* CC17 was able to co-ferment xylose and cellobiose without the need of exogenous β-glucosidase in SSF, though addition of xylanase enhanced the LA concentration. Fed-batch SSF achieved LA production up to 110 g/L with yield of 0.72 g/g cellulose, indicating that SSF has significant advantage over SHF. Similarly, Berlowska and colleagues have also compared SHF and SSF for LA production by lactic acid bacteria (LAB) using sugar beet pulp as feedstock (Berlowska et al., 2018). In SHF process, LAB can only partially utilize the soluble sugar released from the sugar beet pulp. The results showed that SSF achieved 80%–90% higher yield compared with SHF, thus giving higher productivity and lower processing costs.

In CBP, the various stages of enzyme production, saccharification and fermentation are integrated into a single step and carried out by a single microorganism or consortium in a single reactor. Therefore, CBP requires a single microorganism or microbial consortium that can secrete enzyme to degrade lignocellulosic biomass while simultaneously producing desired products. Shahab and colleagues have employed a synthetic bacterial-fungal consortium which consists of facultative anaerobic lactic acid bacteria and aerobic fungus *Trichoderma reesei* for CBP of lactic acid (Shahab et al., 2018). The CBP process achieved LA production yields of 34.7 g/L and 19.8 g/L from microcrystalline cellulose and non-detoxified steam-pretreated beech wood, respectively. The research team mentioned challenges such as formation of acetic acid, production of fermentation inhibitors and carbon catabolite repression. Nevertheless, the study on lactic acid production via CBP of lignocellulosic feedstocks is still limited and more research should be performed.

11.2.1.2 Polyhydroxyalkanoates from Lignocellulosic Biomass

For PHA production of lignocellulosic feedstock, dilute sulfuric acid treatment is the most common method employed. Bertrand and coworkers were the first to utilize hydrolysate from lignocellulosic biomass, poplar wood, as a carbon source for *Pseudomonas pseudoflava* for PHA production (Bertrand et al., 1990). Among ten PHB accumulating strains selected, *Bacillus* spp. emerged as the best PHB producer and able to accumulate PHB up to 25%, 37.4%, 51.6%, and 55.6% utilizing low-cost food crop residues, including pretreated banana peel, teff straw, corncob and sugarcane bagasse as substrate (Getachew & Woldesenbet, 2016).

Kourilova and research team have compared the PHA production performance of mesophilic *Burkholderia sacchari*, halophilic *Halomonas halophila* and thermophilic *Schlegelella thermodepolymerans* utilizing model media mimicking hydrolysates of lignocellulosic biomass, such as softwood, hardwood, rice straw, sugarcane bagasse, wheat bran and wheat straw (Kourilova et al., 2021). *B. sacchari* was able to utilize all examined model media, demonstrating robustness against fermentation inhibitors and giving comparable PHA yields. Both extremophiles were more sensitive to fermentation inhibitors compared with *B. sacchari*. *H. halophila* preferred model hydrolysates rich in hexoses, while *S. thermodepolymerans* gave higher PHA yield on pentoses-rich media.

De Souza and colleagues first employed biological pretreatment by white rot fungi (*Pleurotus ostreatus* MTCC 142) on various lignocellulosic waste, including corn husk, newspaper, rice straw, sugarcane bagasse and wheat bran and subsequently utilized them as feedstock for PHA production by *Bacillus megaterium* Ti3 in a nitrogen-deficient medium (de Souza et al., 2020). Corn husk gave the highest sugar yield (6 g/L) and *B. megaterium* Ti3 accumulated PHB up to 57.8 wt/wt% utilizing pretreated corn husk hydrolysate, which was comparable to xylose-based medium and significantly higher than glucose-based medium (45.7 wt/wt%).

Due to the perishable nature of fruits and their inedible portions, a large quantity of fruit waste (including bagasse, core, empty fruit bunch, peel, pulp, pomace, rind, seed, stem and others) is produced during fruit production and consumption, and some has high fermentable sugar content. In comparison of five media derived from fruit peel residues of watermelon, papaya, orange, melon and banana, *Klebsiella pneumoniae* accumulated the highest PHB content of 8.4×10^{-10} g/CFU in watermelon peel medium (Valdez-Calderón et al., 2020). Matos and colleagues have carried out pilot-scale PHA production by mixed-microbial cultures (MMC) utilizing fruit waste as feedstock in a three-stage pilot plant (100 L upflow anaerobic sludge blanket reactor as acidogenic reactor + 100 L aerobic sequential for enrichment of MCC with PHA-producing culture + 60 L aerobic accumulation reactor for PHA production; Matos et al., 2021). In the first stage, pH and organic loading rate were employed as tuning parameters to obtain butyrate-rich fermentate with a high yield of 0.74 g COD/g COD. The combined process achieved high overall process yield and productivity of 0.45 g COD/g COD and 8.1 g PHA/L/d, the highest value reported for MMC employing a real feedstock at pilot scale.

In order to achieve high cell density cultivation as well as enhanced yield and productivity, researchers have carried out many studies on fed-batch fermentation of *Pseudomonas resinovorans, Burkholderia sacchari, Burkholderia cepacia* and *Bacillus megaterium* for PHA production utilizing lignocellulosic biomass hydrolysate. Lopez and research team have employed *Burkholderia* sp. F24, which has high tolerance to fermentation inhibitors for PHA production utilizing undetoxified sugarcane bagasse hydrolysate as carbon source (Lopes et al., 2014). The high cell density cultivation achieved high PHB content of 49% and productivity of 0.28 g/L/h, compared with 48% and 0.1 g/L/h of batch fermentation.

Lignocellulosic biomass has been confirmed to be a promising feedstock for bioplastic production, though large-scale production still faces challenges. Some major challenges for utilization of lignocellulosic biomass as feedstock for bioplastic production are their low mechanical properties, high hydrophilicity and structural complexity (Raj et al., 2022). Not only the compositions of different types of lignocellulosic biomass are different, but the quality of the same lignocellulosic wastes may also vary greatly between different batches due to influence from environmental factors.

Biotechnology tools such as genetic engineering, metabolic engineering and others can be employed to create strains which have higher sugar consumption efficiency, the ability to utilize both pentoses and hexoses, higher bioplastics yield and productivity, low nutrient requirements, tolerance to fermentation inhibitors, the ability to utilize a diverse range of substrates and other qualities. The concept of a biorefinery-based circular economy should be employed for the valorization of lignocellulosic biomass to various products. For example, the pentoses from hydrolysate can be used for bioplastic production by pentose-utilizing microbes while hexoses, on the other hand, can be utilized for bioethanol production by yeast.

11.2.2 Whey

As the primary by-product of the dairy industry, whey is produced by removal and precipitation of milk casein during the cheese-making process. With 9–10 L of whey generated per 1 kg of cheese produced, the annual production of whey reaches approximately 120 million tons (Amaro et al., 2019). Only approximately 50% of the generated cheese whey is being further valorized as feedstock for production of animal feeds, food supplements and ricotta cheese. Mainly consisting of lactose (39–60 g/L), proteins (27–60 g/L), fats (0.99–10.6 g/L) and mineral salts (5–8 g/L), the relatively high organic content (50–100 g COD/L) of whey makes its disposal issue an environmental concern. Therefore researchers have suggested to use whey as substrate of microbial fermentation to produce value-upgraded bioproducts. The utilization of whey as fermentation substrate (carbon or nitrogen source) for bioplastic production has the benefit of not requiring energy-intensive and cost-incurring pretreatment via enzymatic or acid hydrolysis.

11.2.2.1 Lactic Acid Production from Whey

Different lactobacilli cultures (e.g., *L. acidophilus, L. casei, L. delbrueckii* subsp. *bulgaricus, L. helveticus*) have been employed in the production of lactic acid from whey. Kolev et al. screened 24 strains of lactic acid bacteria to determine their β-galactosidase activity in acid whey (Kolev et al., 2022). Although lactobacilli generally have lower growth in acid whey, 18 strains demonstrated higher β-galactosidase activity compared with control medium. Among all, *L. helveticus* strain OSU-PECh-4A demonstrated almost fivefold higher activity than average. Panesar et al. agreed that *L. helveticus* is preferred, as it gives almost twofold higher LA yield (Panesar et al., 2007).

Lopez-Gomez and research team have performed batch and continuous lactic acid fermentations by *Lactobacillus coryniformis* subsp. *torquens*, utilizing acid whey as substrate (López-Gómez et al., 2019). Batch fermentation gave LA yield and productivity of 0.47 g/g and 1.2 g/L/h, while continuous fermentation provided an average LA yield and productivity of 0.92 g/g and 9.2 g/L/h. Utilizing cheese whey as low-cost substrate, Luongo and colleagues have examined the mixed-culture fermentation for lactic acid production at various hydraulic retention time (HRT) under semi-continuous mode and uncontrolled pH conditions (Luongo et al., 2019). Maximum lactic acid yield and concentration of 0.37 g LA/g COD and 20.1 g/L was achieved at HRT of 2 days.

Choi and colleagues have performed continuous mixed-culture lactic acid fermentation utilizing whey as feedstock (Choi et al., 2016). Compared with constantly maintained pH at 5.5, having a low pH start-up at pH 3 and maintaining it for the beginning stage gave a positive impact and produced higher LA yield. *Bacillus coagulans* was found to be the dominating LA producer. The cell recycling strategy can be adopted in continuous fermentation to retain cell and enhance the concentration of biomass in the reactor, allowing reduced retention rate and enhanced lactic acid productivity as well as reducing substrate concentration of effluent (Panesar et al., 2007).

The productivity of lactic acid in whey fermentation can be enhanced by employing a membrane recycling bioreactor. Ramchandran and coworkers employed a novel compact submerged hollow-fiber membrane fermenter to enhance the cell yield and lactic acid production performance of *Lactococcus lactis* ssp. *cremoris* using whey-based medium as substrate (Ramchandran et al., 2012). The membrane allowed for simultaneous spent medium removal and fresh medium replacement. This helped in promoting the cell growth and lactic acid production to more than double, giving lactic acid productivity of 9.7 g/L/h.

11.2.2.2 Polyhydroxyalkanoates Production from Whey

The search for microbes which can effectively utilize whey as substrate and produces PHAs prove challenging as conventional PHA-accumulating microbes (such as *C. necator*) cannot metabolize whey directly (Oliveira et al., 2018). Andrade reported that lactose in cheese whey suppress the glucose consumption by *Burkholderia sacchari* LFM 101, causing lower biomass growth when cheese whey was employed as co-substrate (de Andrade et al., 2019). Therefore researchers have employed microbes with adequate β-galactosidase activity in whey-based PHA fermentation.

Bosco and colleagues were the first to report the PHA production from the *Leuconostoc* genus, where pure culture and consortium of *Leuconostoc mesenteroides* was able to utilize cheese whey for biosynthesis of PHA (Bosco et al., 2021). Bustamante and coworkers have screened PHA-producing microbes which can utilize whey and demonstrated that *Caulobacter segnis* DSM 29236 as a novel PHB producer from whey with concentration of 9.3 g/L achieved in fed-batch cultivation (Bustamante et al., 2019). Stimulated by nitrogen deficiency, halophilic *Paracoccus homiensis* was able to utilize cheese whey mother liquor and produce P(3HB-co-3HV) with 3HV monomer up to 60.6% (Możejko-Ciesielska et al., 2022). Employing whey as feedstock, *Thermus thermophilus* HB8 produced novel PHA heteropolymer with unique composition, which consists of 3-hydroxyvalerate of 38 mol%, 3-hydroxyundecanoate of 35.4 mol%, 3-hydroxynanoate of 16.6 mol% and 3-hydroxyheptanoate of 9.9 mol% (Pantazaki et al., 2009).

The inability of PHA biosynthesis from whey lactose by high PHA-producing microbes can be overcome by two strategies, which are (1) hydrolyzing whey lactose into galactose and glucose enzymatically or chemically and (2) converting whey into organic acids (e.g., carboxylic acids, volatile fatty acids [VFAs]) via anaerobic fermentation prior to PHA production. Subsequently, high PHA-accumulating microbes can utilize the hydrolysate or VFAs for PHA production. Utilizing cheese whey as feedstock, batch cultivation of *Haloferax mediterranei* in bioreactor of 2 L working volume achieved biomass and P(3HB-co-3HV) concentration of 7.54 g/L and 4 g/L (Pais et al., 2016). However, it should be taken note that this could increase the overall PHA production cost due to extra cost incurred unless the high PHA yield is sufficient to compensate it.

Colombo and colleagues showed that the composition of fermented cheese whey (FCW) affect the type of PHAs produced by mixed microbial culture, where PHB was produced from FCW consisting of acetic, butyric and lactic acid, while PHBV was biosynthesized from FCW containing acetic, butyric, lactic, propionic and valeric acid (Colombo et al., 2016). Chang and coworkers have performed two-stage PHA production from cheese whey of low lactose content, where *Acetobacter pasteurianus* C1 was employed to convert cheese whey to acetic acid in the first stage, while *Bacillus* sp. CYR-1 subsequently utilize the acetic acid to produce PHA (Chang et al., 2021). Pretreatment of protein removal further promoted the PHA production by 3.3-fold up to 411 mg/L. Similarly, Domingos and research team have converted cheese whey into carboxylic acids (containing hexanoic acid, octanoic acid and others) via anaerobic fermentation and performed a dialysis process involving seven sequential batches to obtain a concentrated carboxylic acid stream (Domingos et al., 2018). The concentrated stream was then employed as substrate in fed-batch fermentation of *C. necator* DSMZ 545 for PHAs production with yield of approximately 0.6 g PHA/g VFA.

Researchers have co-produced PHAs and biohydrogen via a two- or three-stage process utilizing whey as feedstock. Asunis and colleagues have adopted a three-step bioprocess of dark fermentation, PHA-storing microorganisms selection and PHA accumulation for combined production of PHA and bioH_2 from sheep cheese whey (Asunis et al., 2022). Employing deproteinized whey as substrate, the daily production volume of bioH_2 reached up to 5.07 L/d following the increase of organic loading rate (OLR), while high a PHAs conversion yield of 0.74 mg COD_{PHA}/mg COD was achieved with maximum polymer content of 62% (Colombo et al., 2019).

Environmental life cycle assessment (LCA) has been performed for production of PHAs from cheese whey (CW) employing conventional anaerobic digestion (AD) as a benchmark (Asunis et al., 2021). The results indicated that baseline PHA production scenario has poorer environmental performance compared with AD (climate change indicator of –35.7 and 44.8 kg CO_2 eq./t CW for AD and PHA). By pinpointing the hotspot of the process by LCA and improvise the process, the improved PHA production process demonstrated comparable environmental performance of –50.3 kg CO_2 eq./t CW.

11.2.3 Waste Animal Fats and Cooking Oils

Huge quantities of lipid-rich wastes, such as waste cooking oil (WCO) and waste animal fats, are generated from the edible oil processing industry, food processing industries, rendering industries, oil mills and slaughterhouses (Ganesh Saratale et al., 2021). Waste management of such lipid-rich waste streams has been a great concern as the accidental or intentional discharge of these waste streams into water bodies can cause severe pollution to the environment, threatening the balance of ecosystem and human health. Therefore, these abundant and inexpensive WCO and animal fats residues can be recycled and employed as potential substrate for bioplastic-producing microorganisms to minimize the production cost of bioplastics and promote the sustainability and carbon footprint of the process. With high level of free fatty acids, WCO usually gave significantly higher yields of bioplastics, which give it an edge to be utilized as carbon source for bioplastic production via fermentation. Chanprateep estimated that the theoretical PHA yield from fatty acids is 0.65 g/g, which is more than 0.3–0.4 g/g from glucose (Chanprateep, 2010). *Bacillus, Cupriavidus, Halomonas, Pseudomonas, Ralstonia* sp. and recombinant *Escherichia coli* have been widely studied for the ability to assimilate waste oil as carbon source for PHA production efficiently.

For waste plant oil, *Cupriavidus necator* has been demonstrated to accumulate PHB up to 1.2 g/L from waste rapeseed oil (Verlinden et al., 2011) and 4.6 g/L from waste sesame oil (Taniguchi et al., 2003) as well as poly(3-hydroxybutyrate-co-4-hydroxybutyrate) [P(3HB-co-4HB)] from spent palm oil (Rao et al., 2010). Employing WCO as substrate, Kamilah and coworkers have obtained P(3HB) of 18.03 g/L from wild type *C. necator* H16 as well as poly(3-hydroxybutyrate-co-3-hydroxyhexanoate) [P(3HB-co-3HHx)] of 18.96 g/L from recombinant *C. necator* PHB 4 (Kamilah et al., 2013). Zainab-L and colleagues have explored the potential of underutilized plant oils, including African elemi oil, *Amygdalus pedunculata* oil, bitter apple oil and desert date oil as feedstock for PHAs biosynthesis from *C. necator* (Zainab-L et al., 2018). Both *C. necator* H16 and Re2058/pCB113 can effectively utilize these plant oils to produce P(3HB) up to 6 g/L and P(3HB-co-3HHx) up to 5 g/L with 31 mol% 3HHx. Sludge palm oil, solid by-products from palm oil industry have been used for the biosynthesis of P(3HB-co-3HHx) with productivity up to 1.1 g/L.h from fed-batch cultivation of recombinant *C. necator* Re2058/pCB113 under the aid of surfactant (Thinagaran & Sudesh, 2019).

Waste vegetable oil has been employed as substrate for cultivation of *Pseudomonas* sp. to biosynthesis various type of medium-chain-length (mcl) PHAs (C6–16) (Song et al., 2008). Using waste rapeseed oil as feedstock, Możejko and colleagues have produced mcl-PHAs up to 19.3% and 21.0% dry cell weight (DCW) for *Pseudomonas* sp. GI01 and GI06 (Możejko et al., 2011). It was found that the monomeric composition was strain-dependent, where monomers of 3-hydroxydecanoic acid (3HD) and 3-hydroxyoctanoic acid (3HO) were present in significant amounts in strain GI01, while 3-hydroxyhexanoate was dominant in strain GI06. In the next research, fed-batch strategy was adopted to produce mcl-PHAs (C6–12) up to 44 wt/wt% content from *Pseudomonas* sp. GI01 after 41 h of cultivation period (Możejko & Ciesielski, 2014). Ruiz et al. showed that fed-batch cultivation of *Pseudomonas putida* KT2440 and *Pseudomonas chlororaphis* 555 achieved biomass concentration of 159.4 g/L and 73 g/L with mcl-PHA titer and productivity of 58/13.9 g/L and 1.93/0.29 g/L.h using hydrolyzed WCO as sole carbon source (Ruiz et al., 2019a, 2019b).

Researchers have proposed an integrated process where the waste oils are first hydrolyzed and subsequently employed for biodiesel production. The low-quality biodiesel produced was utilized as feedstock for PHAs synthesis, while the high-quality biodiesel can be employed as fuel.

11.3 CONCLUSIONS AND FUTURE PROSPECTS

PHA and PLA has attracted wide attention in the market of biodegradable polymer as they can potentially replace the fossil-based bioplastics. Nevertheless, the production cost is always the most significant barrier for wide-scale applications, which the feedstocks contributed approximately 50% of the production cost. Therefore it is crucial to search for feedstocks which are abundant, renewable and low-cost to increase the economic competitiveness of bioplastics. Thus various types of by-products or wastes can serve as potential substrate for bioplastic production.

Following the biorefinery concept, the co-production of bioplastic and biofuels or other bioproducts should be performed to fully valorize the feedstock. The establishment of kinetic models, computer simulations and mathematical tools for the whole biotransformation process from upstream (pretreatment, hydrolysis, fermentation) to downstream (extraction/separation, purification) can help to predict and optimize the efficiency and yield of bioplastics production. More pilot-scale studies can be performed to bridge the gap between lab-scale studies and actual-scale process. Furthermore, techno-economic analysis and life cycle assessment can be carried out to evaluate the economic feasibility and environmental friendliness of the process, and can also pinpoint the weak spots to allow for improvement.

REFERENCES

Amaro, T.M.M.M., Rosa, D., Comi, G., Iacumin, L. 2019. Prospects for the use of whey for polyhydroxyalkanoate (PHA) production. *Frontiers in Microbiology*, **10**.

Andler, R., Valdés, C., Urtuvia, V., Andreeßen, C., Díaz-Barrera, A. 2021. Fruit residues as a sustainable feedstock for the production of bacterial polyhydroxyalkanoates. *Journal of Cleaner Production*, **307**, 127236.

Asunis, F., Carucci, A., De Gioannis, G., Farru, G., Muntoni, A., Polettini, A., Pomi, R., Rossi, A., Spiga, D. 2022. Combined biohydrogen and polyhydroxyalkanoates production from sheep cheese whey by a mixed microbial culture. *Journal of Environmental Management*, **322**, 116149.

Asunis, F., De Gioannis, G., Francini, G., Lombardi, L., Muntoni, A., Polettini, A., Pomi, R., Rossi, A., Spiga, D. 2021. Environmental life cycle assessment of polyhydroxyalkanoates production from cheese whey. *Waste Management*, **132**, 31–43.

Barrett, A. 2018. Bioplastic feedstock 1st, 2nd and 3rd generations, Vol. 2022. Bioplastics News.

Berlowska, J., Cieciura-Włoch, W., Kalinowska, H., Kregiel, D., Borowski, S., Pawlikowska, E., Binczarski, M., Witonska, I. 2018. Enzymatic conversion of sugar beet pulp: A comparison of simultaneous saccharification and fermentation and separate hydrolysis and fermentation for lactic acid production. *Food Technology and Biotechnology*, **56**(2), 188.

Bertrand, J.L., Ramsay, B.A., Ramsay, J.A., Chavarie, C. 1990. Biosynthesis of poly-β-hydroxyalkanoates from pentoses by *Pseudomonas pseudoflava*. *Applied and Environmental Microbiology*, **56**(10), 3133–3138.
Bioplastics, E. 2018. What are bioplastics?. Vol. 2022. European Bioplastics.
Bomrungnok, W., Arai, T., Yoshihashi, T., Sudesh, K., Hatta, T., Kosugi, A. 2020. Direct production of polyhydroxybutyrate from waste starch by newly-isolated *Bacillus aryabhattai* T34-N4. *Environmental Technology*, **41**(25), 3318–3328.
Bosco, F., Cirrincione, S., Carletto, R., Marmo, L., Chiesa, F., Mazzoli, R., Pessione, E. 2021. PHA production from cheese whey and "scotta": Comparison between a consortium and a pure culture of *Leuconostoc mesenteroides*. *Microorganisms*, **9**(12), 2426.
Bustamante, D., Segarra, S., Tortajada, M., Ramón, D., del Cerro, C., Auxiliadora Prieto, M., Iglesias, J.R., Rojas, A. 2019. In silico prospection of microorganisms to produce polyhydroxyalkanoate from whey: *Caulobacter segnis* DSM 29236 as a suitable industrial strain. *Microbial Biotechnology*, **12**(3), 487–501.
Chang, Y.-C., Reddy, M.V., Imura, K., Onodera, R., Kamada, N., Sano, Y. 2021. Two-stage polyhydroxyalkanoates (PHA) production from cheese whey using *Acetobacter pasteurianus* C1 and *Bacillus* sp. CYR1. *Bioengineering*, **8**(11), 157.
Chanprateep, S. 2010. Current trends in biodegradable polyhydroxyalkanoates. *Journal of Bioscience and Bioengineering*, **110**(6), 621–632.
Choi, G., Kim, J., Lee, C. 2016. Effect of low pH start-up on continuous mixed-culture lactic acid fermentation of dairy effluent. *Applied Microbiology and Biotechnology*, **100**(23), 10179–10191.
Colombo, B., Pepè Sciarria, T., Reis, M., Scaglia, B., Adani, F. 2016. Polyhydroxyalkanoates (PHAs) production from fermented cheese whey by using a mixed microbial culture. *Bioresource Technology*, **218**, 692–699.
Colombo, B., Villegas Calvo, M., Pepè Sciarria, T., Scaglia, B., Savio Kizito, S., D'Imporzano, G., Adani, F. 2019. Biohydrogen and polyhydroxyalkanoates (PHA) as products of a two-steps bioprocess from deproteinized dairy wastes. *Waste Management*, **95**, 22–31.
de Andrade, C.S., Nascimento, V.M., Cortez-Vega, W.R., Fakhouri, F.M., Silva, L.F., Gomez, J.G.C., Fonseca, G.G. 2019. Exploiting cheese whey as co-substrate for polyhydroxyalkanoates synthesis from *Burkholderia sacchari* and as raw material for the development of biofilms. *Waste and Biomass Valorization*, **10**(6), 1609–1616.
de Souza, L., Manasa, Y., Shivakumar, S. 2020. Bioconversion of lignocellulosic substrates for the production of polyhydroxyalkanoates. *Biocatalysis and Agricultural Biotechnology*, **28**, 101754.
Domingos, J.M.B., Puccio, S., Martinez, G.A., Amaral, N., Reis, M.A.M., Bandini, S., Fava, F., Bertin, L. 2018. Cheese whey integrated valorisation: Production, concentration and exploitation of carboxylic acids for the production of polyhydroxyalkanoates by a fed-batch culture. *Chemical Engineering Journal*, **336**, 47–53.
Ganesh Saratale, R., Cho, S.-K., Dattatraya Saratale, G., Kadam, A.A., Ghodake, G.S., Kumar, M., Naresh Bharagava, R., Kumar, G., Su Kim, D., Mulla, S.I., Seung Shin, H. 2021. A comprehensive overview and recent advances on polyhydroxyalkanoates (PHA) production using various organic waste streams. *Bioresource Technology*, **325**, 124685.
Getachew, A., Woldesenbet, F. 2016. Production of biodegradable plastic by polyhydroxybutyrate (PHB) accumulating bacteria using low cost agricultural waste material. *BMC Research Notes*, **9**(1), 509.
Kamilah, H., Tsuge, T., Yang, T., Sudesh, K. 2013. Waste cooking oil as substrate for biosynthesis of poly(3-hydroxybutyrate) and poly(3-hydroxybutyrate-co-3-hydroxyhexanoate): Turning waste into a value-added product. *Malaysian Journal of Microbiology*, **9**(1), 51–59.
Kolev, P., Rocha-Mendoza, D., Ruiz-Ramírez, S., Ortega-Anaya, J., Jiménez-Flores, R., García-Cano, I. 2022. Screening and characterization of β-galactosidase activity in lactic acid bacteria for the valorization of acid whey. *JDS Communications*, **3**(1), 1–6.
Kourilova, X., Novackova, I., Koller, M., Obruca, S. 2021. Evaluation of mesophilic *Burkholderia sacchari*, thermophilic *Schlegelella thermodepolymerans* and halophilic *Halomonas halophila* for polyhydroxyalkanoates production on model media mimicking lignocellulose hydrolysates. *Bioresource Technology*, **325**, 124704.
Lopes, M.S.G., Gomez, J.G.C., Taciro, M.K., Mendonça, T.T., Silva, L.F. 2014. Polyhydroxyalkanoate biosynthesis and simultaneous remotion of organic inhibitors from sugarcane bagasse hydrolysate by *Burkholderia* sp. *Journal of Industrial Microbiology and Biotechnology*, **41**(9), 1353–1363.
López-Gómez, J.P., Alexandri, M., Schneider, R., Venus, J. 2019. A review on the current developments in continuous lactic acid fermentations and case studies utilising inexpensive raw materials. *Process Biochemistry*, **79**, 1–10.

Luongo, V., Policastro, G., Ghimire, A., Pirozzi, F., Fabbricino, M. 2019. Repeated-batch fermentation of cheese whey for semi-continuous lactic acid production using mixed cultures at uncontrolled pH. *Sustainability*, **11**(12), 3330.

Matos, M., Cruz, R.A.P., Cardoso, P., Silva, F., Freitas, E.B., Carvalho, G., Reis, M.A.M. 2021. Combined Strategies to Boost Polyhydroxyalkanoate Production from Fruit Waste in a Three-Stage Pilot Plant. *ACS Sustainable Chemistry & Engineering*, **9**(24), 8270–8279.

Możejko, J., Ciesielski, S. 2014. Pulsed feeding strategy is more favorable to medium-chain-length polyhydroxyalkanoates production from waste rapeseed oil. *Biotechnology Progress*, **30**(5), 1243–1246.

Możejko, J., Przybyłek, G., Ciesielski, S. 2011. Waste rapeseed oil as a substrate for medium-chain-length polyhydroxyalkanoates production. *European Journal of Lipid Science and Technology*, **113**(12), 1550–1557.

Możejko-Ciesielska, J., Marciniak, P., Moraczewski, K., Rytlewski, P., Czaplicki, S., Zadernowska, A. 2022. Cheese whey mother liquor as dairy waste with potential value for polyhydroxyalkanoate production by extremophilic *Paracoccus homiensis*. *Sustainable Materials and Technologies*, **33**, e00449.

Oliveira, C.S.S., Silva, M.O.D., Silva, C.E., Carvalho, G., Reis, M.A.M. 2018. Assessment of protein-rich cheese whey waste stream as a nutrients source for low-cost mixed microbial PHA production. *Applied Sciences*, **8**(10), 1817.

Pais, J., Serafim, L.S., Freitas, F., Reis, M.A.M. 2016. Conversion of cheese whey into poly(3-hydroxybutyrate-co-3-hydroxyvalerate) by *Haloferax mediterranei*. *New Biotechnology*, **33**(1), 224–230.

Panesar, P.S., Kaur, S. 2015. Bioutilisation of agro-industrial waste for lactic acid production. *International Journal of Food Science & Technology*, **50**(10), 2143–2151.

Panesar, P.S., Kennedy, J.F., Gandhi, D.N., Bunko, K. 2007. Bioutilisation of whey for lactic acid production. *Food Chemistry*, **105**(1), 1–14.

Pantazaki, A.A., Papaneophytou, C.P., Pritsa, A.G., Liakopoulou-Kyriakides, M., Kyriakidis, D.A. 2009. Production of polyhydroxyalkanoates from whey by *Thermus thermophilus* HB8. *Process Biochemistry*, **44**(8), 847–853.

Rahman, M.H., Bhoi, P.R. 2021. An overview of non-biodegradable bioplastics. *Journal of Cleaner Production*, **294**, 126218.

Raj, T., Chandrasekhar, K., Naresh Kumar, A., Kim, S.-H. 2022. Lignocellulosic biomass as renewable feedstock for biodegradable and recyclable plastics production: A sustainable approach. *Renewable and Sustainable Energy Reviews*, **158**, 112130.

Ramchandran, L., Sanciolo, P., Vasiljevic, T., Broome, M., Powell, I., Duke, M. 2012. Improving cell yield and lactic acid production of *Lactoccocus lactis* ssp. *cremoris* by a novel submerged membrane fermentation process. *Journal of Membrane Science*, **403–404**, 179–187.

Rao, U., Sridhar, R., Sehgal, P.K. 2010. Biosynthesis and biocompatibility of poly(3-hydroxybutyrate-co-4-hydroxybutyrate) produced by *Cupriavidus necator* from spent palm oil. *Biochemical Engineering Journal*, **49**(1), 13–20.

Ruiz, C., Kenny, S.T., Babu P, R., Walsh, M., Narancic, T., O'Connor, K.E. 2019a. High cell density conversion of hydrolysed waste cooking oil fatty acids into medium chain length polyhydroxyalkanoate using *Pseudomonas putida* KT2440. *Catalysts*, **9**(5).

Ruiz, C., Kenny, S.T., Narancic, T., Babu, R., Connor, K.O. 2019b. Conversion of waste cooking oil into medium chain polyhydroxyalkanoates in a high cell density fermentation. *Journal of Biotechnology*, **306**, 9–15.

Saito, K., Hasa, Y., Abe, H. 2012. Production of lactic acid from xylose and wheat straw by *Rhizopus oryzae*. *Journal of Bioscience and Bioengineering*, **114**(2), 166–169.

Shahab, R.L., Luterbacher, J.S., Brethauer, S., Studer, M.H. 2018. Consolidated bioprocessing of lignocellulosic biomass to lactic acid by a synthetic fungal-bacterial consortium. *Biotechnology and Bioengineering*, **115**(5), 1207–1215.

Song, J.-H., Jeon, C.-O., Choi, M.-H., Yoon, S.-C., Park, W.-J. 2008. Polyhydroxyalkanoate (PHA) production using waste vegetable oil by *Pseudomonas* sp. strain DR2. *Journal of Microbiology and Biotechnology*, **18**(8), 1408–1415.

Taniguchi, I., Kagotani, K., Kimura, Y. 2003. Microbial production of poly(hydroxyalkanoate)s from waste edible oils. *Green Chemistry*, **5**(5), 545–548.

Thinagaran, L., Sudesh, K. 2019. Evaluation of sludge palm oil as feedstock and development of efficient method for its utilization to produce polyhydroxyalkanoate. *Waste and Biomass Valorization*, **10**(3), 709–720.

Valdez-Calderón, A., Barraza-Salas, M., Quezada-Cruz, M., Islas-Ponce, M.A., Angeles-Padilla, A.F., Carrillo-Ibarra, S., Rodríguez, M., Rojas-Avelizapa, N.G., Garrido-Hernández, A., Rivas-Castillo, A.M. 2020. Production of polyhydroxybutyrate (PHB) by a novel *Klebsiella pneumoniae* strain using low-cost media from fruit peel residues. *Biomass Conversion and Biorefinery*, **12**, 4925–4938.

Verlinden, R.A.J., Hill, D.J., Kenward, M.A., Williams, C.D., Piotrowska-Seget, Z., Radecka, I.K. 2011. Production of polyhydroxyalkanoates from waste frying oil by *Cupriavidus necator. AMB Express*, **1**(1), 11.

Zainab-L, I., Uyama, H., Li, C., Shen, Y., Sudesh, K. 2018. Production of polyhydroxyalkanoates from under-utilized plant oils by *Cupriavidus necator. CLEAN—Soil, Air, Water*, **46**(11), 1700542.

Zhou, J., Ouyang, J., Xu, Q., Zheng, Z. 2016. Cost-effective simultaneous saccharification and fermentation of l-lactic acid from bagasse sulfite pulp by *Bacillus coagulans* CC17. *Bioresource Technology*, **222**, 431–438.

12 Business Models for Innovative Bioplastic Feedstocks

Niklas Mathias Döhler and André Wolf

CONTENTS

12.1 INTRODUCTION

Over the years, a tremendous variety of process routes for the production of bio-based plastics has emerged. First, this is due to the development of ever new polymers and non-polymer additives, suitable for an increasingly broad field of applications. Second, the investigation and elaboration of new biological feedstock sources for these polymers has further widened the spectrum of resources, production steps and material requirements along the supply chain. The process routes not only differ in the quantity of raw materials needed but can also place different demands on the use of process energy and chemical auxiliaries. Therefore any investigation of the economic competitiveness of bio-based plastics must carefully distinguish between characteristic groups.

A group of so-called drop-in solutions (e.g., bio-PET, bio-PE) shares the chemical features of conventional fossil-based materials, with the only exception that the fossil feedstocks are replaced by biological alternatives. In 2021, this group of materials accounted for about 35.1% of global bioplastic production capacities (European Bioplastics, 2021). As these bio-based solutions are designed to copy the pre-existing fossil solutions, they can be processed along the same long-established refinery stages. The only difference in process routes is the initial phase of feedstock extraction. From a cost perspective, this entails the basic advantage that producers can benefit from past experience gathered in the refinery of fossil-based plastics. Rather, a challenge might lie in insufficiently exploited static economies of scale in the form of shrinking average costs with increasing

DOI: 10.1201/9781003344018-12

production quantity, because of the comparatively low production volumes in comparison to fossil-based competitors. This, in turn, could be aggravated by the existence of higher feedstock costs.

By contrast, in the group of bio-based and biodegradable plastic materials, production patterns are considerably more complex and heterogeneous. They largely represent newly developed chemical compounds with unique features. Deviations from conventional plastic production chains thus do not only concern feedstock acquisition but also the refinery stage. Procedures of fermentation and polymerization, supplemented by several pre- and after-treatment steps are common to most bio-based biodegradable plastics. However, detailed procedures can differ substantially between polymers concerning feedstock requirements, intensity of energy consumption (e.g., electricity, heat) and use of auxiliary chemicals.

At the moment, PLA has the largest market share and promises the greatest growth potential compared to other biodegradable bio-based polymers (Wellenreuther et al., 2022). For instance, feedstock choice is one of the central issues for PLA production also from an economic perspective. It not only influences costs related to the primary stage of the life cycle but also has cost implications for the refinery stage. In particular, this involves costs related to the cultivation of agricultural land, such as expenditures for fertilizers, pesticides, energy used by farm machinery and irrigation.

In general, cost superiority of PLA production from first-generation feedstocks compared to more innovative second- and third-generation feedstock solutions is still said to represent a barrier for a widespread feedstock switch. In view of the complexity of the production processes, the reasons for this cannot be attributed from the outset to a particular input or technology. To better understand the market-based obstacles, representative estimates are required. Unfortunately, relevant data is currently very limited in this respect. For example, no public database exists on the development of the average production costs of certain polymers at industry level. Existing estimates are usually based on individual technoeconomic case studies, the results of which are to a certain extent dependent on local production conditions and decisions regarding the use of specific feedstocks and technological setups.

This chapter attempts to summarize current knowledge on the economic potentials and barriers for using innovative bioplastic feedstocks. Section 12.2 outlines the technological implications of a feedstock switch, including a literature review on recent innovative feedstock solutions and input requirements for different feedstock technologies. Section 12.3 presents the results of a cradle-to-gate cost assessment in the case of corn stover. Section 12.4 provides an overview on relevant factors for the competitiveness of bioplastics from innovative feedstocks, distinguishing between four dimensions: political, economic, social and technological. Section 12.5 draws a conclusion.

12.2 TECHNOLOGICAL IMPLICATIONS OF THE FEEDSTOCK SWITCH

12.2.1 Feedstock Generations

In general, a distinction can be made between three different generations of bioplastic feedstocks. However, there still exists no official definition of how exactly these three generations are to be distinguished from each other. In this chapter, we follow the definition of Vandenberghe et al. (2021). According to them, first-generation bioplastics are defined as bioplastics produced from food plants, such as sugarcane and corn. In contrast, the second-generation includes the feedstock lignocellulosic biomass, namely waste materials from agriculture or wood. Moreover, non-food vegetable oils are also considered to be second-generation feedstocks. Sugars or oils produced by microorganisms, such as bacteria, mushrooms and micro-algae, are part of the third-generation. They also include municipal waste materials, such as wastewater or organic waste. Different types of polymers can be derived from the above-mentioned feedstocks. However, in this chapter we only focus on the polymers polylactic acid (PLA), polyhydroxyalkanoates (PHAs) and polybutylene succinate (PBS), as they make up the most significant market shares within the segment of biodegradable and bio-based plastic polymers (European Bioplastics, 2021).

12.2.2 Feedstocks for the Production of PLA

PLA is a bio-based and biodegradable polymer built from lactic acid molecules. It is a thermoplastic polyester that softens when heated and hardens when cooled. It can be cooled and heated several times. This allows it to be shaped and processed using liquefaction and molding techniques, and the material can then be recycled using the same processes. Dependent on the specific process, the quality of the recycled material can vary. This flexibility and other technical properties make PLA technically suitable for a wide range of applications, from disposable packaging to durable consumer goods. In the case of starch, corn kernels are still the dominant biological source. After harvesting, the corn starch is extracted from the grains. The starch molecules are then broken down into glucose molecules by a hydrolysis process. The glucose is dried and then enters an industrial fermentation process using bacteria. The resulting lactic acid solutions are converted into lactide, purified by crystallization and finally polymerized into PLA. In the case of sugarcane, the general process is similar, with the difference that the sucrose extracted from the sugarcane stalks is the input for the fermentation processes (Wellenreuther & Wolf, 2020).

From an economic point of view, the production of PLA is currently more cost-efficient than the production of other biodegradable polymers, as technical knowledge and production capacities have reached a comparatively high level (Ögmundarson et al., 2020). In terms of market share, European Bioplastics estimates that PLA has the second largest share of global production capacity of all bio-based and/or biodegradable polymers in 2021, at 18.9%. However, over the next 5 years, this share is expected to decline as a consequence of the more extensive growth of other polymers like PHA (European Bioplastics, 2021). In terms of end products, rigid and flexible packaging, medical articles and textiles are the dominant application areas.

A variety of resources for the production of lactic acid are now being discussed. The main purpose of these development processes has been to overcome the environmental problems related to land use and competition with food production associated with the current production of raw materials from food crops. The use of plant residues from the cultivation (stover) and processing (e.g., sugarcane bagasse) of these plants as cellulose-based feedstock has been discussed for some time (Kim & Dale, 2005). The fact that these materials arise as a by-product prevents or at least alleviates reservations about first-generation feedstocks. In addition, several proposals have been made in recent years to completely decouple PLA production from agricultural land use. One strand of this literature focuses on the use of by-products and waste from the food industry that otherwise have little or no economic value. Harbec (2010) and Broeren et al. (2017) analyze the use of wastewater generated from the industrial processing of potatoes. Liu et al. (2018) investigate the production of lactic acid from cheese whey, using lactose and proteins as feedstock. Nguyen et al. (2013) investigate a scenario in which waste from industrial extraction of curcuminoid, used in medicinal applications from *Curcuma longa* root, is fermented into lactic acid by simultaneous saccharification and fermentation. De la Torre et al. (2018) consider a mixture of orange peel waste and corn steep liquor as substrates. Pleissner et al. (2016) investigate waste from coffee production, coffee pulp, as a substrate. Alves de Oliviera et al. (2020) propose the use of sugar beet pulp as a feedstock, which is obtained as a by-product of extracting sugar from sugar beet.

In another area, the potential for a shift from land-based to sea-based resources is being explored. The cultivation and fermentation of carbohydrate-rich marine plants is considered to be a way to create new production pathways from scratch, saving existing food production chains from disruption by plastic manufacturing. In this vein, Helmes et al. (2018) investigate the use of the marine plant *Ulva* spp. for lactic acid production. Ögmundarson et al. (2020) are experimenting with the cultivation of brown algae of the species *Laminaria* sp. as a raw material source.

12.2.3 Feedstocks for the Production of PHA

PHAs are not a single polymer but a whole family of polyesters. Unlike other bio-based plastics, they are not exclusively artificially produced, but also occur in nature. They are bio-based and biodegradable polymers that have the properties of thermoplasticity and UV stability. Due to their heterogeneity, the family of PHAs is potentially suitable for a wide range of applications, from packaging to medical articles. Polyhydroxybutyrate (PHB) is currently the most discussed polymer within the family. Its mechanical and physical properties are similar to those of fossil-based polypropylene (PP), making it a potential substitute for applications such as food packaging and agricultural films (Wellenreuther & Wolf, 2020). The same biomaterials as for PLA are currently used as feedstocks. The first steps of production are therefore the same: extraction of carbohydrates from cultivated food crops and (in the case of starch) breakdown of the polymers into glucose monomers by hydrolysis. An important difference, however, is the subsequent fermentation phase. Unlike PLA, the biochemical processes inside the bacteria already produce the polymer in its final chemical form, so that no artificial polymerization is required. This puts more emphasis on the fermentation phase. The remaining task is to isolate the PHB produced (i.e., to extract it from the bacterial cells), a process step for which various techniques are discussed.

Although the PHA family is considered by many researchers to be particularly promising materials and its technical potential has been discussed for some time, its overall market share is still very modest. European Bioplastics records a market share of only 1.8% of the total segment of bio-based and/or biodegradable plastics for 2021. For the future, this share is expected to increase. For 2026, a share of 6.4% is forecasted, which is still significantly lower than what is expected for PLA or PBAT (European Bioplastics, 2021). There is consensus in the literature that the reasons for this are mainly to be found on the economic side. Primarily due to the energy- and material-intensive fermentation and extraction steps, the production costs are comparatively high compared to both PLA and competing fossil-based plastics. Mass production with its economies of scale and standardization effects could remedy this situation but would require a significant demand stimulus.

For the PHA family, the literature on feedstock solutions is particularly extensive and diverse. Most research focuses on PHB as a concrete example. In addition, a considerable number of papers are dedicated to third-generation feedstocks. Among these, substrates related to waste generation have attracted particular attention in recent years. These are, on the one hand, solid wastes or effluents generated as residues from production processes in the food industry. Shahzad et al. (2013), for example, consider the use of slaughterhouse waste as a source of fatty acids as a feedstock for PHA production. Similarly, Bhatia et al. (2018) investigate coffee waste as a substrate whose oil is extracted and the fatty acids it contains are used as feedstock. Zarroli (2020) examines the recyclability of starch from wastewater generated during potato processing. Another body of literature investigates the suitability of municipal waste. Morgan-Sagastume et al. (2016) investigate the possibilities of integrating PHA production into a municipal wastewater treatment plant. Sangkharak et al. (2020) explore the potential of using used cooking oil as a substrate for PHA production, using the free fatty acids it contains as feedstock. Kendall (2012) looks at the use of organic waste produced as residues in material recovery plants.

Another strand of literature investigates the feedstock potential of various forms of by-products in the bioeconomy. Koller at al. (2013) experiment with cheese whey as a by-product in milk processing. Thinagaran and Sudesh (2019) are testing the use of palm oil sludge, which is a by-product of palm oil milling. Penkhrue et al. (2020) are testing the use of pineapple peel solution as a substrate. Heng et al. (2017) are investigating rice husks as a substrate, using their cellulose content as feedstock. Pernicova et al. (2019) show the possibility of obtaining PHA from the controlled degradation of chicken feather waste. Finally, the use of biogas has also been investigated, with the possibility of a closed carbon cycle if the gas comes from landfill emissions. Rostkowski et al. (2012) analyze the environmental impact of such a cycle. The company Mango Materials has already put such a cycle into practice (Pieja et al., 2016).

12.2.4 Feedstocks for the Production of PBS

PBS is a polyester which is particularly interesting due to its extreme flexibility and its strong heat resistance. In line with the other polymers discussed, it is potentially suitable for a wide range of applications from packaging to consumer goods. It is also biodegradable. Its bio-content, however, varies with the production method. It is produced by polycondensation of two materials: succinic acid and 1,4-butanediol. Traditionally, both materials are obtained from fossil sources. However, in recent years new production technologies have emerged that allow for a partial or full substitution by biological feedstocks. In the former case, only one of the two materials is obtained from biological sources, in the latter case both. The degree of variation in potential process routes is thus higher than in the case of the other polymers considered. After the succinic acid is filtrated, it undergoes a chemical reaction with 1,4-butanediol in an esterification process, which yields an oligomer. In the last step, the 1,4-butanediol is then separated via polycondensation so that a high-molecular-weight PBS is obtained (Luyt & Malik, 2019).

Existing market statistics do not differentiate between fossil-based and (partly) bio-based PBS. Therefore the specific contribution of bio-based PBS to the segment of bio-based plastic production is unclear. For PBS in total, European Bioplastics records in 2021 a global market share of 3.5% within the segment of bio-based and/or biodegradable plastics. Over the next 5 years, this share is expected to increase to 16.0% (European Bioplastics, 2021).

Given the fact that the production of PBS was until recently almost exclusively based on fossil sources, the literature on innovative biological raw materials is still comparatively thin. A major focus is on the production of the precursor succinic acid from biological sources. Zhang et al. (2013) investigate the use of mixed bakery waste as a substrate for the production of succinic acid by using the glucose contained in pastries and cakes as an input to the fermentation process. Dessie et al. (2018) use fruit and vegetable waste that was first subjected to solid-state fermentation and then hydrolyzed to obtain glucose and fructose monomers and finally used as input for bacterial fermentation. Khoshnevisan et al. (2020) used the organic fraction of municipal solid waste to produce several high-value bioproducts, including succinic acid. Another part of the literature focuses on by-products. Li et al. (2010) produced succinic acid from wheat straw and orange peel by first hydrolyzing the material with a strong acid and then subjecting it to bacterial fermentation. Luthfi et al. (2016) used bagasse from oil palm fruits, a residue from oil extraction from oil palm fruits, as raw material. González-García et al. (2018) investigated the performance of apple pomace, a residue from apple juice extraction, as a feedstock.

12.3 CRADLE-TO-GATE COST ASSESSMENT: THE CASE OF CORN STOVER

12.3.1 Methodology

The sustainability of product systems can be measured in terms of three aspects: environmental protection, economic performance and social responsibility (Blanc et al., 2019).

In a life cycle cost analysis (LCC), the product system is evaluated from an economic perspective. Here the system's boundaries are the narrowest ones. In an LCC, all costs incurred on each stage of the process system are recorded. In addition, all revenues per stage are included in the evaluation of the system. This approach allows for an assessment of the economic viability of new systems. An alternative and more prominent method, environmental life cycle assessment (ELCA), evaluates the product system from an environmental point of view. In an ELCA, all relevant flows associated with each phase of a product's life cycle are recorded and evaluated. This includes the assessment of input flows such as energy, water and raw materials as well as output flows such as emissions to air, soil and water. The analysis of the occurring costs does not fall within the system boundaries of an ELCA. Finally, the assessment of the social impacts of a product system can be carried out with the help of an externality assessment (ExA). This assessment method extends the LCA and additionally evaluates the external effects of the product system's emissions. Figure 12.1 illustrates the boundaries between the three forms of life cycle analysis. In this chapter, to document the nature of the existing cost barriers for innovative feedstocks, we perform a cradle-to-gate LCC for the case of PLA derived from corn grain vs. PLA from corn stover.

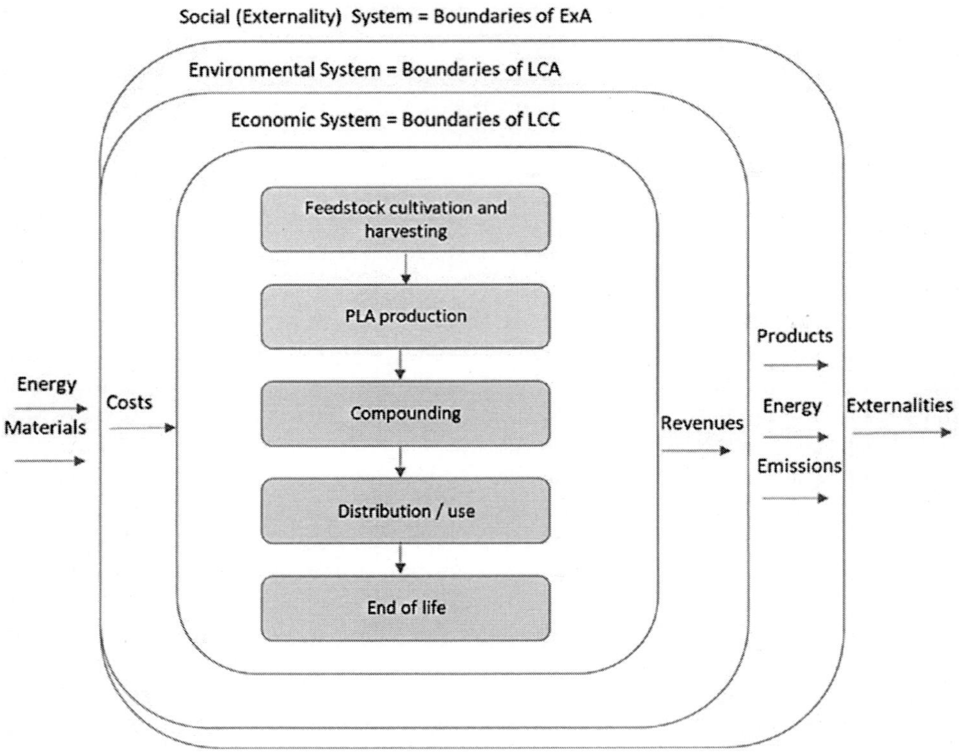

FIGURE 12.1 System boundaries in life cycle methods.

Source: Author's representation.

12.3.2 Results from the LCC Literature

The choice of feedstock is one of the key issues in the production of PLA from an economic point of view, not only because of its impact on the costs of the primary stage of the life cycle but also because of its technological impact on the refinery stage. Depending on the feedstock, there may be costs associated with the management of agricultural land in the primary stage, such as expenditure on fertilizers, pesticides, energy for agricultural machinery and irrigation. Costs associated with the extraction of the chemical feedstock from the biological source may also depend on the type of marketable by-products produced as financial compensation.

The level of production costs is therefore strongly dependent on raw material prices and technological progress in bioplastics production. In addition, production capacities and the associated economies of scale have an influence on unit costs. Political measures to promote sustainable alternatives to fossil-based plastics can support the expansion of production capacities for PLA. Furthermore, the development of the crude oil price plays an important role in the development of demand for bio-based plastics and thus also in the expansion of production capacities through its influence on the prices of fossil-based plastics.

Despite the growing interest in the production of bio-based plastics such as PLA, there are only a few studies in the literature that evaluate the cost structure and techno-economic feasibility of the commercial production of PLA.

In this section, the most relevant studies are presented in detail. These five studies examine the costs of the individual process steps of PLA production. Due to the different raw materials considered for the production of PLA, the process routes investigated also differ between the studies. For this reason, the results are not directly comparable with each other, but they do provide information about the cost structure and the approximate cost range for the production of PLA.

In an older study by Jim Lunt and Associates (2010), the manufacturing process of PLA from locally grown feedstock is investigated to test its economic viability. By applying an engineering cost model based on technical expertise and publicly available information, the study estimates the cost of manufacturing PLA in Maine (US). Two different process routes are investigated, which differ in the choice of feedstock. One production route involves the production of PLA from potato starch, the other the production of PLA from sugar extracted from wood. The results of the cost analysis are compared to the current cost of PLA production from midwestern corn by NatureWorks.

The study identifies four main factors for the cost of PLA. Firstly, raw material costs, which in the first case are the costs of dextrose, which is mainly obtained from potato processing waste, and in the second case the costs of wood sugar. Secondly, the cost of additives and waste disposal, in particular the cost of chemicals, nutrients and gypsum waste associated with bacterial fermentation. The third cost driver relates to process yield and the fourth cost driver includes utility costs (e.g., electricity costs).

The process routes are examined in three scenarios, which differ in capital costs and the technology used to produce PLA. In all scenarios, a plant capacity of 50,000 t PLA is considered. In the first scenario, the "greenfield" scenario, it is assumed that the production plant needs to be built from scratch. This involves higher capital costs, mainly resulting from the provision of electricity, steam and water systems. In addition, this scenario assumes that the traditional bacterial fermentation process is used to produce lactic acid, which is a more expensive technology. The second scenario leads to significant cost savings (~22%) compared to the first scenario, as it assumes a "brownfield" construction strategy: the PLA plant is co-located at existing industrial sites. Compared to the greenfield scenario, this strategy reduces not only capital costs but also annual operating costs, including taxes, insurance, maintenance, return on investment and other expenses. Similar to the first scenario, this uses traditional bacterial fermentation technology. The third scenario, the preferred one, achieves further significant cost reductions by using the innovative yeast fermentation technology while maintaining a brownfield construction strategy. Costs are 8% lower than in the other brownfield scenario and 29% lower than in the greenfield scenario. The cost structure of these processes is compared to the cost structure of common processes producing PLA from midwestern corn. The study concludes that the scenario considered can compete with corn-based PLA if advanced fermentation technologies and existing industrial infrastructure are used.

Chiarakorn et al. (2014) investigate the production of PLA from cassava starch. In contrast to the aforementioned study by Jim Lunt and Associates (2010), this study assesses environmental costs (indirect costs) in addition to financial costs (or direct costs such as production and investment costs). Two production scenarios for PLA are analyzed. In the first scenario, cassava roots are used as raw material for the production of PLA. Cassava starch is extracted from the cassava roots, which is then converted into glucose and finally into PLA. The second scenario assumes that the cassava starch is not obtained from own cassava plants but bought from external sources, otherwise it corresponds to the first scenario. The calculations include the potential direct costs of PLA production, raw material costs (cassava roots and chemicals), capital costs, labor costs, operating costs and waste treatment costs. The environmental costs (indirect costs) of PLA production include two main cost items: the cost of CH_4 emissions from wastewater and the cost of CO_2 emissions from electricity and fuel consumption. The total cost of PLA production from cassava starch to PLA resin (Scenario 2) is USD 2890/t PLA, which is higher than the total cost of PLA resin production from cassava roots in the first scenario (i.e., USD 2710/t PLA). The differences in the results of the two scenarios arise from the fact that the production of PLA from cassava roots in the first scenario generates two by-products in the starch extraction process: cassava flour and gypsum. The costs and benefits of these by-products only accrue in the first scenario.

In a study by Kwan et al. (2018), a techno-economic assessment was carried out to investigate the technical feasibility, profitability and level of investment risk between the production of lactic acid (LA), lactide and PLA using food waste powder as feedstock in one plant. The economic performance of the three scenarios was assessed by estimating the capital costs, operating costs and revenue generated. The total capital costs include the fixed investment costs and the operating capital costs. The fixed capital costs refer to the expenditure for the construction of the plant, including the cost of equipment purchase, installation, piping and other related costs. The estimated operating costs include the total variable production costs, fixed costs, plant overheads and general costs. Revenues come from the sale of products and from the food waste treatment fee. Different profitability indicators were used to assess the economic performance of the three scenarios. Kwan et al. (2018) conclude that all scenarios examined in this study are economically feasible, as evidenced by the application of a range of profitability indicators, with LA production (scenario 1) being the most profitable option. The minimum selling prices for one tonne of LA, lactide and PLA are USD 943, USD 2073, and USD 3330, respectively.

Sanaei and Stuart (2018) investigate the cost performance of producing PLA using an innovative feedstock called triticale. Triticale (X Triticosecale Wittmack) is a crop that the authors believe to have the potential to become a preferred industrial energy crop for biorefineries. Compared to existing cereals such as wheat, the plant grows on marginal land and provides higher yields. Another advantage is that this plant does not compete with food crops. The aim of the work is to identify an economically promising strategy for the production of PLA based on this new raw material. In addition to several economic indicators used to evaluate the economic performance of different production scenarios, the study also includes cost estimates. The baseline scenario assumes the application of established commercial technologies. Therefore it has the lowest technological risk while maximizing the production capacity of the product. The alternative scenarios feature less conventional and well-tested technology choices. They have higher technological risks compared to the baseline scenario but can potentially lead to a better return on investment. Total production costs include the costs of raw materials (biomass and chemicals), energy and operating materials, maintenance and repair, labor, operating materials, insurance and overhead, administration, distribution and sales. The total costs estimated in this study range from USD 911/t PLA to USD 1496/t PLA, with the benchmark scenario having the highest costs.

Manandhar and Shah (2020) investigate the techno-economic feasibility of producing 100,000 t of lactic acid per year from maize grain in a biorefinery. In doing so, the study estimates the resource requirements (equipment, raw materials, energy and labor) and costs of producing lactic acid from bacterial, fungal and yeast-based fermentation pathways. The study found that lactic acid production costs are highly dependent on the conversion rate of sugar to lactic acid, maize prices, plant size, annual operating hours and the required use of gypsum. The minimum selling price for lactic acid produced from maize grains using different fermentation processes was comparable to the market price for lactic acid. It was found that the fermentation processes using microorganisms such as yeast, which tolerate low pH and have high lactic acid yields, had the lowest production costs, estimated at USD 844 per tonne of lactic acid. The total production costs for lactic acid from maize grains were USD 1181/t and USD 1251/t for the bacteria- and fungus-based fermentation processes, respectively. The authors point out that improvements in process efficiency and lower costs for raw materials, equipment and chemicals could further reduce production costs and improve the techno-economic feasibility of lactic acid production.

Table 12.1 compares the various results of the studies presented and gives the maximum and minimum production costs for 1 t of PLA or, in the case of Manandhar and Shah (2020), for 1 t of lactic acid in USD. The values show a wide range from USD 844 to USD 3558 per t of PLA/LA. The main cost drivers identified in the studies were costs for raw materials, energy costs, labor costs and capital costs.

The wide range of results is due to the different process routes that were analyzed. The processes differ in the selection of feedstocks and in the assumptions made regarding the production process. Therefore the results are not directly comparable. A significant difference is the choice of feedstock, which has a great influence on the results. Feedstock choice not only affects the costs directly

TABLE 12.1
Comparison of Literature Results on the Unit Costs of PLA Production

Study	Feedstock(s)	System Boundaries	Range of Results: Costs per t PLA (USD)	
			Min.	Max.
Chiarakorn et al. (2014)	Cassava	Feedstock—PLA polymerization	2410	2620
Jim Lunt and Associates (2010)*	Potato; Wood	Feedstock—PLA polymerization	1808	2977
Kwan et al. (2018)	Food waste	Feedstock—PLA polymerization	3558	3558
Manandhar and Shah (2020)**	Corn grain	Feedstock—Fermentation	844	1251
Sanaei and Stuart (2018)	Triticale	Feedstock—PLA polymerization	911	1496

Note: * Converted from USD/lb to USD/t. ** Costs per t lactic acid.

Source: Author's representation.

associated with the raw material input but also changes the subsequent process steps: different feedstocks require different chemical conversion steps, associated with energy and material needs.

Another very relevant factor for the cost of PLA production is the energy use (i.e., electricity, heat) incurred for the individual process steps, particularly in the PLA refinery process. When innovative raw materials are used, the technology, which is not yet fully developed, is usually associated with high energy intensity, representing a cost driver. The costs for additives and waste disposal also depend on the choice of feedstock and the subsequent technological process steps (Wellenreuther et al., 2022).

12.3.3 Specific Simulation Results: Deterministic Scenarios

In what follows, we present in detail the results of a cost assessment on the cradle-to-gate costs of producing PLA from two alternative feedstocks: corn grain (first generation) and corn stover (second generation).

In a first simulation exercise, a deterministic case is considered. This means all parameter values are fixed (i.e., existing uncertainty in prices and production technologies is ignored). For each production input, prices and quantities are set equal to the mean values obtained from a meta-analysis of the current ELCA literature (see section 12.2.2) on PLA and from price information on market platforms. The resulting cost estimates can be interpreted as reflecting the status quo observed for production technologies and input markets according to the available information sources. Table 12.2 shows the estimates for variable unit costs and their components for both product systems. The first important result is that the variable unit costs of PLA production based on corn stover are already at a similar level as those of PLA production based on corn grains. At the same time, the cost composition between the two systems differs considerably in some cases. This already holds for the stage of raw material extraction, as the prices per kg differ between the two raw materials. While the underlying average price for corn grain is USD 0.142/kg, the cost of using corn stover is estimated at only USD 0.052/kg. In the latter estimate, harvesting costs are the main source (USD 0.034/kg straw). The opportunity cost of additional fertilizer use plays only a minor role (USD 0.018/kg straw). However, this cost advantage in the initial phase is offset by significantly higher process costs for the corn stover–based system in both pre-treatment and fermentation. This in turn is partly a consequence of the more energy-intensive processing. In particular, the electricity demand for fermentation is higher in the stover-based system, a finding shared by the two existing literature sources (Ögmundarson et al., 2020; Adom & Dunn, 2017). Another reason is the use of the costly enzyme cellulase in the pre-treatment of maize straw, which alone contributes USD 0.04 to the cost per kg PLA.

TABLE 12.2
Variable Unit Costs of PLA Production in System Comparison

	USD/kg PLA	
By Production Stage	**System Corn Grain**	**System Corn Stover**
Feedstock	USD 0.28	USD 0.10
Pre-treatment	USD 0.01	USD 0.08
Fermentation	USD 0.38	USD 0.50
Polymerization	USD 0.08	USD 0.08
Total	**USD 0.75**	**USD 0.76**
Share of energy costs	**11.1%**	**19.7%**

Source: Author's calculations.

Regarding the results for capital costs (see Table 12.3), the fact that only one literature source was available to quantify the capital requirements for the maize straw-based system represents a serious limitation of this analysis. Nevertheless, it is noteworthy that our estimates of capital costs per kg PLA based on this source are well above the range of capital cost estimates derived from the results of the maize grain literature. Overall, this creates a gap in total unit costs between the two systems of about USD 0.2/kg PLA.

12.3.4 Specific Simulation Scenarios: Stochastic Case

An essential task for a meta-analysis of PLA production costs is to make the extent of input-specific cost variations transparent. Then implications for overall cost uncertainty can be assessed. In the following, we apply Monte Carlo simulation as a common tool for such a task. It has been used for cost estimations by various studies, including the cost structure of corn-based biofuel production (Petter & Tyner, 2014). The concept behind the Monte Carlo approach is to capture the output uncertainty of a system by means of specifying probability distributions for the relevant input parameters. Conducting a very large number of draws from these distributions and computing for each draw the resulting output leads to a distribution of output values, in our case a distribution of the unit costs of PLA production.

TABLE 12.3
Fixed and Total Unit Costs of PLA Production in System Comparison

	USD/kg PLA	
By Component	**System Corn Grain**	**System Corn Stover**
Labor costs	USD 0.05	USD 0.05
Maintenance and repair	USD 0.11	USD 0.17
Depreciation	USD 0.12	USD 0.16
Return on investment	USD 0.16	USD 0.24
Fixed unit costs	**USD 0.44**	**USD 0.62**
Total costs	**USD 1.18**	**USD 1.38**

Source: Author's calculations.

Table 12.4 presents statistics summarizing the results of the Monte Carlo simulations for corn grain-based PLA production. Comparing the production steps, the fermentation costs show the highest level of uncertainty, followed by the feedstock costs. The variations in fermentation costs are to some extent due to observable differences in the estimated energy intensities, especially with regard to the extent of heat utilization, between the available studies. Variations in the level of nutrient costs (Kwan et al., 2018) are also relevant here. Variations in feedstock costs are less a consequence of volatile corn prices but are more due to discrepancies in estimated input quantities between the data sources. Uncertainty in capital costs concerns both the annual loss in value of the capital employed and the level of annual expenditures for maintenance.

Table 12.5 shows the same statistics for the Monte Carlo simulations of the stover-based system. In this case, the range of results obtained from the drawings is somewhat more pronounced. This is mainly due to the greater uncertainty in the cost of the pre-treatment stage. In particular, the price of the enzyme cellulase varies considerably in the literature (Liu et al., 2018). Likewise, the not insignificant fluctuations in fermentation costs are more strongly due to fluctuations in electricity use than in the maize grain scenario. In the case of the raw material costs, the uncertainty about the extent of harvesting costs is in the foreground. In total, the probability of total unit costs to be less than USD 1.19/kg, the mean of the corn grain scenario, is less than 1% for the corn stover scenario.

A direct comparison of the range of our results with the unit costs reported in the bioplastics literature is presented in Table 12.6. The results of our corn stover scenario are roughly in the range of the scenario results reported by Manandhar and Shah (2020). A quantitative comparison of the maize straw scenario with literature results for other second- or third-generation feedstocks is

TABLE 12.4
Distribution of Unit Costs for Corn Grain–Based PLA Production

System Corn Grain	USD/kg PLA		
	Mean	Max.	Min.
Feedstock	USD 0.28	USD 0.32	USD 0.21
Pre-treatment	USD 0.01	USD 0.01	USD 0.01
Fermentation	USD 0.38	USD 0.55	USD 0.26
Polymerization	USD 0.08	USD 0.12	USD 0.05
Variable unit costs	**USD 0.75**	**USD 0.91**	**USD 0.57**
Fixed unit costs	USD 0.44	USD 0.51	USD 0.36
Total unit costs	**USD 1.19**	**USD 1.37**	**USD 1.00**

Source: Author's calculations.

TABLE 12.5
Distribution of Unit Costs for Corn Stover–Based PLA Production

System Corn Stover	USD/kg PLA		
	Mean	Max.	Min.
Feedstock	USD 0.10	USD 0.15	USD 0.06
Pre-treatment	USD 0.08	USD 0.58	USD 0.03
Fermentation	USD 0.50	USD 0.65	USD 0.37
Polymerization	USD 0.08	USD 0.12	USD 0.06
Variable unit costs	**USD 0.77**	**USD 1.30**	**USD 0.59**
Fixed unit costs	USD 0.63	USD 0.74	USD 0.51
Total unit costs	**USD 1.40**	**USD 1.97**	**USD 1.13**

Source: Author's representations.

TABLE 12.6
Comparison of Simulation Results with the Literature

Study	Feedstock(s)	Cost Estimates: USD/t PLA Min.	Cost Estimates: USD/t PLA Max.
Own	Corn grain	1004	1374
Own	Corn stover	1130	1972
Chiarakorn et al. (2014)	Cassava	2410	2620
Jim Lunt and Associates (2010)*	Potato, wood	1808	2977
Kwan et al. (2018)	Food waste	3558	3558
Manandhar and Shah (2020)**	Corn grain	844	1251
Sanaei and Stuart (2018)	Triticale	911	1496

Notes: * Converted from USD/lb to USD/t. ** Costs per t lactic acid.

Source: Author's representation.

difficult, as these studies differ significantly in their scenario specifications. At least it can be said that the higher maturity of stover-based production seems to lead to cost advantages compared to third-generation feedstocks such as food waste. However, to isolate the role of technology, a consistent comparative analysis of different innovative feedstocks under the same framework conditions would be needed.

12.4 FACTORS FOR COMPETITIVENESS OF SUSTAINABLE BIOPLASTICS

Technology-based cost spreads are only one group of influencing factors producers of bioplastics from innovative feedstocks need to cope with. Factors from the social and political dimension that affect that implementation of business models for sustainable bioplastics are also an important part of the equation. In what follows, an overview is given on different types of barriers discussed in the literature. Applying the basic concept of a PEST analysis (Sammut-Bonnici & Galea, 2015), we distinguish four different dimensions: economic, technological, political, social.

12.4.1 Economic Factors

- *Feedstock costs*: The production costs of bioplastics depend on the development of raw material prices. Currently, bioplastics are mainly produced from corn starch or sucrose from sugarcane (European Bioplastics, 2022). If the prices for raw materials rise, the production costs also rise and thus the prices for bioplastic products. Since the prices for maize and sugar are determined on the world markets, this means a considerable dependence on general agro-economic factors, such as the occurrence of droughts in important growing regions, the price development of seeds and fertilizers, or the labor shortage during harvest. Some of these factors are difficult to predict and exhibit a high degree of volatility, leading to uncertainties in the cost of bioplastic raw materials. In addition, other long-term factors such as increasing land scarcity, but also improvements in agricultural productivity, can have an impact on production costs in the feedstock phase (Döhler et al., 2022). In this regard, the switch to alternative feedstocks, especially from the most recent third generation, could imply a reduction in exposure toward agro-economic or market-related risks. On the other hand, the absence of formal markets for some of these feedstocks (e.g., food waste) can also entail new supply risks producers need to manage when scaling up capacities.

- *Energy costs*: Similar to fossil plastics, the steps involved in extracting plastics from biological resources require significant amounts of energy in the form of electricity and process heat. This is especially true for the fermentation stage with its stringent process temperature requirements (Manandhar & Shah, 2020). Therefore, trends and fluctuations in industrial prices for electricity and natural gas can have a significant impact on the overall production costs of bioplastic polymers. In this respect, the comparatively still low level of technological maturity of many innovative feedstocks often implies that energy intensities are higher, as is also in the case in our simulation exercise in the previous section. Hence at least for the scale-up phase, fluctuating energy prices can represent a significant risk factor for companies promoting plastic products from new materials.

12.4.2 Technological Factors

- *Productivity increases through learning effects*: The experience gained over time in the production of innovative materials reveals optimization potential in processing. The introduction of more efficient production processes in turn lowers average costs and thus improves competitiveness. Existing evidence confirms the relevance of such experience curves in the bioeconomy sector (Junqueira et al., 2017). For bioplastic production from third-generation feedstocks in particular, rapid learning rates will be crucial to eliminate cost-related competitive drawbacks.
- *Economies of scale*: By expanding the production volume of bioplastics, companies can take advantage of cost benefits by spreading existing fixed costs over a larger volume and thus producing more output at lower average (unit) costs. Currently, the production volume for bioplastics is still comparatively low. At the same time, fixed capital costs for buildings and equipment account for a sizeable share of total production costs (Manandhar & Shah, 2020), suggesting significant potential for untapped economies of scale. The cost development in more advanced biotechnologies documents the realization of such potentials over time (de Jong et al., 2017). The existence of significant scale economies hampers the dissemination of innovative feedstock solutions: their less mature technologies cause cost disadvantages compared to established first-generation feedstocks On the other hand, this means that the effect of exogenous impulses for a feedstock switch (e.g., the policy instruments discussed in the next subsection) could be magnified: if certain policies are able to raise production volumes for bioplastic producers in the innovative feedstock segment, the resulting cost reduction could further increase their competitiveness.

12.4.3 Political Factors

- *Taxes*: A targeted plastic tax on products made from fossil fuels could lead to a price increase for conventional plastics. As a result, the prices of bioplastics would decrease in relative terms, thus increasing the demand for bioplastics. However, to stimulate the most–environmentally friendly alternatives, tax solutions should ideally be implemented in a more differentiated manner. For instance, one option would be to limit taxes not just to fossil origin, but to introduce a general tax on plastic products, whose extent is however differentiated according to the (actual or approximated) environmental footprint of its production process. In this way, the environmental benefits of a feedstock switch could be translated into a reduction of the tax burden, providing a market-based incentive for producers. At the same time, the impact of environmental taxes with less comprehensive targets could also be detrimental. For example, a general tax on non-recyclable plastic products, regardless of their origin, could hinder the market introduction of innovative biodegradable plastics destined for composting as an alternative treatment option (Döhler et al., 2022).

- *Subsidies*: Government subsidies would allow bioplastics producers to offer their products at lower prices, which in turn would increase demand. In principle, subsidies can take many different forms. They can consist of support for research and development activities, investment support for capacity building through public grants or access to favorable credit, as well as tax exemption schemes and direct forms of production subsidy. Again, a differentiated approach linking subsidies to the environmental superiority of specific process routes and feedstock alternatives could be favorable for the market penetration of second- and third-generation feedstocks.
- *Bans*: Government bans on certain types of fossil-based plastic products would increase demand for bioplastic products. However, if the bans do not differentiate according to the origin of the raw materials, but according to other characteristics such as reusability or the recycled share of production, certain types of bio-based and/or biodegradable plastic products would also be excluded from the market, as is the case with taxes (Döhler et al., 2022).
- *Labeling of products*: Another way that policy makers can stimulate demand for bioplastics is to encourage the development of public labeling schemes. The rationale behind labeling is to reduce the information gap between sellers and buyers, which is caused by the fact that buyers usually do not have the time to identify the environmental impacts of products (Galarraga Gallastegui, 2002). If plastic products are certified as bio-based and/or biodegradable and the certification is documented by meaningful labels, this can influence buyers' decision-making, as empirical studies on consumers' willingness to pay have shown (Yokessa & Marette, 2019). However, the effectiveness of labels depends crucially on their transparency and credibility in the eyes of customers. It is therefore a prerequisite that they are linked to effective and clear environmental standards, compliance with which is regularly monitored by competent bodies. This requires an active role of public authorities in the development of labeling schemes. By adding the requirement of using sustainable feedstocks in such schemes, policymakers could help to boost the market penetration of second- and third-generation feedstocks from the demand side.

12.4.4 Social Factors

- *Preferences for sustainability*: A general increase in environmental awareness could create new market opportunities for sustainable products through its influence on consumer preferences. Increased media coverage of plastic pollution of the oceans, combined with a widespread realization of the scarcity of fossil resources, could also have a specific positive effect on demand for bioplastics. The prerequisite for this is that existing information gaps and public sustainability concerns about bioplastics are addressed (Leal Filho et al., 2021). This fact represents an important chance for innovative feedstocks: by communicating their sustainability potential, producers could try to create greater leeway in price setting.
- *Caveats regarding land use conflicts*: A particular consumer reservation relates to the first production phase of commodity extraction. The currently predominant first generation of feedstocks consists of agricultural products such as maize and sugarcane, whose land requirements mean direct competition with food production. Studies of public opinion on bioenergy, a sector facing the same feedstock issue, show that concerns about competition with the food system are relevant to consumer perceptions (Radics et al., 2015). In this respect, future feedstock generations in the form of agricultural by-products or waste could bring about a long-term change in consumer sentiment.

12.5 CONCLUSION

This chapter has provided a summary of current knowledge on the economic implications of a feedstock switch in bioplastic production, focusing on the aspect of cost competitiveness. For the central polymers/polymer families PLA, PHA and PBS, a wide range of feedstock solutions has recently been proposed and tested. However, given that technological readiness is mostly still in an experimental stage, information on the expected cost structure of industrial-scale production is still very sparse. The existing life cycle cost studies focus on PLA. While differing in their assumptions and quantitative predictions, they all point to the relevance of fixed costs for the overall level of unit costs in production. Hence production volumes and associated economies of scale are crucial factors for cost competitiveness. To further illustrate this point, detailed results of a simulation exercise comparing costs of corn grain–based PLA with corn stover–based PLA were presented. They document that cost disadvantages of the second-generation solution are largely restricted to the fixed cost component. Finally, barriers to future market success of bioplastics from innovative feedstocks are discussed, grouped into four segments. We argue that besides technological questions, the regulatory framework and social attitudes toward bioplastics also play an important role. By implementing a comprehensive and differentiated regulatory framework, which acknowledges the implications of a feedstock switch for the overall environmental life cycle balance of bioplastics, policymakers can stimulate the dissemination of innovative feedstocks in at least two ways. First, they can create cost incentives for a feedstock switch on the side of producers. Second, they can signal the environmental potentials of such a feedstock switch to consumers, thereby addressing the rather skeptical stance toward bioplastics of the wider public. In turn, this simultaneous stimulation of supply and demand can help to support innovative producers in exploiting scale economies, thereby also improving the economic performance of second- and third-generation feedstocks. The road ahead will undoubtedly be long, but the global conditions are more favorable than ever.

REFERENCES

Adom, F. K., & Dunn, J. B. (2017). Life cycle analysis of corn-stover-derived polymer-grade l-lactic acid and ethyl lactate: greenhouse gas emissions and fossil energy consumption. Biofuels, Bioproducts and Biorefining, 11(2), 258–268.

Alves de Oliveira, R., Schneider, R., Hoss Lunelli, B., Vaz Rossell, C. E., Maciel Filho, R., & Venus, J. (2020). A simple biorefinery concept to produce 2G-lactic acid from sugar beet pulp (SBP): a high-value target approach to valorize a waste stream. Molecules, 25(9), 2113.

Bhatia, S. K., Kim, J. H., Kim, M. S., Kim, J., Hong, J. W., Hong, Y. G., Kim, H. J., Kim, Jeon, J. M., Kim, S. H., Ahn, J., Lee, H., & Yang, Y. H. (2018). Production of (3-hydroxy- butyrate-co-3-hydroxyhexanoate) copolymer from coffee waste oil using engineered *Ralstonia eutropha*. Bioprocess and Biosystems Engineering, 41(2), 229–235.

Blanc, S., Massaglia, S., Brun, F., Peano, C., Mosso, A., & Giuggioli, N. R. (2019). Use of bio-based plastics in the fruit supply chain: an integrated approach to assess environmental, economic, and social Sustainability. Sustainability, 11(9), 2475.

Broeren, M. L. M., Kuling, L, Worrell, E., & Shen, L. (2017). Environmental impact assessment of six starch plastics focusing on wastewater-derived starch and additives resources. Conservation and Recycling, 127, 246–255.

Chiarakorn, S., Permpoonwiwat, C. K., & Nanthachatchavankul, P. (2014). Financial and economic viability of bioplastic production in Thailand; WorldFish (ICLARM)—Economy and Environment Program for Southeast Asia (EEPSEA), Research Report 7, Philippines.

de Jong, S., Hoefnagels, R., Wetterlund, E., Pettersson, K., Faaij, A., & Junginger, M. (2017). Cost optimization of biofuel production—the impact of scale, integration, transport and supply chain configurations. Applied Energy, 195, 1055–1070.

De la Torre, I., Ladero, M., & Santos, V. E. (2018). Production of d-lactic acid by *Lactobacillus delbrueckii* ssp. *delbrueckii* from orange peel waste: techno-economical assessment of nitrogen sources. Applied Microbiology and Biotechnology, 102(24), 10511–10521.

de Souza Vandenberghe, L. P., de Oliveira, P. Z., Bittencourt, G. A., de Mello, A. F. M., Vásquez, Z. S., Karp, S. G., & Soccol, C. R. (2021). The 2G and 3G bioplastics: an overview. Biotechnology Research and Innovation, 5(1).

Dessie, W., Zhang, W., Xin, F., Dong, W., Zhang, M., Ma, J., & Jiang, M. (2018). Succinic acid production from fruit and vegetable wastes hydrolyzed by on-site enzyme mixtures through solid state fermentation. Bioresource Technology, 247, 1177–1180.

Döhler, N., Wellenreuther, C., & Wolf, A. (2022). Market dynamics of biodegradable bio-based plastics: projections and linkages to European policies. EFB Bioeconomy Journal, 2.

European Bioplastics (2021). Bioplastics market update 2021. https://docs.european-bioplastics.org/publications/market_data/Report_Bioplastics_Market_Data_2021_short_version.pdf

Galarraga Gallastegui, I. (2002). The use of eco-labels: a review of the literature. European Environment, 12(6), 316–331.

González-García, S., Argiz, L., Míguez, P., & Gullón, B. (2018). Exploring the production of bio-succinic acid from apple pomace using an environmental approach. Chemical Engineering Journal, 350, 982–991.

Harbec, A. (2010). Lactic acid production from agribusiness waste starch fermentation with *Lactobacillus amylophilus* and its cradle-to-gate Life Cycle Assessment as a precursor to Poly-L-Lactide; Université De Montréal, Montréal, Canada.

Helmes, R. J. K., López-Contreras, A. M., Benoit, M., Abreu, H., Maguire, J., Moejes, F., & Burg, S. W. K. (2018). Environmental impacts of experimental production of lactic acid for bioplastics from *Ulva* spp. Sustainability, 10, 2462, 1–15.

Heng, K. S., Hatti-Kaul, R., Adam, F., Fukui, T., & Sudesh, K. (2017). Conversion of rice husks to polyhydroxyalkanoates (PHA) via a three-step process: optimized alkaline pretreatment, enzymatic hydrolysis, and biosynthesis by *Burkholderia cepacia* USM (JCM 15050). Journal of Chemical Technology & Biotechnology, 92(1), 100–108.

Jim Lunt & Associates (2010). The business case for commercial production of bioplastics in Maine. Preliminary report to the Maine Technology Institute, Jim Lunt & Associates LLC, Waxhaw.

Junqueira, T. L., Chagas, M. F., Gouveia, V. L., Rezende, M. C., Watanabe, M. D., Jesus, C. D., Cavalett, O., Milanez, A. Y., Bonomi, A. (2017). Techno-economic analysis and climate change impacts of sugarcane biorefineries considering different time horizons. Biotechnology for Biofuels and Bioproducts, 10(1), 1–12.

Kendall, A. (2012). A life cycle assessment of biopolymer production from material recovery facility residuals. Resources, Conservation and Recycling, 61, 69–74.

Khoshnevisan, B., Tabatabaei, M., Tsapekos, P., Rafiee, S., Aghbashlo, M., Lindeneg, S., & Angelidaki I. (2020). Environmental life cycle assessment of different biorefinery platforms valorizing municipal solid waste to bioenergy, microbial protein, lactic and succinic acid. Renewable and Sustainable Energy Reviews, 117.

Kim, S., & Dale, B. E. (2005). Life cycle assessment of various cropping systems utilized for producing biofuels: bioethanol and biodiesel. Biomass and Bioenergy, 29(6), 426–439.

Koller, M., Sandholzer, D., Salerno, A., Braunegg, G., & Narodoslawsky, M. (2013). Biopolymer from industrial residues: life cycle assessment of poly(hydroxyalkanoates) from whey. Resources, Conservation and Recycling, 73, 64–71.

Kwan, T. H., Hu, Y., & Lin, C. S. K. (2018). Techno-economic analysis of a food waste valorisation process for lactic acid, lactide and poly (lactic acid) production. Journal of Cleaner Production, 181, 72–87.

Leal Filho, W., Salvia, A. L., Bonoli, A., Saari, U. A., Voronova, V., Klõga, M., Kumbhar, S. S., Olszewski, K., Müller de Quevedo, D., Barbir, J. (2021). An assessment of attitudes towards plastics and bioplastics in Europe. Science of the Total Environment, 755, 142732.

Li, Q., Siles, J. A., & Thompson, I. P. (2010). Succinic acid production from orange peel and wheat straw by batch fermentations of *Fibrobacter succinogenes* S85. Applied Microbiology and Biotechnology, 88(3), 671–678.

Liu, P., Zheng, Z., Xu, Q., Qian, Z., Liu, J., & Ouyang, J. (2018). Valorization of dairy waste for enhanced D-lactic acid production at low cost. Process Biochemistry, 71, 18–22.

Luthfi, A. A. I., Jahim, J. M., Harun, S., Tan, J. P., & Mohammad, A. W. (2016). Biorefinery approach towards greener succinic acid production from oil palm frond bagasse. Process Biochemistry, 51(10), 1527–1537.

Luyt, A. S., & Malik, S. S. (2019). Can biodegradable plastics solve plastic solid waste accumulation? In Plastics to Energy (pp. 403–423). William Andrew Publishing, Norwich.

Manandhar, A., & Shah, A. (2020). Techno-economic analysis of bio-based lactic acid production utilizing corn grain as feedstock. Processes, 8(2), 199.

Morgan-Sagastume, F., Heimersson, S., Laera, G., Werker, A., & Svanström, M. (2016). Techno-environmental assessment of integrating polyhydroxyalkanoate (PHA) production with services of municipal wastewater treatment. Journal of Cleaner Production, 137, 1368–1381.

Nguyen, C. M., Kim, J. S., Nguyen, T. N., Kim, S. K., Choi, G. J., Choi, Y. H., Jang, K. S., Kim, J. C. (2013). Production of L-and D-lactic acid from waste *Curcuma longa* biomass through simultaneous saccharification and cofermentation. Bioresource Technology, 146, 35–43.

Ögmundarson, Ó., Sukumara, S., Laurent, A., & Fantke, P. (2020). Environmental hotspots of lactic acid production systems. GCB Bioenergy, 12(1), 19–38.

Penkhrue, W., Jendrossek, D., Khanongnuch, C., Pathom-Aree, W., Aizawa, T., Behrens, R. L., & Lumyong, S. (2020). Response surface method for polyhydroxybutyrate (PHB) bioplastic accumulation in *Bacillus drentensis* BP17 using pineapple peel. PloS one, 15(3), e0230443.

Pernicova, I., Enev, V., Marova, I., & Obruca, S. (2019). Interconnection of waste chicken feather biodegradation and keratinase and mcl-PHA production employing *Pseudomonas putida* KT2440. Applied Food Biotechnology, 6(1), 83–90.

Petter, R., & Tyner, W. E. (2014). Technoeconomic and policy analysis for corn stover biofuels. International Scholarly Research Notices, 2014.

Pieja, A., Schauer-Gimenez, A., Oakenfull, A., & Morse, M. (2016). Biorenewable materials at Mango. In Industrial Biorenewables: A Practical Viewpoint. John Wiley & Sons, Hoboken, NJ.

Pleissner, D., Neu, A. K., Mehlmann, K., Schneider, R., Puerta-Quintero, G. I., & Venus, J. (2016). Fermentative lactic acid production from coffee pulp hydrolysate using *Bacillus coagulans* at laboratory and pilot scales. Bioresource Technology, 218, 167–173.

Radics, R., Dasmohapatra, S., & Kelley, S. (2015). Systematic review of bioenergy perception studies. BioResources, 10(4), 8770–8794.

Rostkowski, K. H., Criddle, C. S., & Lepech, M. D. (2012). Cradle-to-gate life cycle assessment for a cradle-to-cradle cycle: biogas-to-bioplastic (and back). Environmental Science & Technology, 46(18), 9822–9829.

Sammut-Bonnici, T., & Galea, D. (2015). PEST analysis. Wiley Encyclopedia of Management, 12 – Strategic Management, 2312–2319.

Sanaei, S., & Stuart, P. R. (2018). Systematic assessment of triticale-based biorefinery strategies: techno-economic analysis to identify investment opportunities. Biofuels, Bioproducts and Biorefining, 12, S46–S59.

Sangkharak, K., Khaithongkaeo, P., Chuaikhunupakarn, T., Choonut, A., & Prasertsan, P. (2020). The production of polyhydroxyalkanoate from waste cooking oil and its application in biofuel production. Biomass Conversion and Biorefinery, 11, 1651–1664.

Shahzad, K., Kettl, K., Titz, M., Koller, M., Schnitzer, H., & Narodoslawsky, M. (2013). Comparison of ecological footprint for biobased PHA production from animal residues utilizing different energy resources. Clean Technologies and Environmental Policy, 15, 525–536.

Thinagaran, L., & Sudesh, K. (2019). Evaluation of sludge palm oil as feedstock and development of efficient method for its utilization to produce polyhydroxyalkanoate. Waste and Biomass Valorization, 10(3), 709–720.

Wellenreuther, C., & Wolf, A. (2020). Innovative feedstocks in biodegradable bio-based plastics: a literature review. HWWI Research Papers, 194. Hamburg Institute of International Economics, Hamburg, Germany.

Wellenreuther, C., Wolf, A., & Zander, N. (2022). Cost competitiveness of sustainable bioplastic feedstocks—A Monte Carlo analysis for polylactic acid. Cleaner Engineering and Technology, 6.

Yokessa, M., & Marette, S. (2019). A review of eco-labels and their economic impact. International Review of Environmental and Resource Economics, 13(1–2), 119–163.

Zarroli, A. (2020). Assessment of the production of bioplastics from industrial wastewater from fish canning industry. Politecnico di Torino, Torino, Italy.

Zhang, A. Y., Sun, Z., Leung, C. C. J., Han, W., Lau, K. Y., Li, M., & Lin, C. S. K. (2013). Valorisation of bakery waste for succinic acid production. Green Chemistry, 15(3), 690.

13 Techno-Economic Aspects and Life Cycle Assessment of Second- and Third-Generation Bioplastics

Susan Grace Karp, Priscilla Zwiercheczewski de Oliveira, Leonardo Wedderhoff Herrmann, Luiz Alberto Junior Letti, Walter José Martínez-Burgos, Gabriel Rossignol Frassetto, Natália Rodrigues Nitsch, Jéssica Aparecida Viesser, and Carlos Ricardo Soccol

CONTENTS

13.1 INTRODUCTION

Bioplastics represent a group of materials obtained from renewable resources or from oil-based resources that are biodegradable. They can be classified in three categories, namely (1) bio-based and biodegradable, like polylactic acid (PLA), polyhydroxyalkanoates (PHAs), and thermoplastic starch (TPS); (2) bio-based non-biodegradable, like bio-polyethylene (bio-PE) and bio-polyethylene terephthalate (bio-PET); and (3) petroleum-based and biodegradable, like polybutylene succinate (PBS) and polybutylene adipate terephthalate (PBAT; Zimmermann et al. 2020; Van Roijen and Miller 2022).

Among bio-based bioplastics, those obtained from food resources, like sugarcane juice or corn starch, are classified as first-generation (1G) bioplastics. The use of food resources as feedstocks to produce biomaterials brings concerns about the security of food supply. In this scenario emerged the second and third generation (2G and 3G) bioplastics (i.e., those obtained from non-food resources). The 2G bioplastics are those derived from lignocellulosic biomass, including agricultural wastes and wood, and from non-edible vegetable oils, like castor bean oil. The 3G bioplastics are obtained from sugars or oils produced by microorganisms and from municipal wastes (organic wastes and wastewater; Vandenberghe et al. 2021).

The production of 2G and 3G bioplastics is chiefly based on fermentation technologies, that is, the conversion of substrates (usually sugars and oils) into polyesters or their precursors by microorganisms. There is substantial scientific literature reporting the microbial production of, for example,

DOI: 10.1201/9781003344018-13

lactic acid to be further polymerized into polylactic acid and polyhydroxyalkanoates, which are polymerized inside the microbial cell. Most scientific papers on this topic along the last decades are related to laboratory-scale bioprocesses. After optimizing a bioprocess and confirming its feasibility on a laboratory scale, it is necessary to analyze the advantages of the bioprocess developed on a pilot and commercial scale. Through the techno-economic analysis, it is possible to obtain some answers based on the data collected from the performance on a laboratory scale and thus design the bioprocess under analysis for an industrial scale (Wellenreuther et al. 2022).

The techno-economic analysis can be carried out for several purposes, such as mass and energy balances, the definition of the production capacity of an industrial plant, scaling projection and economic aspects such as substrate costs, machinery costs, production costs and the definition of the product price. The main application of this technique is to identify whether the process has a minimum economic index.

There are many economic indices that can be considered, the most common are the internal rate of return (IRR) and the net present value (NPV). Through economic analysis, it is possible to assess the risks associated to the process. Other objectives that may be related to the techno-economic analysis are the identification of the costliest step in the process, the comparison between two technologies known as benchmarking and the definition of the minimum product price (Kwan et al. 2018).

Another fundamental aspect concerning the production of 2G and 3G bioplastics is related to the analysis of sustainability and environmental impacts, representing the bioeconomy metrics that have gained prominence in recent years. Environmental legislations and certificates were developed and improved, and analytical tools have been proposed and validated. One of the most important tools is the life cycle assessment (LCA) of products and processes, which is the most well-established bioeconomy metric to date (Cristóbal et al. 2016).

The main objective of an LCA is the quantitative evaluation of environmental impacts generated by a product, process or service. The first stage of an LCA consists in the definition of the productive chain to be analyzed and its main objectives being, among other elements, the delimitation of the boundaries, definition of the functional units and the choice of environmental indicators to be evaluated. Next, the life cycle inventory defines the interconnections between the subsystems of the chain, as well as all the mass and energy flows. Finally, the life cycle impact assessment defines the classes of environmental impacts to be evaluated, followed by their classification and characterization (Curran 2006).

This chapter will present techno-economic and life cycle assessment data related to the most important 2G and 3G bioplastics, namely polylactic acid, polyhydroxyalkanoates and bioplastics derived from polysaccharides.

13.2 TECHNO-ECONOMIC ASPECTS IN THE PRODUCTION OF SECOND- AND THIRD-GENERATION BIOPLASTICS

13.2.1 Polylactic Acid

This section describes a global techno-economic analysis of polylactic acid (PLA). Considerations and information are found in the literature according to the uniqueness of the process, following simple steps as shown in Figure 13.1.

PLA is the major bioplastic produced around the world and the most researched in the literature. In this way, according to Wellenreuther et al. (2022), PLA exhibits the largest market share and a larger potential growth when compared to other biodegradable polymers. It could be seen in the European Bioplastics (2022) reports, which said that in 2017, 2.05 million tons of bioplastics were produced, while PLA represented a fraction of 10.3%. Furthermore, as the bioplastic market grows, PLA grows at the same pace.

European Bioplastics (2022) expects global production capacities to grow by 36% over the period 2020–2025 and 79% in the period 2018–2030. For the PLA, the production should be increased from 300 kt in 2019 to 328 kt in 2024 by the entry of new producers into this market. Nowadays, the

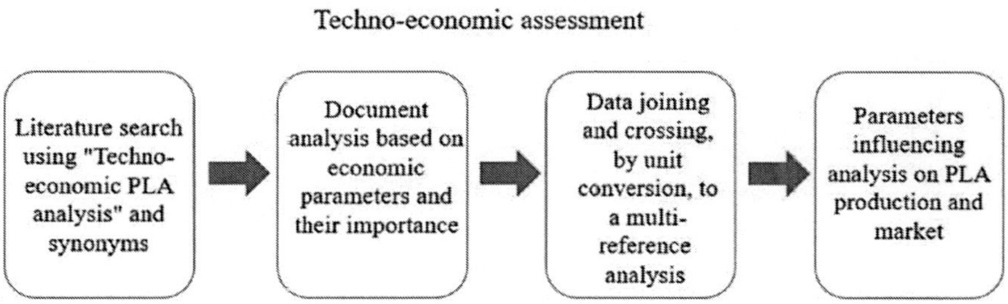

FIGURE 13.1 Schematic process for collecting literature data for a techno-economic analysis.

principal PLA market players are Corbion, NatureWorks LLC, Cargill, and BASF SE (Ratshoshi et al. 2021). These producers, according to Zwiercheczewski de Oliveira et al. (2022), will be installed in North America and Asia-Pacific. However, this growth will only be possible if the PLA price becomes more competitive in the market. Data gathered by Ioannidou et al. (2022) show that PLA costs from USD 1.91 to USD 2.64 per kg, while biaxially oriented polypropylene (BOPP) costs from USD 1.08 to USD 2.00 per kg, which is a large difference for a low-added-value product.

Bioplastics face financial hurdles as sustainable substitutes for conventional plastics, as they impose "premium green" prices on end users compared with conventional plastics derived from abundant, cheap oil (Rastogi and Shrivastava 2017). An alternative for replacing 1G-based bioplastics is the 2G and 3G bioplastics, produced from agroindustry residues, that at the same time can boost the circular economy of the agroindustry. The crescent number of techno-economic studies about PLA production using different substrates, operations and processes, in different industrial scales and markets, is remarkable in the literature. The substrate choice is guided by the social and economic environment where the industrial plant is established, available material and logistics. These aspects are crucial to defining the production organization chart, the pre-treatment method and the purity composition of the PLA end product. For example, production lines that use lignocellulosic wastes had the additional cost of removing lignin from the celluloses (Fahim et al. 2019; Ioannidou et al. 2022; Ratshoshi et al. 2021).

Another variable that is important to analyze is the productive capacity of the industry. A high substrate rate means high production of PLA; however, it also means high fixed and variable costs. Then, the best choice of production volume is not the higher or the lower but the most economically profitable. Table 13.1 shows examples in this regard. Ratshoshi et al. (2021) achieved the best substrate rate (t/h) at more than 20 times the larger number reported by Basticas-Oyanedel and Schmidt (2018), and more than four times higher than Kwan et al. (2018). This happens because sugarcane bagasse and harvesting residues are more widespread and accessible than food waste, which has more complicated logistics.

Furthermore, the techno-economic studies present in the literature allow comparing different economic parameters among PLA production using different feedstocks, production capacities, and rates. First, the total capital investment (TCI) in Table 13.2 remains equivalent to the substrate and the substrate rate previously presented in Table 13.1. It is due not only to the lignocellulosic feedstock providing a superior production but also superior initial investments and operating costs.

Meantime, Ratshoshi et al. (2021) reported that the use of 1G/2G feedstock reduced by 31% the TCI under the same conditions of 2G feedstock (bagasse and harvesting residues). According to the authors, bagasse, waste and molasses were used as feedstock to produce PLA, which resulted in the lowest operational cost of production (OPEX) of USD 1043/t when compared to 2G (USD 1069/t), due to increased production volumes.

A discordant value is observed when comparing the reports of Kwan et al. (2018) and Rajendran and Han (2022). Despite using the same substrate, the fivefold higher substrate rate (t/h) used by Kwan et al. (2018) resulted in an approximately eightfold higher TCI. However, this difference became positive when this data was analyzed together with the minimum selling price (MSP),

TABLE 13.1
Literature Techno-Economic Studies by Authors, Substrate, Substrate Rate (t/h) and Annual Operating Time (h)

Author	Substrate	Substrate Rate (t/h)	Annual Operating Time (h)
Ratshoshi et al. (2021)	2G bagasse, harvesting residues	35.6	5000
	1G/2G feedstock	49.3	5000
Bastidas-Oyanedel and Schmidt (2018)	Food waste	2.09	NE
Rajendran and Han (2022)	Food waste	2.09	7920
Ioannidou et al. (2022)	Corn stover, sugar beet pulp	5.71	7920
Fahim et al. (2019)	Coffee and cotton waste	0.51	NE
Kwan et al. (2018)	Food waste	10	NE

Abbreviation: NE = not evaluated.

TABLE 13.2
Principal Parameters Found in Techno-Economic Studies

Author	Income Tax Rate (%)	TCI[a] (USD t/year)	MSP[b] (USD/t)	ROI[c] (%)	Payback (years)	Obs.
Ratshoshi et al. (2021)	28	5247	7138	NE[d]	NE	
		3613	3920			
Bastidas-Oyanedel and Schmidt (2018)	NE	4148	NE	98	7.8	
Rajendran and Han (2022)	25	838	6530	8.35	11.98	China
		901	5350	12.42	8.05	India
		964	4750	15.12	6.62	Brazil
		946	4290	17.11	5.84	US
Ioannidou et al. (2022)	NE	186	2000	NE	12	
		210	1100		6	
Fahim et al. (2019)	NE	1.82	NE	56.4	NE	Egypt
Kwan et al. (2018)	16.5	6937	3300	17.2	6.6	Hong Kong

Notes: [a] = total capital investment; [b] = minimum selling price; [c] = return over investment; [d] = not evaluated.

which is lower in the report by Kwan et al. (2018). MSP means the commodity price where total revenues are equal to costs (i.e., where the profit is equal to zero). According to Ratshoshi et al. (2021), the market value for this parameter is USD 5140/t, so, among the techno-economic studies analyzed, 3 of 11 were overpriced. It was noted by Ratshoshi et al. (2021) that using 2G substrates not only raised the TCI but also the MSP when compared to 1G/2G substrates. According to the authors, this outcome also implies that a 1G/2G PLA biorefinery annexed to a standard sugar mill could be economically competitive in the market, but this does not occur when using only 2G feedstock. Moreover, MSP depends on where the industrial plant is installed.

Rajendran and Han (2022) evaluated the PLA production in different countries and on different continents and demonstrated that location is highly connected to profitability. In the report, China and India had higher MSP values than the market due to higher costs concerning production and logistics in these countries. In comparison, the US and Brazil had lower MSP prices, making the plants profitable. Brazil had a higher PLA production value than the US, but economically, the US showed more feasibility than Brazil due to factors such as high capital investment and operating costs.

Other parameters (Table 13.2) that are important in a techno-economic report are return over investment (ROI) and payback. ROI is the rate of a cash return without the consideration of cash discount in the plant's lifetime, while payback refers to the time required to recover the investment cost (Kwan et al. 2018). A high ROI and a low payback value are essential to the effectiveness of the process. The higher ROI value (98%) was achieved by Bastidas-Oyanedel and Schmidt (2018), followed by Fahim et al. (2019) at 56.4%; these two reports stand out from the others, which showed values in the range of 8.35%–17.2%. The first report considered the feedstock in tons of volatile solids, and in the second the production was directed to PLA 3D printing, which had the highest market value. Considering the other reports, ROI values were strictly related to supplementary parameters, such as TCI and MSP. According to Rajendran and Han (2022), a high-income tax shows a negative impact on ROI, payback period and IRR, while high PLA selling price shows a positive impact on ROI, payback period and IRR.

Recent techno-economic studies show positive and negative aspects concerning the industrial PLA plant type. Feedstock type and accessibility are also connected to the place where the plant is installed. This factor is central when looking at the TCI and MSP, as seen in the study of Rajendran and Han (2022). According to Wellenreuther et al. (2022), feedstock choice may have a significant influence on PLA production feasibility, not only through its impact on the costs of the primary stage of the life cycle but also due to its technological implications at the refinery stage, in numbers, 40%–60% of OPEX.

Nowadays, PLA is not able to compete in price with petroleum-based plastics, not even in its 1G/2G/3G configuration. In the future, depending on technological and scientific advances, the economic profitability of bioplastics compared to their fossil-based counterparts shall be improved. Studies focusing on PLA production optimization and on the pretreatment of 2G and 3G substrates are essential to reduce OPEX and TCI for bioplastics production.

In conclusion, PLA is the flagship of biotechnological advances in bioplastics. In this way, studies related to its optimization and profitability need to be developed, bringing new production parameters, feedstock, processes, and variables that will help in the establishment of its market. Furthermore, the comparison of the techno-economic studies proved that the profitability of production depends on the substrate (type and rate), place and initial investment.

13.2.2 Polyhydroxyalkanoates

The polyhydroxyalkanoates (PHAs) are a family of linear polyesters, which are produced intracellularly mainly by bacteria as a carbon storage material (Chen et al. 2020). These polymers can be obtained from 1G, 2G and 3G sources, and the greatest interest today is focused on the production of 2G and 3G PHAs aiming at circular economy systems (Vandenberghe et al. 2021). The use of waste (solid or liquid), or of sugars and oils produced by microorganisms as substrates reduces the production costs of the biopolymer, since the cost of the raw material represents at least 40% of the total production cost (Sabbagh and Muhamad 2017).

These polyesters have thermoplastic properties, are renewable and are biodegradable in the natural environment (Wang et al. 2022). Among the potential applications of PHAs are general plastic packaging and packaging for food, pharmaceuticals and medical implants. The properties of PHAs, as well as the policies being adopted around the world regarding easily degradable plastics, are encouraging the use and production of PHAs. Thus, according to Statista (Statista 2022a), the PHA market in the years 2019 and 2020 reached USD 57 million and USD 62 million, respectively, and it is expected that by the year 2025 the trade of this polyester will be USD 121 million.

The main bottleneck in the production and use of PHAs on a large scale is the production costs, since these are around three to four times higher with respect to plastics made from petroleum resins (Sabbagh and Muhamad 2017). For example, the price of a metric ton of polyethylene is USD

1200 (Statista 2022b), while just the cost of producing a ton of PHA from a by-product is USD 4000 (Wang et al. 2022), which is a cost 3.33 times higher than the cost of conventional plastic. In this sense, further research and new technologies are still needed to reduce polymer production costs.

In general, PHA is produced by some bacteria of the genera *Burkholderia, Pseudomonas, Bacillus, Escherichia* and *Halomonas*, which need specific nutrients for their growth. According to Chen et al. (2020), the high productivity of PHA largely depends on the substrates of the culture media, so they must be composed of easily fermentable carbon sources (glucose, fructose and sometimes fatty acids), nitrogen and micronutrients, which adds to the costs of the culture media. An alternative is to use hydrolyzed lignocellulosic biomass or effluents rich in carbohydrates, such as cassava processing wastewater, as carbon sources (Martinez-Burgos et al. 2019), and agro-industrial by-products such as corn steep liquor and whey as sources of nitrogen and micronutrients (Martinez-Burgos et al. 2021; Junior et al. 2021).

On the other hand, when the medium used for the production of PHAs is based on pure reagents, the cost of the medium is extremely high and economically unfeasible. For example, Ashby et al. (2022) achieved a yield of 2.6 g PHA/L using the *Burkholderia sacchari* strain and a mixture of carbon sources (glucose, xylose, galactose and arabinose) and salts for medium supplementation (Na_2HPO_4, KH_2PO_4, $MgSO_4.7H_2O$, $CuSO_4.5H_2O$, $NiCl_2.6H_2O$, $CaCl_2.2H_2O$, $(NH_4)_2SO_4$, H_3BO_3, $CoCl_2$, $ZnSO_4.7H_2O$, $MnCl_2.4H_2O$ and $Na_2MoO_4.2H_2O$), however, the cost of the medium was extremely high at approximately USD 600/m^3. Salgaonkar and Bragança (2017) reached a maximum concentration of 1.9 g PHA/L using hydrolyzed lignocellulosic biomass (sugarcane bagasse) as a carbon source. However, the medium was supplemented with yeast extract and salts (KH_2PO_4, $MgCl_2.6H_2O$, $CaCl_2.2H_2O$, $NaHCO_3$, NH_4Cl, $FeCl_3.6H_2O$, NaCl and KCl), so the cost of the medium increased sharply (USD 684/m^3). In contrast, a yield of 2.96 g PHA/L was achieved using a medium composed of rice straw hydrolysate, meat extract and peptone, and the cost of the medium was approximately USD 50/m^3; despite having a low value with respect to media prepared only with pure reagents, this still remains a high cost for the type of polymers being produced (Van Thuoc et al. 2021). Table 13.3 shows the costs of some media used for the production of PHAs.

TABLE 13.3
Costs of Some Pure Reagent-Based Media Used in the Production of Polyhydroxyalkanoates

Carbon Source	Microorganism	Cost/m³ (USD)	Yield (kg PHA/m³)	Cost/kg PHA (USD)	Reference
Palm oil, fructose	*Cupriavidus necator* B-10646	1086	5.12	212	(Volova et al. 2020)
Tamanu oil	*C. necator*	1153	10.6	108	(Arumugam et al. 2018)
Xylose	*Schlegelella thermodepolymerans*	1047	2.85	367	(Kourilova et al. 2020)
Brewer's spent grain hydrolysates	*Burkholderia cepacia, Bacillus cereus* and *C. necator*	624	1.19	525	(Corchado-Lopo et al. 2021)
Glycerol, octanoate	*Pseudomonas putida* KT2440	1340	8.47	158	(Li et al. 2021)
Glucose	*Paracoccus* sp. LL1	1312	44.2	29.7	(Khomlaem et al. 2021)
Waste fish oil, glycerol	*Salinivibrio* sp. M318	1050	35.6	29.5	(Van Thuoc et al. 2019)
Methanol	*Methylobacterium organophilum*	593	130	4.56	(Kim et al. 1996)

Note: All media contained a mixture of salts as supplement.

13.2.3 Bioplastics Derived from Polysaccharides

Starch and cellulose are the main polysaccharides derived from 1G, 2G and 3G sources that are used for the industrial production of bioplastics. In addition, they are often used to produce intermediates for blending with other plastics (Nandakumar et al. 2021). Starch-based bioplastics make up around 50% of the global bioplastics market and constitute many of the thermoplastics in use today (Kaur et al. 2017).

Polymers that contain native or modified starch moieties fall under the category of starch-based polymers (Nandakumar et al. 2021). Several raw materials such as raw starch, modified starch (e.g., thermoplastic starch), and other starch-derived sugars can be employed for the manufacturing of bioplastics (Kaur et al. 2017). Starch is a natural polymer produced by many plant species during photosynthesis and can be found in roots, pulps, seeds and tubers of plants in large concentration (Alrefai et al. 2020). The main sources of starch are cassava, maize, wheat and potatoes (Gadhave et al. 2018). Starch is also present in the biomass of several microalgal species. The life cycle of starch-based 2G and 3G bioplastics involves obtaining the starch source (which can be an agricultural residue or microalgal biomass, for example); extracting the starch; refining and processing the starch; manufacturing the final product; delivering the product to the user; and the final disposal and biodegradation of the product.

Industrially, starch-based bioplastics are often blended with biodegradable polyesters, including PLA, PBAT, PBS, and PHAs (Kadri et al. 2017; Gadhave et al. 2018). According to Nandakumar et al. (2021), starch is considered a good filler for conventional plastics since it is biodegradable and has adequate thermal stability so as to not interfere with the melt flow processing of the polymer. Starch has proved its significant efficiency in many bioplastic applications as it is cheap, easy to extract, abundant, biodegradable and renewable. Starch-based plastic is one of the most common bioplastic types used in packaging applications (Alrefai et al. 2020).

Starch can be utilized in materials in either its native, granular form or after heat processing with a plasticizer into destructurized or thermoplastic starch (TPS). TPS has also been the target of much fundamental research as well as commercialization (Shogren et al. 2019). The process of producing TPS is relatively simple: it does not require long processing times or even several complicated chemical reactions (Alrefai et al. 2020). Traditional processing technologies to manufacture TPS include film casting, compression molding, injection molding and extrusion (RameshKumar et al. 2020; Alrefai et al. 2020). Some producers and processors of TPS bioplastics are Rodenburg Biopolymers of Oosterhut, in the Netherlands, that transforms starch from potato processing waste into bio-based and biodegradable blends with an annual capacity of 130 million pounds per year; and BIOTEC, in Germany, that produces a line of biodegradable starch-polymer blends and has a capacity of 55 million pounds per year. According to Allied Market Research, the global starch-based bioplastics market was estimated to be USD 424 million in 2016 (Shogren et al. 2019).

Bioplastics may also be derived from lignocellulosic biomass such as agro-industrial residues and forestry biomass (Brodin et al. 2017). Cellulose-based bioplastics are derived from cellulose, the most abundant organic compound and the main constituent in plant fibers (Di Bartolo et al. 2021; Kaur et al. 2017). Cellulose-based bioplastics are made from chemically modified plant cellulose such as cellulose acetate (CA; Kaur et al. 2017). Cellulose acetate is a strong and tough material, but its cost is substantially higher than that of commodity polymers. Global production of cellulose esters is around 2 billion pounds per year (Shogren et al. 2019).

The production of bioplastics as a main product from forestry resources is presently not economically viable (Brodin et al. 2017). Cellulose is typically obtained from wood through a pulping process and can be converted to different materials, like cellulose-based plastics that consist of regenerated cellulose and cellulose diacetates (Di Bartolo et al. 2021). The most common methods for cellulose extraction are the kraft method, the caustic soda processing and the sulfite method. All three processes employ a mixture of wood chips with chemicals ranging from sodium hydroxide to sodium sulfite to cleave the bonds between hemicelluloses, cellulose and lignin. However, these processes, especially the kraft process, have been considered controversial due to the environmental impacts, odorous substances and large amounts of liquid waste produced as by-products (Nandakumar et al. 2021). On the other side, Kim et al. (2020) proposed a new approach for the production of 2,5-furandicarboxylic acid (FDCA) as an eco-friendly bio-based plastic monomer, obtained from cellulose via hydroxymethyl furfural (HMF); and 1,5-pentaindiol (1,5-PDO) as a high-value product for the polymer industry, obtained from C5 sugars. This highlights the importance of exploring cellulose and hemicellulose derivatives as precursors of bio-based materials.

13.3 LIFE CYCLE ASSESSMENT OF SECOND- AND THIRD-GENERATION BIOPLASTICS

Regarding bioplastics, there is an intense effort to transform their production to be as sustainable and ecologically friendly as possible. Currently, plastics are among the petroleum-based products with major urgency of being replaced by greener alternatives, due to their large environmental hazard and importance for human use. Approaches such as circular bioeconomy, biorefinery and the life cycle assessment (LCA) studies can provide an overview of the whole production process and how each step can be destined to a more sustainable direction (Talan et al. 2022; Yadav et al. 2020; Vandenberghe et al. 2021). The LCA of bioplastics can also be compared to that of fossil-based plastics in order to verify their ecological performance and economical or energetic costs, wherein the main argument for bioplastic production is the environmental advantage. However, it is important to take into account that LCA studies may diverge in terms of which parameters were used and the interpretation of them, such as global warming potential, acidification potential, carbon footprint or water footprint (Narodoslawsky et al. 2015; Klai et al. 2021). A representation of the LCA of 2G and 3G bioplastics can be seen in Figure 13.2.

The initial evaluations of LCA performed for analyzing PHA production processes were not so positive, especially due to the early stages of developing the technology requiring several

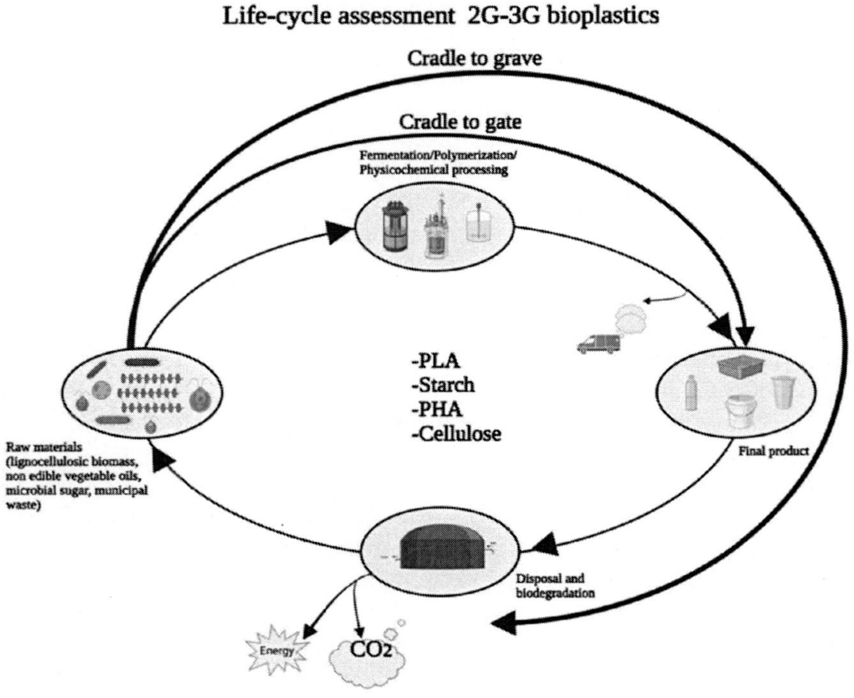

FIGURE 13.2 Representation of the life cycle assessment of 2G and 3G bioplastics.

investments with low productivity (Patel et al. 2005; Narodoslawsky et al. 2015). For instance, in the study of Patel et al. (2005), a comparison of 20 cradle-to-gate studies showed some cases with higher total energy requirement and carbon dioxide emission than fossil-based polyethylene. Another study performed by Tabone et al. (2010) indicated that the majority of the environmental impacts of bio-based plastics were associated with agricultural efforts to produce the raw materials. In their work, they compared two different production methods for PLA and two for PHA with fossil-based polypropylenes, polyethylenes and polycarbonates, reaching equal or even worse global warming potentials, acidification and ozone depletion potentials of the grain-derived biopolymers compared to the petroleum-based ones. However, when comparing corn grain–derived (1G) PHA with corn stover–derived (2G) PHA, the latter always demonstrated a better or equal environmental performance, with a negative global warming potential.

As the technology matures and evolves, more steps of biopolymer production are optimized, and more precise are the LCA results. Besides, the assessments can be refined by increasing the number of factors evaluated, such as adding all energetic and material inputs, all emissions and residues, and providing perspectives of the use, recycle and disposal of the product (cradle-to-grave approach; Harding et al. 2007; Vea et al. 2021; Moretti et al. 2021). The study of Harding et al. (2007) presented a cradle-to-grave analysis with the addition of carbon dioxide generation and other environmental issues, and compared the PHB bioplastic with polypropylene and polyethylene. PHB showed lower environmental impacts than both fossil plastics in terms of global warming, abiotic and ozone depletion, human, aquatic and terrestrial toxicity, being less advantageous than polyethylene only in terms of acidification and eutrophication. A study conducted by Vea et al. (2021) also added the methane production, land use, carcinogenicity, fine particulate matter formation and other factors to the LCA of PHA plastics. The researchers found that, although PHA presented greener environmental impacts compared to polypropylene and polyethylene, the results were highly dependent on the PHA yield, especially as the yield decreased from pilot to industrial scale; the sugar beet molasses used for its production significantly contributed to environmental impacts.

As for the PLA, it is interesting to consider that this bioplastic is one of the few, if not the only, which has been produced in large scale for several years (Groot and Borén 2010; Tabone et al. 2010). The LCA of PLA is similar to the one of PHA, wherein the agricultural phase and the yield of fermentation represent significant steps for the environmental impact. Groot and Borén (2010), for instance, developed a LCA based on the sugarcane production in Thailand with bagasse used to generate energy, evaluating both PLA and lactic acid production separately, and providing perspectives in terms of global warming potential, carbon dioxide emission and energy consumption. They also concluded that agriculture was a hotspot for acidification, photochemical ozone creation, eutrophication and farmland use, although emissions of greenhouse gases, use of material resources and non-renewable energy consumption were lower than those of fossil-based polymers. Morão and Bie (2019) found equal results evaluating a cradle-to-gate LCA with 16 environmental impact categories, pointing the agriculture as the major contributor of the same factors as Groot and Borén with addition of water consumption and particulate matter generation, and highlighting the chemical usage in manufacture. Ghomi et al. (2021) presented similar results, mentioning that the PLA conversion during manufacture was the most energy intensive stage, and most of the products end up in landfills and composting. For polylactide, a slightly different polymer derived from lactic acid as well, almost identical results were found (Moretti et al. 2021), indicating that the energy consumption for conversion and thermoforming polylactide cups was critical, coupled with the climate changes of agriculture. These data highlight the importance of developing 2G and 3G technologies for bioplastics production, in order to minimize the environmental impacts associated to the agricultural activities.

13.4 SUMMARY AND CONCLUSIONS

This chapter presented techno-economic and life cycle assessment data related to the most important 2G and 3G bioplastics, specifically PLA, PHAs and bioplastics derived from polysaccharides.

PLA, which exhibits the largest market share and a larger potential growth among biodegradable polymers, has a reported production cost of USD 1.91 to USD 2.64 per kg. Other reported economic indices were income tax rates (16.5%–28%), TCI (USD 1.82–6937 per t/year), MSP (USD 1100–7138/t), ROI (8.35%–98%) and payback time (5.84–11.98 years). According to the LCA of PLA, the agricultural phase and the fermentation represent significant steps that influence on the environmental impact, demonstrating the importance of developing 2G and 3G technologies for PLA production.

PHAs represent another green alternative for plastics still under scientific development worldwide. The main bottleneck in the production and use of PHAs on a large scale is the production costs, since these are around three to four times higher with respect to plastics prepared from petroleum resins. Reported production cost values ranged from USD 593 to USD 1340/m^3. Although the initial evaluations of LCA performed for PHA production processes were not so positive, the comparison between 1G and 2G technologies demonstrated significant environmental advantages of the 2G technology.

Finally, the techno-economic and environmental perspectives for the bioplastics derived from polysaccharides such as starch and cellulose are promising since these are abundant, cheap and worldwide available substrates, though there are still challenges related to their expansion in the market.

REFERENCES

Alrefai, Raid, Alla M. Alrefai, Khaled Benyounis, and Joseph Stokes. 2020. "A Review on the Production of Thermo-Plastic Starch from the Wastes of Starchy Fruits and Vegetables." *Encyclopedia of Materials: Plastics and Polymers* 3: 89–104. doi:10.1016/b978-0-12-820352-1.00054-7.

Arumugam, A., S. G. Senthamizhan, V. Ponnusami, and S. Sudalai. 2018. "Production and Optimization of Polyhydroxyalkanoates from Non-Edible *Calophyllum inophyllum* Oil Using *Cupriavidus necator*." *International Journal of Biological Macromolecules* 112: 598–607. doi:10.1016/j.ijbiomac.2018.02.012.

Ashby, Richard D., Nasib Qureshi, Gary D. Strahan, David B. Johnston, Joseph Msanne, and Xiaoqing Lin. 2022. "Biocatalysis and Agricultural Biotechnology Corn Stover Hydrolysate and Levulinic Acid: Mixed Substrates for Short-Chain Polyhydroxyalkanoate Production." *Biocatalysis and Agricultural Biotechnology* 43 (100): 102391. doi:10.1016/j.bcab.2022.102391.

Bastidas-Oyanedel, Juan Rodrigo, and Jens Ejbye Schmidt. 2018. "Increasing Profits in Food Waste Biorefinery—A Techno-Economic Analysis." *Energies* 11 (6): 1551. doi:10.3390/EN11061551.

Brodin, Malin, María Vallejos, Mihaela Tanase Opedal, María Cristina Area, and Gary Chinga-Carrasco. 2017. "Lignocellulosics as Sustainable Resources for Production of Bioplastics—A Review." *Journal of Cleaner Production* 162: 646–64. doi:10.1016/j.jclepro.2017.05.209.

Chen, Guo Qiang, Xin Yu Chen, Fu Qing Wu, and Jin Chun Chen. 2020. "Polyhydroxyalkanoates (PHA) toward Cost Competitiveness and Functionality." *Advanced Industrial and Engineering Polymer Research* 3 (1): 1–7. doi:10.1016/j.aiepr.2019.11.001.

Corchado-Lopo, Carlos, Oscar Martínez-Avila, Elisabet Marti, Jordi Llimós, Anna María Busquets, Dan Kucera, Stanislav Obruca, Laia Llenas, and Sergio Ponsá. 2021. "Brewer's Spent Grain as a No-Cost Substrate for Polyhydroxyalkanoates Production: Assessment of Pretreatment Strategies and Different Bacterial Strains." *New Biotechnology* 62: 60–67. doi:10.1016/j.nbt.2021.01.009.

Cristóbal, Jorge, Cristina T. Matos, Jean Philippe Aurambout, Simone Manfredi, and Boyan Kavalov. 2016. "Environmental Sustainability Assessment of Bioeconomy Value Chains." *Biomass and Bioenergy* 89 (June): 159–71. doi:10.1016/J.BIOMBIOE.2016.02.002.

Curran, M. A. 2006. *Life-Cycle Assessment: Principles and Practice*. Scientific Applications International Corporation (SAIC). https://onlinebooks.library.upenn.edu/webbin/book/lookupid?key=ha102328785.

Di Bartolo, Alberto, Giulia Infurna, and Nadka Tzankova Dintcheva. 2021. "A Review of Bioplastics and Their Adoption in the Circular Economy." *Polymers* 13 (8): doi:10.3390/polym13081229.

European Bioplastics. 2022. "European Bioplastics." www.european-bioplastics.org/.

Fahim, I. S., H. Chbib, and Hamada Mohamed Mahmoud. 2019. "The Synthesis, Production & Economic Feasibility of Manufacturing PLA from Agricultural Waste." *Sustainable Chemistry and Pharmacy* 12 (June): 100142. doi:10.1016/J.SCP.2019.100142.

Gadhave, Ravindra V., Abhijit Das, Prakash A. Mahanwar, and Pradeep T. Gadekar. 2018. "Starch Based Bio-Plastics: The Future of Sustainable Packaging." *Open Journal of Polymer Chemistry* 8 (2): 21–33. doi:10.4236/ojpchem.2018.82003.

Ghomi, Erfan Rezvani, Fatemeh Khosravi, Ali Saedi Ardahaei, Yunqian Dai, Rasoul Esmaeely Neisiany, Firoozeh Foroughi, Min Wu, Oisik Das, and Seeram Ramakrishna. 2021. "The Life Cycle Assessment for Polylactic Acid (PLA) to Make It a Low-Carbon Material." *Polymers* 13 (11): 1854. doi:10.3390/POLYM13111854.

Groot, Wim J., and Tobias Borén. 2010. "Life Cycle Assessment of the Manufacture of Lactide and PLA Biopolymers from Sugarcane in Thailand." *International Journal of Life Cycle Assessment* 15 (9): 970–84. doi:10.1007/S11367-010-0225-Y/FIGURES/11.

Harding, K. G., J. S. Dennis, H. von Blottnitz, and S. T. L. Harrison. 2007. "Environmental Analysis of Plastic Production Processes: Comparing Petroleum-Based Polypropylene and Polyethylene with Biologically-Based Poly-β-Hydroxybutyric Acid Using Life Cycle Analysis." *Journal of Biotechnology* 130 (1): 57–66. doi:10.1016/J.JBIOTEC.2007.02.012.

Ioannidou, Sofia Maria, Dimitrios Ladakis, Eleni Moutousidi, Endrit Dheskali, Ioannis K. Kookos, Iana Câmara-Salim, María Teresa Moreira, and Apostolis Koutinas. 2022. "Techno-Economic Risk Assessment, Life Cycle Analysis and Life Cycle Costing for Poly(Butylene Succinate) and Poly(Lactic Acid) Production Using Renewable Resources." *Science of the Total Environment* 806 (February): 150594. doi:10.1016/J.SCITOTENV.2021.150594.

Junior, Jair Rosaario do Nascimento, Luis Alberto Zevallos Torres, Adriane Bianchi Pedroni Medeiros, Adenise Lorenci Woiciechowski, Walter José Martínez-Burgos, and Carlos Ricardo Soccol. 2021. "Enhancement of Biohydrogen Production in Industrial Wastewaters with Vinasse Pond Consortium Using Lignin-Mediated Iron Nanoparticles." *International Journal of Hydrogen Energy* 6: 27431–43. doi:10.1016/j.ijhydene.2021.06.009.

Kadri, Tayssir, Tarek Rouissi, Satinder Brar Kaur, Maximiliano Cledon, Saurabhjyoti Sarma, and Mausam Verma. 2017. "Biodegradation of Polycyclic Aromatic Hydrocarbons (PAHs) by Fungal Enzymes: A Review." *Journal of Environmental Sciences* 51: 52–74. doi:10.1016/j.jes.2016.08.023.

Kaur, Loveleen, Robinka Khajuria, Leena Parihar, and G. Dimpal Singh. 2017. "Polyhydroxyalkanoates: Biosynthesis to Commercial Production—A Review." *Journal of Microbiology, Biotechnology and Food Sciences* 6 (4): 1098–1106. doi:10.15414/jmbfs.2017.6.4.1098-1106.

Khomlaem, Chanin, Hajer Aloui, Won Gyun Oh, and Beom Soo Kim. 2021. "High Cell Density Culture of *Paracoccus* sp. LL1 in Membrane Bioreactor for Enhanced Co-Production of Polyhydroxyalkanoates and Astaxanthin." *International Journal of Biological Macromolecules* 192: 289–97. doi:10.1016/j.ijbiomac.2021.09.180.

Kim, Hyunwoo, Shinje Lee, Yuchan Ahn, Jinwon Lee, and Wangyun Won. 2020. "Sustainable Production of Bioplastics from Lignocellulosic Biomass: Technoeconomic Analysis and Life-Cycle Assessment." *ACS Sustainable Chemistry and Engineering* 8 (33): 12419–29. doi:10.1021/acssuschemeng.0c02872.

Kim, Seon Won, Pil Kim, Hyun S. Lee, and Jung H. Kim. 1996. "High Production of Poly-β-Hydroxybutyrate (PHB) from *Methylobacterium organophilum* under Potassium Limitation." *Biotechnology Letters* 18 (1): 25–30. doi:10.1007/BF00137805.

Klai, Nouha, Bhoomika Yadav, Oumaima El Hachimi, Aishwarya Pandey, Balasubramanian Sellamuthu, and Rajeshwar Dayal Tyagi. 2021. "Agro-Industrial Waste Valorization for Biopolymer Production and Life-Cycle Assessment toward Circular Bioeconomy." In *Biomass, Biofuels, Biochemicals*, edited by Ashok Pandey, Rajeshwar Dayal Tyagi, and Sunita Varjani, 515–55. Elsevier. doi:10.1016/B978-0-12-821878-5.00007-6.

Kourilova, Xenie, Iva Pernicova, Karel Sedlar, Jana Musilova, Petr Sedlacek, Michal Kalina, Martin Koller, and Stanislav Obruca. 2020. "Production of Polyhydroxyalkanoates (PHA) by a Thermophilic Strain of *Schlegelella thermodepolymerans* from Xylose Rich Substrates." *Bioresource Technology* 315: 123885. doi:10.1016/j.biortech.2020.123885.

Kwan, Tsz Him, Yunzi Hu, and Carol Sze Ki Lin. 2018. "Techno-Economic Analysis of a Food Waste Valorisation Process for Lactic Acid, Lactide and Poly(Lactic Acid) Production." *Journal of Cleaner Production* 181 (April): 72–87. doi:10.1016/J.JCLEPRO.2018.01.179.

Li, Ying, Songyuan Yang, Dayao Jin, and Xiaoqiang Jia. 2021. "Optimization of Medium-Chain-Length Polyhydroxyalkanoate Production by *Pseudomonas putida* KT2440 from Co-Metabolism of Glycerol and Octanoate." *Canadian Journal of Chemical Engineering* 99 (3): 657–66. doi:10.1002/cjce.23894.

Martinez-Burgos, Walter Jose, Eduardo Bittencourt Sydney, Satinder Kaur Brar, Valcineide Oliveira de Andrade Tanobe, Adriane Bianchi Pedroni Medeiros, Julio Cesar Carvalho, and Carlos Ricardo Soccol. 2019. "The Effect of Hydrolysis and Sterilization in Biohydrogen Production from Cassava Processing

Wastewater Medium Using Anaerobic Bacterial Consortia." *International Journal of Hydrogen Energy* 44: 25551–64. doi:10.1016/j.ijhydene.2019.08.085.

Martinez-Burgos, Walter José, Eduardo Bittencourt Sydney, Dieggo Rodrigues de Paula, Adriane Bianchi Pedroni Medeiros, Júlio Cesar de Carvalho, Denisse Molina, and Carlos Ricardo Soccol. 2021. "Hydrogen Production by Dark Fermentation Using a New Low-Cost Culture Medium Composed of Corn Steep Liquor and Cassava Processing Water: Process Optimization and Scale-Up." *Bioresource Technology* 320: 124370. doi:10.1016/j.biortech.2020.124370.

Morão, Ana, and François de Bie. 2019. "Life Cycle Impact Assessment of Polylactic Acid (PLA) Produced from Sugarcane in Thailand." *Journal of Polymers and the Environment* 27 (11): 2523–39. doi:10.1007/S10924-019-01525-9/TABLES/7.

Moretti, Christian, Lorie Hamelin, Line Geest Jakobsen, Martin H. Junginger, Maria Magnea Steingrimsdottir, Linda Høibye, and Li Shen. 2021. "Cradle-to-Grave Life Cycle Assessment of Single-Use Cups Made from PLA, PP and PET." *Resources, Conservation and Recycling* 169 (June): 105508. doi:10.1016/J.RESCONREC.2021.105508.

Nandakumar, Ardra, Jo Ann Chuah, and Kumar Sudesh. 2021. "Bioplastics: A Boon or Bane?" *Renewable and Sustainable Energy Reviews* 147 (August 2020): 111237. doi:10.1016/j.rser.2021.111237.

Narodoslawsky, M., K. Shazad, R. Kollmann, and H. Schnitzer. 2015. "LCA of PHA Production—Identifying the Ecological Potential of Bio-Plastic." *Chemical and Biochemical Engineering Quarterly* 29 (2): 299–305. doi:10.15255/CABEQ.2014.2262.

Patel, Martin, Catia Bastioli, Luigi Marini, and Eduard Würdinger. 2005. "Life-Cycle Assessment of Bio-Based Polymers and Natural Fiber Composites." *Biopolymers Online*. doi:10.1002/3527600035.BPOLA014.

Rajendran, Naveenkumar, and Jeehoon Han. 2022. "Integrated Polylactic Acid and Biodiesel Production from Food Waste: Process Synthesis and Economics." *Bioresource Technology* 343 (January): 126119. doi:10.1016/J.BIORTECH.2021.126119.

RameshKumar, Saranya, P. Shaiju, Kevin E. O'Connor, and Ramesh Babu P. 2020. "Bio-Based and Biodegradable Polymers—State-of-the-Art, Challenges and Emerging Trends." *Current Opinion in Green and Sustainable Chemistry* 21: 75–81. doi:10.1016/j.cogsc.2019.12.005.

Rastogi, Meenal, and Smriti Shrivastava. 2017. "Recent Advances in Second Generation Bioethanol Production: An Insight to Pretreatment, Saccharification and Fermentation Processes." *Renewable and Sustainable Energy Reviews* 80 (May): 330–40. doi:10.1016/j.rser.2017.05.225.

Ratshoshi, Brankie Karabo, Somayeh Farzad, and Johann F. Görgens. 2021. "Techno-Economic Assessment of Polylactic Acid and Polybutylene Succinate Production in an Integrated Sugarcane Biorefinery." *Biofuels, Bioproducts and Biorefining* 15 (6): 1871–87. doi:10.1002/BBB.2287.

Sabbagh, Farzaneh, and Ida Idayu Muhamad. 2017. "Production of Poly-Hydroxyalkanoate as Secondary Metabolite with Main Focus on Sustainable Energy." *Renewable and Sustainable Energy Reviews* 72: 95–104. doi:10.1016/j.rser.2016.11.012.

Salgaonkar, Bhakti, and Judith Bragança. 2017. "Utilization of Sugarcane Bagasse by *Halogeometricum borinquense* Strain E3 for Biosynthesis of Poly (3-Hydroxybutyrate- Co -3-Hydroxyvalerate)." *Bioengineering* 4: 1–18. doi:10.3390/bioengineering4020050.

Shogren, Randal, Delilah Wood, William Orts, and Gregory Glenn. 2019. "Plant-Based Materials and Transitioning to a Circular Economy." *Sustainable Production and Consumption* 19: 194–215. doi:10.1016/j.spc.2019.04.007.

Statista. 2022a. "Market Value of Polyhydroxyalkanoate (PHA) Worldwide in 2019 and 2020, with a Forecast for 2025." no. 8.5.2017: 2003–5. Retrieved from: https://www.statista.com/statistics/1010383/global-polyhydroxyalkanoate-market-size/ (August 20, 2022).

Statista. 2022b. "Price of High-Density Polyethylene Worldwide from 2017 to 2022." no. 8.5.2017: 2003–5. Retrieved from: https://www.statista.com/statistics/1171074/price-high-density- polyethylene-forecast-globally/#:~:text=in%202020%2c%20the%20average%20global,u.s.%20dollars%20per%20metric.

Tabone, Michaelangelo D., James J. Cregg, Eric J. Beckman, and Amy E. Landis. 2010. "Sustainability Metrics: Life Cycle Assessment and Green Design in Polymers." *Environmental Science and Technology* 44 (21): 8264–69. doi:10.1021/es101640n.

Talan, Anita, Sameer Pokhrel, R. D. Tyagi, and Patrick Drogui. 2022. "Biorefinery Strategies for Microbial Bioplastics Production: Sustainable Pathway towards Circular Bioeconomy." *Bioresource Technology Reports* 17: 100875. doi:10.1016/j.biteb.2021.100875.

Vandenberghe, Luciana Porto de Souza, Priscilla Zwiercheczewski de Oliveira, Gustavo Amaro Bittencourt, Ariane Fátima Murawski de Mello, Zulma Sarmiento Vásquez, Susan Grace Karp, and Carlos Ricardo Soccol. 2021. "The 2G and 3G Bioplastics: An Overview." *Biotechnology Research and Innovation* 5 (1): e2021004. doi:10.4322/biori.202104.

Van Roijen, Elisabeth C., and Sabbie A. Miller. 2022. "A Review of Bioplastics at End-of-Life: Linking Experimental Biodegradation Studies and Life Cycle Impact Assessments." *Resources, Conservation and Recycling* 181 (June): 106236. doi:10.1016/J.RESCONREC.2022.106236.

Van Thuoc, Doan, Nguyen Thi Chung, and Rajni Hatti Kaul. 2021. "Polyhydroxyalkanoate Production from Rice Straw Hydrolysate Obtained by Alkaline Pretreatment and Enzymatic Hydrolysis Using *Bacillus* Strains Isolated from Decomposing Straw." *Bioresources and Bioprocessing* 8 (98): 1–11. doi:10.1186/s40643-021-00454-7.

Van Thuoc, Doan, Dam Ngoc My, Tran Thi Loan, and Kumar Sudesh. 2019. "Utilization of Waste Fish Oil and Glycerol as Carbon Sources for Polyhydroxyalkanoate Production by *Salinivibrio* sp. M318." *International Journal of Biological Macromolecules* 141: 885–92. doi:10.1016/j.ijbiomac.2019.09.063.

Vea, Eldbjørg Blikra, Serena Fabbri, Sebastian Spierling, and Mikołaj Owsianiak. 2021. "Inclusion of Multiple Climate Tipping as a New Impact Category in Life Cycle Assessment of Polyhydroxyalkanoate (PHA)-Based Plastics." *Science of the Total Environment* 788 (September): 147544. doi:10.1016/J.SCITOTENV.2021.147544.

Volova, Tatiana, Kristina Sapozhnikova, and Natalia Zhila. 2020. "*Cupriavidus necator* B-10646 Growth and Polyhydroxyalkanoates Production on Different Plant Oils." *International Journal of Biological Macromolecules* 164: 121–30. doi:10.1016/j.ijbiomac.2020.07.095.

Wang, Ke, Alex Michael Hobby, Yike Chen, Allan Chio, Bryan Martin Jenkins, and Ruihong Zhang. 2022. "Techno-Economic Analysis on an Industrial-Scale Production." *Processes* 10: 17.

Wellenreuther, Claudia, André Wolf, and Nils Zander. 2022. "Cost Competitiveness of Sustainable Bioplastic Feedstocks—A Monte Carlo Analysis for Polylactic Acid." *Cleaner Engineering and Technology* 6 (February): 100411. doi:10.1016/J.CLET.2022.100411.

Yadav, Bhoomika, Aishwarya Pandey, Lalit R. Kumar, and R. D. Tyagi. 2020. "Bioconversion of Waste (Water)/Residues to Bioplastics—A Circular Bioeconomy Approach." *Bioresource Technology* 298: 122584. doi:10.1016/j.biortech.2019.122584.

Zimmermann, Lisa, Andrea Dombrowski, Carolin Völker, and Martin Wagner. 2020. "Are Bioplastics and Plant-Based Materials Safer than Conventional Plastics? In Vitro Toxicity and Chemical Composition." *Environment International* 145 (December): 106066. doi:10.1016/J.ENVINT.2020.106066.

Zwiercheczewski de Oliveira, Priscilla, Luciana Porto de Souza Vandenberghe, Ariane Fátima Murawski de Mello, and Carlos Ricardo Soccol. 2022. "A Concise Update on Major Poly-Lactic Acid Bioprocessing Barriers." *Bioresource Technology Reports* 18 (June): 101094. doi:10.1016/J.BITEB.2022.101094.

14 Emerging Technologies, Recent Developed Processes, Patents, and Innovation About Second- and Third-Generation Bioplastics

Ariane Fátima Murawski de Mello,
Luciana Porto de Souza Vandenberghe,
Clara Matte Borges Machado,
Manuela Mendonça Cardozo Ribeiro Ravazoli,
Gustavo Amaro Bittencourt, Leonardo Wedderhoff Herrmann,
Priscilla Zwiercheczewski de Oliveira, and Carlos Ricardo Soccol

CONTENTS

14.1 INTRODUCTION

Plastics are versatile materials applied in multiple areas, varying from daily-use disposables to materials for construction, automobile and medical use. However, they carry an environmental burden due to the lack of degradability coupled to a poor disposal system. On the other hand, bioplastics are long-term sustainable alternatives that usually present an equal performance to conventional petroleum-based plastics. Among various definitions, bioplastics are here described as polymers that are biodegradable while also possessing a biological origin (Mehta et al., 2020). As well, their biocompatibility allows versatile industrial applications, mainly in human needs areas such as medicine, biomedicine, pharmacy and agriculture. The most common bioplastics are polylactic acid (PLA) and polyhydroxyalkanoates (PHAs). The feedstock for bioplastics synthesis can be used to classify them: first-generation (1G) substrates comprise food crops (e.g., wheat, corn and rice), second-generation (2G) feedstocks are non-edible materials like lignocellulosic wastes and third-generation (3G) feedstocks relate to microalgal biomass and/or wastewater (Pandit et al., 2021).

DOI: 10.1201/9781003344018-14

TABLE 14.1
Major Bioplastic Producers

Company	Country	Bioplastic	Production Capacity (kton/year)
Newlight Technologies	US	PHA	23
Danimer Scientific	US	PHA	13.6
Tianjin GreenBio Material	China	PHA	10
Bio-On	Italy	PHA	10
Kaneka	Japan	PHA	1–3.5
PHB Industrial	Brazil	PHA	0.5–3
NatureWorks	US	PLA	150
Total Corbion PLA JV	Thailand	PLA	75
Hisun China 45	China	PLA	45
BBCA & Galactic	China	PLA	4

Source: Adapted from Kourmentza et al. (2017), Vandi et al. (2018) and Jem and Tan (2020).

The market of bioplastics has grown in recent years as the concern about plastic waste in the environment has also increased. While 2.42 million tons of bioplastics were produced in 2021, it is expected that 7.59 million tons will be produced by 2026. In 2021, PLA held almost 19% of this production, while PHA had a share of 1.8%. However, the PHA share is expected to grow to 6.4% until 2026, while PLA is expected to have 10.4% of market share (European Bioplastics, 2022a). The main producers of PHAs and PLA with their production capacities are presented in Table 14.1.

Bioplastics production processes involve multiple stages until their commercialization, generally consisting of substrate processing, production by microorganisms, recovery and polymerization when necessary, each one having their bottlenecks which are being surpassed by the development of innovative technologies. Second- and third-generation feedstocks also help to overcome one of the main issues of bioplastics production: their cost when compared to conventional plastics. Using substrates that would be discarded and have low cost when compared to food crops helps to reduce overall processes expenses and allows higher competition of bioplastics.

14.2 BIOPLASTICS GENERAL CHARACTERISTICS AND PROPERTIES

Analysis of bioplastics structure is crucial to optimizing processing conditions and end-use performance. Bioformulations require even more rigorous testing to ensure biocompatibility and biodegradability. Biocompatibility refers to the biopolymer structure being compatible with tissues, and also with some biologically active molecules. Biodegradability is strictly related to how biopolymer degradation can occur in vivo, without posterior removal, in which its versatility is determined by some factors such as pKa, hydrophobic structure, temperature, blend percentage and water solubility (Oliveira et al., 2022). Bioplastics may be submitted to known techniques by the conventional plastic industry as follows:

- *Thermal analysis*: measurements of dimension changing and mechanical properties as a function of temperature, by heat flow and weight loss. Analysis can be performed using a differential scanning calorimeter (DSC), for thermal stability and phase transitions measurements. Thermogravimetric analysis (TGA) is applied to determine structure composition including volatile or solvent content, decomposition temperature and its residual products (Otoni et al., 2021; Akhlaq et al., 2022).
- *Mechanical analysis*: measurements of mechanical properties concerning the structure's response to force. Dynamic mechanical analysis (DMA) is employed in order to predict the bioplastic response to forces and strains at defined temperatures through parameters such as

storage modulus, loss modulus, tan delta and glass transition (T_g). Mechanical strength analyses define the final product behavior, through the Young modulus, yield strength, ultimate strength, span elongation, fatigue and durability (Otoni et al., 2021; Akhlaq et al., 2022).
- *Rheology*: measurements of viscosity flow and bioplastics deformation. Rheometers analyze bioplastics uniformity and durability, and are important tools when a new blend is developed (Akhlaq et al., 2022).

The characterization of these properties facilitates every step of bioplastic manufacturing, from assessing the quality of a given raw material to testing the properties of a final product. As new bioplastic formulations are developed, manufacturers will need systematic approaches to incorporate innovative blends while maintaining efficiency, meeting customer expectations and maximizing sustainability. Second- and third-generation bioplastics obtained via fermentation process have exceptional advantages, such as high purity, crystallinity, optimal polymerization degree, water binding capacity, tensile strength and biocompatibility (Chandel et al., 2018). On the other hand, the final product usually presents an opaque aspect due to the 2G/3G biomass characteristics, bounding the manufacturing procedure, as most of the bioplastics are processed via wet routes. Water causes interference during thermal and resistance tests; therefore, it may be removed before the injection molding and thermoforming process (Otoni et al., 2021). Additives must be avoided, since they inhibit the biodegradation properties; also bioplastics must not induce in vivo toxicity, which means not promoting an inflammatory response by the immune system (Narancic et al., 2020). These ideas could facilitate innovation in bioplastic production, offering multiple options in replacement to petroleum-based products.

14.3 INNOVATION IN SUBSTRATE PROCESSING AND IN COMPOSITES AND BLENDED MATERIALS

Agroindustrial wastes are one of the best sources of lignocellulose for researchers and industries to develop bioplastics. Lignocellulosic materials have a complex structure consisting of lignin, cellulose and hemicellulose, which configure their recalcitrance to decomposition (Haghighi Mood et al., 2013). Cellulose is a polysaccharide made of D-glucose monomers linked by β-(1,4)-glycosidic bonds to each other. It is the main constituent of lignocellulosic materials. Because of that, it can be used as industrial feedstocks when it is converted into glucose by chemical or enzymatic hydrolysis (Haghighi Mood et al., 2013; Salameh, 2009). To overpass the recalcitrance to decomposition, pretreatment is needed so the fractions of hemicellulose and lignin can be solubilized. Lignin is an intricate matrix made by polymers synthesized by phenylpropanoid precursors, and it is mostly acid-insoluble. Hemicellulose consists of branched heteropolymer of hexoses, pentoses (mainly xylose) and sugar acids. It is advisable that such structures must be removed before enzymatic hydrolysis as they may limit the availability of cellulosic substrate (Kumar and Sharma, 2017; Magalhães et al., 2019).

There are four classes of pretreatment methods: physical, chemical, physicochemical and biological. Chemical methods are widely used and are based on acid or alkaline substances acting on the lignocellulosic matrix. Dilute acid pretreatment is widely used in different sectors because it is a fast and low-cost method, which produces lower amounts of inhibitors to fermentation, however, the use of temperatures above 120°C make this method not highly accessible for industrial scale (Haghighi Mood et al., 2013; Kumar and Sharma, 2017; Onsrud, 2021). Alkaline methods using ammonia, sodium, calcium or potassium hydroxides do not require complex reactors because the pretreatment happens in lower temperatures than the acidic ones, thus requiring a higher reaction time (Haghighi Mood et al., 2013; Onsrud, 2021). The synthesis and application of deep eutectic solvents (DES) have been recently used as an option for chemical biomass pretreatment, taking advantage of their main characteristics: biocompatible, non-toxic, non-corrosive, recyclable and optimizable (Tomé et al., 2018). However, there are no references mentioning the use of DES as a pretreatment of biomass for the subsequent production of PHA or PLA, which is an opportunity to explore the junction of these technologies in the process.

Physical methods exist in the extraction of cellulose by increasing the surface area of the material, leading to a higher degree of polymerization and reducing cellulose crystallinity. The most conventional method is milling for the reduction of the particle size to at least 50 times the original one. Recent studies on physical pretreatment have focused on radiation. There is plenty of research using ultrasound as a method of pretreatment to lignocellulosic materials prior to produce bioplastics, concluding that it is an effective approach that contributes to further hydrolysis (Onsrud, 2021; Silvello et al., 2022). Combination of physical and chemical methods can also be applied for higher yields. As an example, alkali pretreatment can be coupled to milling since the increase on the surface area increases material exposure to the applied substance, enhancing the reaction. Another example is the steam explosion method, which consists of applying elevated pressures of steam followed by a sudden reduction of pressure. It uses less energy compared to physical pretreatments, and hydrolysis occurs because of the sudden change in temperature (Haghighi Mood et al., 2013). Biological pretreatments take advantage of lignocellulolytic microorganisms, mostly white rot fungi, for the disruption of the recalcitrant matrix. Although being a greener method for pretreatment as it applies mild conditions for reaction and no chemicals, the time consumed for these processes is far longer when compared to physicochemical pretreatments, which narrows its industrial applications.

After pretreatment, enzymatic hydrolysis can then be performed to break down the carbohydrates. Cellulases are used to hydrolyze the cellulose into monomers, while hemicellulases depolymerize their corresponding molecule (Jørgensen et al., 2007). As hemicellulose is the most complex molecule in the lignocellulosic material, it needs more than one enzyme to degrade it and this can be expensive to develop. One way to overpass that is to produce enzymes on the target of lignocellulosic material which could be employed in biorefineries later to produce enzymes on-site (Jørgensen et al., 2007).

The use of starch as a polymer is a viable alternative due to its composting and film-forming abilities, especially in the production of edible films and in short-time products, such as single-use packaging (Tapia-Blácido et al., 2022). Modification processes are needed to reach the disruption of native starch granules, which have strong inter- and intramolecular hydrogen bonds (Tabasum et al., 2019). Because of its poor mechanical properties and hydrophilic characteristics, starch is commonly blended with fossil-based plastics to improve the final polymer performance, resulting in thermoplastic starch (TPS; Tabasum et al., 2019). Different modification configurations, in terms of additives, temperature, pressure and shear, among others, can lead to different amylose/amylopectin semi-crystalline matrixes with thermoplastic characteristics that can be processed using conventional plastic techniques, such as extrusion, blow and injection molding (Debiagi et al., 2017). Tabasum et al. (2019) reviewed blending processes between corn starch and different polymers. Given that starch sources are highly variable in relation to granule shape and size, amylose/amylopectin ratio and enlargement, the modification step results in TPS with different characteristics. Basiak et al. (2017) found significant differences on the thickness of the films prepared from corn (112 μm), potato (55 μm) and wheat starch (74 μm), which can result in polymers with different mechanical performances. Temperature and process time are also both important factors to the modification of starch. The disruptions of its crystalline regions, as the complete gelatinization intends to do, is reached between 90°C and 119°C, depending on starch source and concentration of ingredients (Chen et al., 2017). Other process configurations related to changes on plasticizers, co-polymers and additives can modify the final biopolymer properties, in relation to the thickness, density, tensile strength, elongation and water absorption capacity.

It is worth mentioning that the worldwide extraction of starch is made mainly from food crops such as corn, potato and cassava, raising the discussion about using food for biomaterials production, which could threaten food security, therefore increasing cost production due to competition for arable lands. Novel feedstocks are being explored to develop completely biodegradable 2G bioblends, such as in the case of starch extracted from rice bran, a waste from the rice milling process, for bioplastic production through injection molding at 80–110°C, using glycerol and water as plasticizers (Alonso-González et al., 2021). Although the final optimized material showed high values of water uptake capacity, they exhibited poor physical integrities, which suggests that blending with other biopolymer with the proper compatibilizer could improve its physical properties (Debiagi et al., 2017). Thus other novel feedstocks for potential 2G bioplastics include, for instance, protein-rich biomasses and seaweeds.

Seed oil cake (SOC) is a by-product from vegetable oil industries and accounts for about 50% of the total weight of the original seeds, containing high amounts of fiber, polysaccharides and proteins. The proteins extracted from SOCs have been evaluated as raw materials for films for the food packaging sector. Also, wheat gluten can be explored to produce protein-based bioplastic through the extrusion processes, in which the protein concentrates are added to a solution with solvent and plasticizer to create a film-forming solution (Mirpoor et al., 2020). Many strategies are being tested to improve protein-based films' mechanical and water vapor barrier properties, which are the major drawbacks in comparison with synthetic films. These approaches include controlled protein denaturation, cross-linking, different plasticization and process configuration, as well as blending with other polymers. Fetzer et al. (2021) extracted a protein concentrate from rapeseed by dissolving its meal in a saline solution under alkaline conditions, adding lauryl and oleoyl chloride to the resulted rapeseed protein concentrate, which was then dialyzed and lyophilized. The film-forming solution was prepared using glycerol as plasticizer and presented an increased hydrophobicity, which shows the potential of modified rapeseed proteins to be used as packaging layers, coatings and adhesives.

Representing the 3G feedstock for the production of bioplastics, marine biomasses have great potential in environment protection and climate change mitigation because of its autotrophic complex, absorbing greenhouse gasses (GHGs; Elrayies, 2018). Seaweeds (macroalgae up to 60 m) and microalgae (0.02–2000 μm) have great capacity for biomass productivity—5 to 10 times faster than conventional crops. *Chlorella* sp. and *Spirulina* sp. are the most important microalgal species for bioplastic production, where they are a source of film-forming material, such as alginate, carrageenan and agar. Their biomasses are harvested directly through thermal-mechanical methods into biopolymers, by compression molding and application of different additives (Vandenberghe et al., 2021). *Kappaphycus alvarezii* and *Gelidium sesquipedale* (red algae species) were explored, respectively, for carrageenan and agar extraction and film production, showing higher viscosity, elasticity and water vapor permeability properties (Martínez-Sanz et al., 2019; Mostafavi and Zaeim, 2020). In another approach, seaweed-based biopolymers can be blended with other materials to improve water vapor barrier properties and mechanical strength. *Ulva* spp. were evaluated for PLA film production, through the fermentation of their starch sugars to lactic acid (Helmes et al., 2018). Madera-Santana et al. (2015) evaluated seaweed waste from agar extraction process as an inexpensive and effective filler for incorporation into PLA matrix by melt blending. The resulting biocomposites presented a slight increase in the tensile modulus, as well as an enhancement in its rigid amorphous phase content.

An alternative and promising approach is based on the cultivation of biopolymers intracellularly within microalgal cells, such as PHB, with its subsequent extraction. In this case, biopolymers are synthesized naturally in microalgal metabolism when applying media conditioning strategies with phosphate and nitrogen depletion and light variations, which increases their concentration. Apart from different production systems, which include raceway ponds as open systems and photobioreactors as closed ones, the harvesting step for substrate processing can account for more than 30% of microalgae production costs (Salim et al., 2011). The best harvesting method to be applied is related to the targeted final product and the morphological features of the microalgal cells, such as cell size, density or specific surface charges. Rahman et al. (2015) reported an *Escherichia coli* fermentation applying wastewater from microalgal wet lipid extracted media, reaching a maximum of 31% PHB of the *E. coli* dry cell weight.

14.4 NEW FERMENTATION STRATEGIES FOR 2G AND 3G BIOPLASTICS PRODUCTION

Feedstocks can be used to produce bioplastics in different fermentation strategies, such as traditional batch fermentation, simultaneous saccharification and fermentation (SSF), or consolidated bioprocessing (CBP). In SSF, the enzymatic hydrolysis of the feedstock occurs alongside the fermentation step. This process reduces product inhibition in hydrolysis, enhances product yield and eliminates the need of separated reactors for hydrolysis and fermentation (Soccol et al., 2011). However, the major drawbacks are the need to balance favorable conditions (such as temperature and pH) for both processes, there is also the difficulty to recycle the enzymes and microbial cells. As an alternative, there is the CBP type of fermentation. In this fermentation strategy, the microorganism expresses the hydrolytic enzyme capable of breaking the feedstock, liberating fermenting sugars which are then metabolized by the microorganism (Cripwell et al., 2020). The CBP strategy reduces the overall operational and input cost, eliminating the need of enzyme addition and reducing the risk of contaminations. However, CPB has longer fermentation periods and only engineered strains can produce high product yields, limiting its territory application since some countries restrict the use of recombinant microorganisms.

Another fermentation strategy for reducing production costs is the utilization of undefined microbial consortia—mixed microbial cultures (MMC). In this case, the process occurs in three steps, described by Tu et al. (2019):

1. Acidogenic fermentation, where the carbon source is converted into volatile fatty acids (VFAs).
2. Culture selection, the desired microorganisms are selected usually through a feast and famine regime.
3. Fermentation, where the MMC selected in (2) is fed in the feedstock produced in (1).

Other fermentation processes have been developed and employed, such as the one proposed by Corsino et al. (2022) that used the fed-on-demand strategy, that is, feeding substrate into the reactor only when the organic substrate was already completely degraded, in this case, citrus fruits processing wastewater. Besides the innovation in the fermentation processes, this study also elucidated that the quality of the wastewater determined the PHA production yield and the biopolymer mechanical characteristics. The process proposed by Rakkan et al. (2022) applied a simultaneous wastewater treatment and biopolymer production process utilizing *Enterobacter* TS3 cultivated in textile wastewater. This strain produced 0.34 gPHA/L and was able to reduce the wastewater chemical oxygen demand (COD) and total Kjeldahl nitrogen (TKN), making the water quality meet the criteria of local water quality standard.

As stated above, genetic engineered microorganisms are the most promising in CBP, such as the strain developed by Jeong et al. (2021). They constructed an alginolytic enzyme complex capable of degrading alginate, a major compound in brown algae. In this study, the enzyme complex was immobilized on the *Ralstonia eutropha* cell surface, enabling both substrate degradation and PHB production (2.58 g/L). In another study developed by Brojanigo et al. (2022), an amylolytic *Cupriavidus necator* DSM 545 strain was constructed for one-step PHA production from starchy residues (broken rice and purple sweet potato waste). A PHA yield of 5.78 g/L from broken rice and 3.65 g/L from purple sweet potato waste was observed, indicating that this strain is suitable for low-cost PHA production from starchy waste streams. Bae et al. (2018) developed a similar process to produce optically pure lactic acid from genetic engineered *Kluyveromyces marxianus*. This strain produced 130 g/L of L-lactic acid and 122 g/L of D-lactic acid from *Helianthus tuberosus* crop in one step without pretreatment or nutrient supplementation.

In the context of metabolic engineering, the overexpression of the *dr1558* gene (*Deinococcus radiodurans*) in *E. coli*–harboring PHB biosynthesis genes (*Ralstonia eutropha*) was proved to enhance PHB production (Park et al., 2019). The engineered strain produced 5.31 g/L of PHB, whereas the wild type strain produced only 1.52 g/L of PHB in 48 hour shake flasks (30 mL) cultivation. Genetic engineering has also been used to produce co-polymers of different bioplastics. For example, Tran and Charles (2020) constructed a recombinant *Pseudomonas putida* capable of accumulating a quaterpolymer P(3HB-co-LA-co-3HHx-co-3HO) up to 42% of its dry cell weight.

Biorefineries—industrial plants utilizing renewable feedstocks producing value-added bioproducts—are required for the implementation of the bioeconomy. Emerging areas for waste biorefinery are simultaneous production of bioplastics, bioenergy, wastewater treatment and soil amendment. As an example of process in these emerging areas is the simultaneous organic municipal waste treatment and PHA microbial production claimed by Yu et al. (2020, WO2020252582-A1) and LUX-ON, an industrial plant funded by Bio-on and Hera Group. This plant aims to produce PHA with atmospheric CO_2 and energy generation with solar power, different from traditional Bio-on plants that utilize sugar beet and sugarcane molasses, fruit and potato wastes, waste glycerol and waste frying oil (Bio-on, 2018). Additionally, Cuervo (2020, WO2020021346) demanded a method to produce and recover 90% of PLA obtained through cheese whey fermentation. In the context of LA production, Chen et al. (2020) showed that using immobilized cells could be beneficial, since they could be used for 4 months without loss of activity, and this system also allowed continuous LA removal which enhanced the process productivity by avoiding product inhibition. PHB is another promising bioplastic in the biorefinery context, as represented by the 3G process proposed by Arun et al. (2022) which is based in the cultivation of algae in wastewater to produce biodiesel via transesterification and the remaining algal biomass (de-oiled *Chlorella vulgaris* cake) is used to produce PHB (0.41 g PHB/g cake).

14.5 INNOVATION ON RECOVERY PROCESSES OF BIOPLASTICS

Bioplastics downstream processes depend on the fermentation characteristics, on the compound being intracellular or extracellular, whether it naturally polymerizes or not, and on the final product characteristics (Koller et al., 2013). Usually, the downstream processes for PHAs begin by removing or harvesting the cells from the culture media, which can be performed through centrifugation, filtration, sedimentation, or flocculation. Most of the time, the biomass is dried after those steps, either by thermal treatment or lyophilizing. As the material is produced intracellularly, the next process is the cell lysis to liberate the PHA, one of the major purification steps. Two main strategies are performed in this step, the first one is the use of organic chemical solvents to solubilize the cell membrane, resulting in a PHA with higher purity, and the second one is the use of enzymatic, mechanic or osmotic disruption to break the cell, wherein the lipids and cell content remain attached to the polymer. The last steps are associated with the purification of the polyester in order to reach enough quality for its application, such as the removal of excessive chemical agents or toxins before medical applications (Koller et al., 2013).

Recent developments are intensively searching for candidates to replace the mostly applied solvents for PHAs extraction, that are usually considered hazardous. Non-halogenated alternatives, for instance bio-based lactic acid esters (ethyl and methyl lactate), ketones (methyl isobutyl and methyl ethyl ketone), and acetates (butyl and ethyl acetate; Riedel et al., 2013), are being tested with promising results, reaching up to 99% purity using a PHA-to-solvent ratio of 2% (w/v). Alternatives to organic treatment comprehend techniques to disrupt the cells through enzymatic digestion, mechanical forces, different chemicals and other interesting ideas. Enzymes applied in the extraction include pepsin, trypsin, papain, alcalase, lysozyme and other nucleases and phospholipases, reaching around 95% purity. However, enzymatic processes demand costly resources to achieve high efficiency and additional bleaching with hydrogen peroxide or recovery with other chemical agents is required in some cases (Yasotha et al., 2006). Mechanical disruption can be performed through high-pressure homogenization, ultrasonication, and bead milling, requiring sequential steps for gathering the liberated PHA. Other chemicals include supercritical carbon dioxide and sodium dodecyl sulfate, each of them conferring slightly different properties, efficiencies and costs to the downstream process, though being rather similar techniques.

Different from PHAs, PLA is a polymer formed by extracellularly produced lactic acid monomer units that do not polymerize naturally. After the lactic acid becomes available in the liquid, an intense purification should be performed in order to reach at least food-grade quality, as impurities severely affect the molecular weight and yield of the polymer. Finally, the polymerization occurs, complexing the lactic acid units into a plastic-like biopolymer (Jem and Tan, 2020). As PLA is industrially produced through a relatively classic method, there is reserved information about innovations in the area of downstream. Similar to PHA, supercritical carbon dioxide has been tested with PLA in order to reduce polymerization temperature and facilitate post-polymerization purification, as it removes impurities and can be recycled. Lee (2020, KR2020122045-A) describes a process for inserting bioactive compounds in an amorphous mass of PLA by submitting it to a solution of the compounds and pressing it through rolls at 60°C and 150°C.

Genetic engineering is also a powerful tool for improving both PHAs and PLA recovery. In the case of PHAs, the goal is to obtain engineered strains capable of secreting the biopolymer in the extracellular media while still maintaining cellular integrity for cell recycling strategies. Being intracellular polymers, PHAs cannot be directly targeted, but their phasins—which are proteins related to their synthesis, bound to the PHA granules (Mezzina and Pettinari, 2016)—could. As an example, in the study conducted by Rahman et al. (2013), an engineered *E. coli* was able to secrete phasins and PHB in the media by type 1 secretion using the HlyA signal. Thirty-six percent of the phasin-bound PHB was secreted in the media, while 64% remained in the intracellular environment. However, through scanning electron microscopy (SEM) analysis, it was found that the extracellular PHB was an amorphous material, rather than the usual granules that are obtained. Therefore although being a promising method for PHAs recuperation with reduced costs, the technology of exporting the intact granules to the media with higher yields still needs to be developed.

On the other hand, the goal with PLA polymers is to achieve polymerization of lactic acid during the fermentation process, excluding the need of this step after monomer synthesis and recovery. This can be advantageous since PLA obtained by ring-opening polymerization (ROP) can have trace metal catalysts, restraining their use for medical applications. Therefore lactic acid synthetizing microorganisms are usually engineered to express LA-polymerizing-enzyme (LPE), which is a type of PHA synthase, responsible for the polymerization of HA monomers to PHA. In the study developed by Jung et al. (2010), an engineered *E. coli* was able to synthetize up to 11% of its dry cell weight of PLA homopolymer with glucose as carbon source and 56% of dry cell weight (DCW) of P(3HB-co-LA) copolymer when using glucose and 3HB as substrates.

14.6 NEW APPLICATIONS OF 2G AND 3G BIOPLASTICS

At the present moment, no worldwide regimentation was found concerning specifically rules to meet the safety of bioplastics, or standard methods to bioplastics composition. Although, some regions are engaged in making bioplastics support strategies. The European association European Bioplastics (2022b) gathers all the details for the data presented below:

- *Minimal biomass percentage*: termed renewable material; 25% was set by an industrial commitment in Japan, while the USDA Biopreferred Programme sets a range of 7%–95% depending on product class.
- *Bio-based carbon content*; measurement through the isotope 14C, European norm EN 16640 and the standard EN 16785–1. Certifications are offered by the certifiers TÜV AUSTRIA Belgium and DIN CERTCO, based on EN 16640 norm or ISO 16620–2 (or ASTM D 6866).
- *Industrial composting*: the European standard EN 13432 requires, after 12 weeks, at least 90% fragmentation, including ecotoxicity test and the heavy metal content.

- *Home composting*: the Australian norm AS 5810 and the French standard NF T 51–800 require, after 12 months, at least 90% degradation at ambient conditions.
- *Biodegradable/compostable products*: certifications are offered by TÜV AUSTRIA Belgium, DIN CERTCO (Germany), AfOR (United Kingdom), and COBRO (Poland).

Bioplastics final quality depends on the initial substrate, the production technique, if chemicals are employed during any treatment method, and avoid using the common plasticizer additives. The physical-chemical properties will define the bioplastic application. PLA is already used in packaging applications, shopping bags, and cups. The recent focus is widely applied in tissue engineering, due to the range of the melting temperature (T_m) and the glass transition temperature (T_g) presented by the copolymers, with some of them being commercially available (Narancic et al., 2020). Due to high biocompatibility and biodegradability, PHA polymers are researched for temporary in vivo applications, such as antimicrobial releasing sutures, drug release capsules and bone scaffold applications inducing neural regeneration. Since P3HB is a natural metabolic component of human blood, this biopolymer has been applied in surgical implants (Lamberti et al., 2020).

Medical PLA disposables are single-use products, such as gloves, face masks, needles, and blood bags. These products can replace equipment that normally needs sterilization procedures, so bio-based disposals can help to prevent the spread of health care–acquired infection (HAI), common in hospitals. Also, PLA is a candidate for applications in medical implants, orthopedic devices, and drug delivery systems. PLA was important during the COVID-19 pandemic, used as a 3D printable biopolymer for ventilator components and also as a filament to manufacture personal protective equipment. PHA is under experimentation to be used as medical consumables for surgical sutures and meshes. In this sense, research and development teams are performing and validating tests to safely replace petroleum-based plastics and clear the eventual regulatory hurdles (DeStefano et al., 2020).

Bioplastics can also be applied in other areas, like electronics, agriculture, and the environment. Highly porous foam materials are directed for application as filters, adsorbents, and insulators, such materials are able to improve fertilizer's quality and remove organic pollutants from wastewater (Otoni et al., 2021). Some industrial leaders are developing other practical uses and updating their technologies. Innovative applications are presented below:

- Corbion PLA from the Netherlands produces high quality lactic acid and lactide using 2G substrates via fermentation process. PLA is converted into resin by its partners, which is used to produce foam, film, molded plastic parts and fibers. The company also follows the life cycle assessment (Corbion, 2022).
- PHAXTEC from the US produces rigid and moldable PHA pellets derived from methane, used in lids, jars, and bottles. Other applications include fibers (used in shoes and rope), 3D printing and film (PHAXTEC, 2022).
- Mushroom Packaging from the US developed a sustainable, high-performance packaging solution grown from mushrooms mycelium. It replaces petroleum-based plastics in protective packaging, certified by C2C Gold Certification for sustainable goods (Mushroom Packaging, 2022).
- Notpla from the UK is a packaging material made from brown algae that naturally biodegrades in weeks contributing to de-acidify oceans (Notpla, 2022).

14.7 PATENT ANALYSIS ABOUT 2G AND 3G BIOPLASTICS

With the rising concern about human activity on the environment, highly innovative new technologies classified as green alternatives are being constantly researched and developed. Second- and third-generation bioplastics, being biodegradable and derived from biological sources, are potential substitutes to conventional petroleum-based materials that are harmful to the environment. These bioplastics also have the advantage of being produced from renewable resources that do not compete with the food chain, creating opportunities of using substrates that would be discarded as residues and reducing production costs for these materials. Therefore, new processes of substrate preparation, bioplastics production and recovery, besides brand new applications of these materials are being developed and transferred to the industrial sector for implementation, commercialization and spread of new technologies.

In order to unravel the scenario of research, development and innovation (RDI) in this topic, a patent search and analysis was conducted. Although presenting limitations, the patent system is important for the protection—granting exclusivity of technology exploitation to the applicant—and for the dissemination of new processes. Thus a search was conducted in the Derwent Innovations Index (DII) database combining selected terms with Boolean language referring to bioplastics, substrates and processes. The International Patent Classification (IPC) codes C08 (which refers to organic macromolecular compounds, their preparation and applications) and C12 (involving biochemistry) were also included for search refining. The time period was set from 2000 to August 2022; outliers that did not fit in the theme were discarded and 400 documents were sorted for classification and data analysis.

While analyzing the year of publication of the patent documents, it is noticeable that since 2006 the technology development of 2G and 3G bioplastics is growing constantly, with a more expressive number of published documents in the last 4 years, since 2019 (Figure 14.1(a)). These dates are directly related to the years that the European Commission launched the document "Knowledge-Based Bio-Economy" in 2005 and an update of their bioeconomy strategy in 2018. The first document was important for triggering the development of new alternative technologies and policies

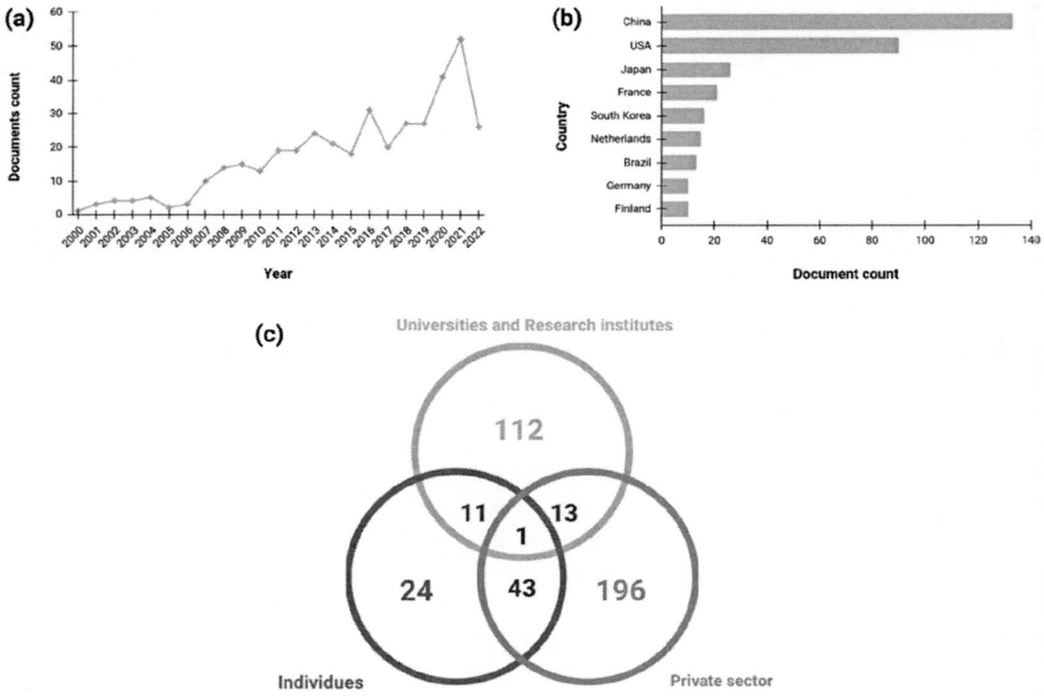

FIGURE 14.1 Number of published patent documents regarding the search of 2G and 3G bioplastics in Derwent Innovations Index (DII) database per (a) year of publication, (b) country of origin and (c) applicant profile.

regarding bioeconomy and including the schedule of sustainable development around the world (Patermann and Aguilar, 2021). Besides, the release of the Sustainable Development Goals (SDGs) Agenda in 2015, which has to be achieved by 2030 by the United Nations (UN), could have impacted politics on various levels and institutions and on the communication among society actors when regarding technology development and implementation (Biermann et al., 2022). The 2018 update on the EU's bioeconomy strategy has as a priority to strengthen and scale up bio-based sectors, thereby assisting new investments and market growth (European Commission, 2019).

The profile of countries that possess the higher number of published patent documents is directly related to the largest producers of bioplastics in the world (Figure 14.1(b)). Besides holding several industries of PLA and PHA production, China surpassed the US in 2019, becoming the largest patent applicant in the world (WIFO, 2020), mainly due to governmental subsidies for this activity. However, the quality of these documents has already been questioned since not all of the patents are in fact industrialized or commercialized. Besides, only 33% of the Chinese residential patents applied in 2020 were granted (WIPO, 2022), demonstrating the gap that may exist in this country. On the other hand, the United States and Japan (the two countries that follow China being significant patent applicants in general and during this search) had approximately 61% of their patents being granted in 2020, showing a higher rate of commercialization of the developed technologies (WIPO, 2022).

Other countries that stand out in the search about 2G and 3G bioplastics are France, which holds in its territory the enterprise Veolia, which is developing research in PHA production integrated to wastewater treatment plants (WWTP) as described in the patent documents WO2016020884 and WO2014108878; the Netherlands, which has the headquarters of both DSM (responsible for the development of new enzymatic compositions for substrate processing) and Paques Biomaterials (technology involving the production of PHAs, mainly polyhydroxybutyrate-co-valerate or PHBV, in pilot scale WWTP plants as described in the documents EP3760591 and WO2015181083); and Germany, which holds BASF and CureVac, the latter one being responsible for application of PLAs and their copolymers in medical and pharmaceutical systems. Brazil is also highlighted for being an agroindustrial country that generates large quantities of lignocellulosic biomass and therefore holds a great number of patent applications, most of them assigned by PHB Industrial, BioCelere (a research center of GranBio enterprise responsible for the application of patents that involve genetic engineering of microorganisms for the utilization of xylose in fermentation, as described in the documents BR102015012691 and BR102015008841), and federal universities that are constantly developing research in the use of 2G biomass for the production of value-added products. The profile of the private sector being the main technology holder (Figure 14.1c) demonstrates that the majority of developed technologies are in fact being commercialized. The academy is also responsible for a large number of documents that evidence the interest of achieving sustainable development from different actors in society. It is possible to elucidate that, in the next few years, a greater cooperation among academia and the private sectors is likely to be observed mainly because cooperation is one of the cornerstones of bioeconomy and sustainable development (Schütte, 2018).

Through manual classification, the focus of the document, the type of bioplastic produced or applied and their generation (2G or 3G) were retrieved. PLA was the main bioplastic researched and developed and this impacted directly on the focus classification, holding the largest number of application registers. This shows that the PLA market is more consolidated, mainly due to historic events. While PHA was first described in 1920 by the French microbiologist Maurice Lemoigne (Keshavarz and Roy, 2010), Lactic acid was produced for the first time in 1770 and commercialized in 1880, and its polymerization to PLA began in 1932 (Jem and Tan, 2020). Besides, as already stated, PLA has a larger share of the bioplastics market in general. However, technology development on PHAs has been increasing in recent years (Figure 14.2(a)), and its market share of bioplastics is also projected to grow in the next few years as more processes are implemented in the industry. The main areas of application found for these bioplastics were packaging (46 registers),

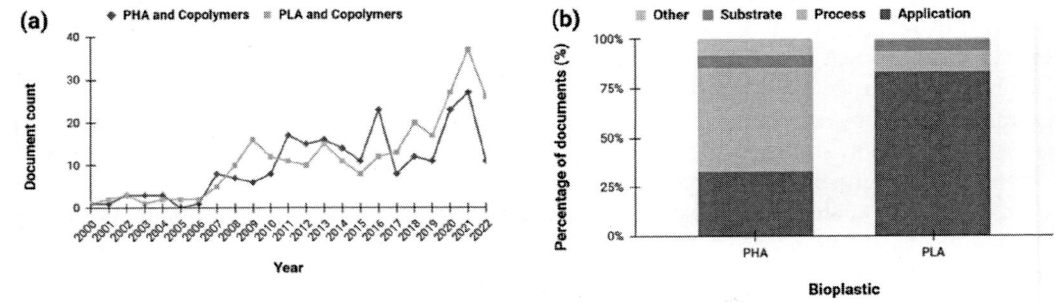

FIGURE 14.2 Number of registers regarding the search of 2G and 3G bioplastics in Derwent Innovations Index (DII) database: (a) evolution of applications of different bioplastics in the last 22 years; (b) classification of technology focus of different bioplastics.

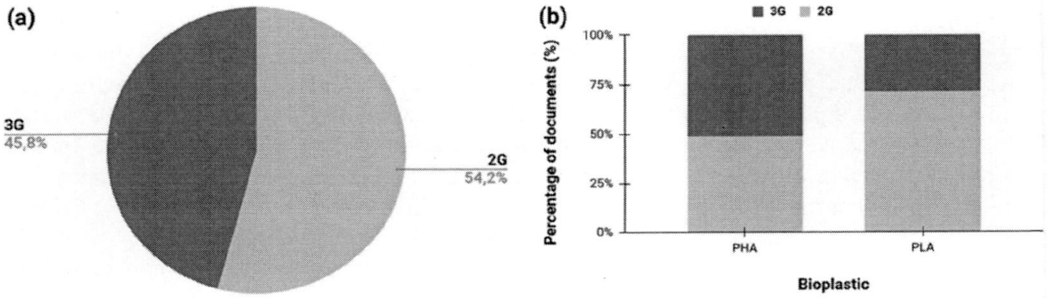

FIGURE 14.3 Percentage of registers regarding the search of 2G and 3G bioplastics in Derwent Innovations Index (DII) database: (a) comparison of documents regarding 2G and 3G feedstocks; (b) classification of generation of different bioplastics.

agricultural (32), pharmaceutical (24), disposables (23) and medicine (22), which reiterates the versatility of these materials for substituting conventional petroleum-based plastics.

Regarding the generation of bioplastics produced, 2G holds a larger share of documents (Figure 14.3(a)), but the 3G technology is growing mainly due the development of PHAs deriving from mixed microbial cultures in WWTP (Figure 14.3(b)). The 2G processes are more consolidated mainly due to the fact that biofuels derived from lignocellulosic biomass have been researched in depth in recent years and technology developed for this area regarding substrate processing can be adapted to the obtainment of other bioproducts. The 3G processes, however, are still incipient for the majority of biomolecules and more research needs to be conducted for the implementation of these products in the industry. At least 96 registers of composites with PHA or PLA were found in the database, mainly applying lignocellulosic material for the manufacture of the desired material. This shows that 2G feedstocks can be applied not only to the recovery of sugars to posterior fermentation but also for enhancing bioplastics properties.

14.8 CONCLUSIONS

Bioplastics are great candidates for substituting conventional petroleum-based plastics, yet their cost of production restrains their applications. Second- and third-generation feedstocks are promising substrates for the obtainment of these materials as they help to reduce production costs and do not compete with the food chain. Second- and third-generation bioplastics are surrounded by innovation in all stages of manufacturing and commercialization. The main technology tendencies are the use of greener pretreatments for substrate processing, as steam explosion and ultrasound, different forms of fermentation, as SSF and CPB, and the development of engineered strains for higher yields of bioplastics. The bioplastics obtained can be used in several areas, but mainly in health care–related applications due to their biocompatibility. It is expected that, in the next few years, technology development in this area will grow to solve the gaps found and reduce production costs, making bioplastics competitive with the conventional ones. Innovations and increasing applications are essential for bioplastics impact. As these bioplastics become more advanced and adapted to new applications, validating tests performed by scientists will serve as a proving ground for regimentations, evaluating physical properties and even discovering potential unknown uses.

REFERENCES

Akhlaq, Sumaiya, Dhananjay Singh, Nishu Mittal, Gaurav Srivastava, Saba Siddiqui, Soban Ahmad Faridi, and Mohammed Haris Siddiqui. 2022. "Polyhydroxybutyrate Biosynthesis from Different Waste Materials, Degradation, and Analytic Methods: A Short Review." *Polymer Bulletin*. doi:10.1007/s00289-022-04406-9.

Alonso-González, María, Manuel Felix, Antonio Guerrero, and Alberto Romero. 2021. "Rice Bran-Based Bioplastics: Effects of the Mixing Temperature on Starch Plastification and Final Properties." *International Journal of Biological Macromolecules* 188 (July): 932–40. doi:10.1016/j.ijbiomac.2021.08.043.

Arun, Jayaseelan, Sivakumar Shri Vigneshwar, Authilingam Swetha, Kannappan Panchamoorthy Gopinath, Sakeenabi Basha, Kathirvel Brindhadevi, and Arivalagan Pugazhendhi. 2022. "Bio-Based Algal (*Chlorella vulgaris*) Refinery on de-Oiled Algae Biomass Cake: A Study on Biopolymer and Biodiesel Production." *Science of the Total Environment* 816: 151579. doi:10.1016/j.scitotenv.2021.151579.

Bae, Jung Hoon, Hyun Jin Kim, Mi Jin Kim, Bong Hyun Sung, Jae Heung Jeon, Hyun Soon Kim, Yong Su Jin, Dae Hyuk Kweon, and Jung Hoon Sohn. 2018. "Direct Fermentation of Jerusalem Artichoke Tuber Powder for Production of L-Lactic Acid and D-Lactic Acid by Metabolically Engineered Kluyveromyces Marxianus." *Journal of Biotechnology* 266 (March 2017): 27–33. doi:10.1016/j.jbiotec.2017.12.001.

Basiak, Ewelina, Andrzej Lenart, and Frédéric Debeaufort. 2017. "Effect of Starch Type on the Physico-Chemical Properties of Edible Films." *International Journal of Biological Macromolecules* 98: 348–56. doi:10.1016/j.ijbiomac.2017.01.122.

Biermann, Frank, Thomas Hickmann, Carole Anne Sénit, Marianne Beisheim, Steven Bernstein, Pamela Chasek, Leonie Grob, Rakhyun E. Kim, Louis J. Kotzé, Måns Nilsson, Andrea Ordóñez Llanos, Chukwumerije Okereke, Prajal Pradhan, Rob Raven, Yixian Sun, Marjanneke J. Vijge, Detlef van Vuuren, and Birka Wicke. 2022. "Scientific Evidence on the Political Impact of the Sustainable Development Goals." *Nature Sustainability* 5 (9): 795–800. doi:10.1038/s41893-022-00909-5.

Bio-On. 2018. "Bio-on and Hera Create Lux-on the New Challenge to Produce Bioplastic from CO 2." *Bio-On*. www.bio-on.it/project.php?lin=inglese.

Brojanigo, Silvia, Nicoletta Gronchi, Tiziano Cazzorla, Tuck Seng Wong, Marina Basaglia, Lorenzo Favaro, and Sergio Casella. 2022. "Engineering *Cupriavidus necator* DSM 545 for the One-Step Conversion of Starchy Waste into Polyhydroxyalkanoates." *Bioresource Technology* 347 (October 2021): 126383. doi:10.1016/j.biortech.2021.126383.

Chandel, Anuj Kumar, Vijay Kumar Garlapati, Akhilesh Kumar Singh, Felipe Antonio Fernandes Antunes, and Silvio Silvério da Silva. 2018. "The Path Forward for Lignocellulose Biorefineries: Bottlenecks, Solutions, and Perspective on Commercialization." *Bioresource Technology* 264 (April): 370–81. doi:10.1016/j.biortech.2018.06.004.

Chen, Po Ting, Zih Syuan Hong, Chieh Lun Cheng, I. Son Ng, Yung Chung Lo, Dillirani Nagarajan, and Jo Shu Chang. 2020. "Exploring Fermentation Strategies for Enhanced Lactic Acid Production with Polyvinyl Alcohol-Immobilized *Lactobacillus plantarum* 23 Using Microalgae as Feedstock." *Bioresource Technology* 308 (March): 123266. doi:10.1016/j.biortech.2020.123266.

Chen, Xu, Li Guo, Peirong Chen, Yang Xu, Huili Hao, and Xianfeng Du. 2017. "Investigation of the High-Amylose Maize Starch Gelatinization Behaviours in Glycerol-Water Systems." *Journal of Cereal Science* 77: 135–40. doi:10.1016/j.jcs.2017.08.012.

Corbion. 2022. "Corbion—PLA." *Corbion*. www.totalenergies-corbion.com/.

Corsino, Santo Fabio, Daniele Di Trapani, Francesco Traina, Ilenia Cruciata, Laura Scirè Calabrisotto, Francesco Lopresti, Vincenzo La Carrubba, Paola Quatrini, Michele Torregrossa, and Gaspare Viviani. 2022. "Integrated Production of Biopolymers with Industrial Wastewater Treatment: Effects of OLR on Process Yields, Biopolymers Characteristics and Mixed Microbial Community Enrichment." *Journal of Water Process Engineering* 47 (February). doi:10.1016/j.jwpe.2022.102772.

Cripwell, Rosemary A., Lorenzo Favaro, Marinda Viljoen-Bloom, and Willem H. van Zyl. 2020. "Consolidated Bioprocessing of Raw Starch to Ethanol by *Saccharomyces cerevisiae*: Achievements and Challenges." *Biotechnology Advances* 42 (June): 107579. doi:10.1016/j.biotechadv.2020.107579.

Cuervo, Laura Viviana Garcés. 2020. *Method for Obtaining Polylactic Acid (PLA) From Cheese Whey.* WO2020021346. World Intellectual Property Organization, Geneva, Switzerland.

Debiagi, Flávia, Léa Rita P. F. Mello, and Suzana Mali. 2017. Thermoplastic Starch-Based Blends: Processing, Structural, and Final Properties. In *Starch-Based Materials in Food Packaging: Processing, Characterization and Applications.* Elsevier Inc. doi:10.1016/B978-0-12-809439-6.00006-6.

DeStefano, Vincent, Salaar Khan, and Alonzo Tabada. 2020. "Applications of PLA in Modern Medicine." *Engineered Regeneration* 1 (April): 76–87. doi:10.1016/j.engreg.2020.08.002.

Elrayies, Ghada Mohammad. 2018. "Microalgae: Prospects for Greener Future Buildings." *Renewable and Sustainable Energy Reviews* 81 (August): 1175–91. doi:10.1016/j.rser.2017.08.032.

European Bioplastics. 2022a. "Bioplastics Market Data Update." *European Bioplastics.* www.european-bioplastics.org/market/.

European Bioplastics. 2022b. "Welcome to European Bioplastics." *European Bioplastics.* www.european-bioplastics.org.

European Commission. 2019. "Updated Bioeconomy Strategy 2018 | Knowledge for Policy." *European Commission.* https://knowledge4policy.ec.europa.eu/publication/updated-bioeconomy-strategy-2018_en.

Fetzer, Andreas, Cornelia Hintermayr, Markus Schmid, Andreas Stäbler, and Peter Eisner. 2021. "Effect of Acylation of Rapeseed Proteins with Lauroyl and Oleoyl Chloride on Solubility and Film-Forming Properties." *Waste and Biomass Valorization* 12 (2): 745–55. doi:10.1007/s12649-020-01012-6.

Haghighi Mood, Sohrab, Amir Hossein Golfeshan, Meisam Tabatabaei, Gholamreza Salehi Jouzani, Gholam Hassan Najafi, Mehdi Gholami, and Mehdi Ardjmand. 2013. "Lignocellulosic Biomass to Bioethanol, a Comprehensive Review with a Focus on Pretreatment." *Renewable and Sustainable Energy Reviews* 27: 77–93. doi:10.1016/j.rser.2013.06.033.

Helmes, Roel J. K., Ana M. López-Contreras, Maud Benoit, Helena Abreu, Julie Maguire, Fiona Moejes, and Sander W. K. van den Burg. 2018. "Environmental Impacts of Experimental Production of Lactic Acid for Bioplastics from *Ulva* spp." *Sustainability (Switzerland)* 10 (7): 1–15. doi:10.3390/su10072462.

Jem, K. Jim, and Bowen Tan. 2020. "The Development and Challenges of Poly (Lactic Acid) and Poly (Glycolic Acid)." *Advanced Industrial and Engineering Polymer Research* 3 (2): 60–70. doi:10.1016/j.aiepr.2020.01.002.

Jeong, Da Woon, Jeong Eun Hyeon, Myeong Eun Lee, Young Jin Ko, Minhye Kim, and Sung Ok Han. 2021. "Efficient Utilization of Brown Algae for the Production of Polyhydroxybutyrate (PHB) by Using an Enzyme Complex Immobilized on *Ralstonia eutropha*." *International Journal of Biological Macromolecules* 189 (July): 819–25. doi:10.1016/j.ijbiomac.2021.08.149.

Jørgensen, Henning, Jan Bach Kristensen, and Claus Felby. 2007. "Enzymatic Conversion of Lignocellulose into Fermentable Sugars: Challenges and Opportunities." *Biofuels, Bioproducts and Biorefining* 1 (2):119–34. doi: 10.1002/bbb.4.Jung, Yu Kyung, Tae Yong Kim, Si Jae Park, and Sang Yup Lee. 2010. "Metabolic Engineering of *Escherichia coli* for the Production of Polylactic Acid and Its Copolymers." *Biotechnology and Bioengineering* 105 (1): 161–71. doi:10.1002/bit.22548.

Keshavarz, Tajalli, and Ipsita Roy. 2010. "Polyhydroxyalkanoates: Bioplastics with a Green Agenda." *Current Opinion in Microbiology* 13 (3): 321–26. doi:10.1016/j.mib.2010.02.006.

Koller, Martin, Horst Niebelschütz, and Gerhart Braunegg. 2013. "Strategies for Recovery and Purification of Poly[(R)-3-Hydroxyalkanoates] (PHA) Biopolyesters from Surrounding Biomass." *Engineering in Life Sciences* 13 (6): 549–62. doi:10.1002/elsc.201300021.

Kourmentza, Constantina, Jersson Plácido, Nikolaos Venetsaneas, Anna Burniol-Figols, Cristiano Varrone, Hariklia N. Gavala, and Maria A. M. Reis. 2017. "Recent Advances and Challenges towards Sustainable Polyhydroxyalkanoate (PHA) Production." *Bioengineering* 4 (2): 1–43. doi:10.3390/bioengineering4020055.

Kumar, Adepu Kiran, and Shaishav Sharma. 2017. "Recent Updates on Different Methods of Pretreatment of Lignocellulosic Feedstocks: A Review." *Bioresources and Bioprocessing* 4 (1). doi:10.1186/s40643-017-0137-9.

Lamberti, Fabio M., Luis A. Román-Ramírez, and Joseph Wood. 2020. "Recycling of Bioplastics: Routes and Benefits." *Journal of Polymers and the Environment* 28 (10): 2551–71. doi:10.1007/s10924-020-01795-8.

Lee, K. Y. 2020. *Producing Poly Lactic Acid Mask Pack Useful in Cosmetic Packs for Moisturizing and Removing Dead Skin Cells, Comprises e.g. Producing Amorphous Biodegradable Poly Lactic Acid Nonwoven Fabric, and Purified Water Containing Various Components*. KR2020122045. Korean Intellectual Property Office, Daejon, South Korea.

Madera-Santana, Tomás J., Yolanda Freile-Pelegrín, José C. Encinas, Carlos R. Ríos-Soberanis, and Patricia Quintana-Owen. 2015. "Biocomposites Based on Poly(Lactic Acid) and Seaweed Wastes from Agar Extraction: Evaluation of Physicochemical Properties." *Journal of Applied Polymer Science* 132 (31): 1–8. doi:10.1002/app.42320.

Magalhães, Antonio Irineudo, Júlio Cesar de Carvalho, Gilberto Vinícius de Melo Pereira, Susan Grace Karp, Marcela Candido Câmara, Jesus David Coral Medina, and Carlos Ricardo Soccol. 2019. "Lignocellulosic Biomass from Agro-Industrial Residues in South America: Current Developments and Perspectives." *Biofuels, Bioproducts and Biorefining* 13 (6): 1505–19. doi:10.1002/bbb.2048.

Martínez-Sanz, Marta, Antonio Martínez-Abad, and Amparo López-Rubio. 2019. "Cost-Efficient Bio-Based Food Packaging Films from Unpurified Agar-Based Extracts." *Food Packaging and Shelf Life* 21 (July): 100367. doi:10.1016/j.fpsl.2019.100367.

Mehta, Neha, Eoin Cunningham, Deborah Roy, Ashley Cathcart, Martin Dempster, Emma Berry, and Beatrice M. Smyth. 2021. "Exploring Perceptions of Environmental Professionals, Plastic Processors, Students and Consumers of Bio-Based Plastics: Informing the Development of the Sector." *Sustainable Production and Consumption* 26: 574–87. doi:10.1016/j.spc.2020.12.015.

Mezzina, Mariela P., and M. Julia Pettinari. 2016. "Phasins, Multifaceted Polyhydroxyalkanoate Granule-Associated Proteins." *Applied and Environmental Microbiology* 82 (17): 5060–67. doi:10.1128/AEM.01161-16.

Mirpoor, Seyedeh Fatemeh, Concetta Valeria L. Giosafatto, Prospero Di Pierro, Rocco Di Girolamo, Carlos Regalado-González, and Raffaele Porta. 2020. "Valorisation of *Posidonia oceanica* Sea Balls (Egagropili) as a Potential Source of Reinforcement Agents in Protein-Based Biocomposites." *Polymers* 12 (12): 1–13. doi:10.3390/polym12122788.

Mostafavi, Fatemeh Sadat, and Davood Zaeim. 2020. "Agar-Based Edible Films for Food Packaging Applications—A Review." *International Journal of Biological Macromolecules* 159: 1165–76. doi:10.1016/j.ijbiomac.2020.05.123.

Mushroom Packaging. 2022. "Mushroom Packaging®." *Mushroom Packaging*. https://mushroompackaging.com/.

Narancic, Tanja, Federico Cerrone, Niall Beagan, and Kevin E. O'Connor. 2020. "Recent Advances in Bioplastics: Application and Biodegradation." *Polymers* 12 (4). doi:10.3390/POLYM12040920.

Notpla. 2022. "Notpla®." *Notpla*. www.notpla.com/.

Oliveira, Priscilla Zwiercheczewski de, Luciana Porto de Souza Vandenberghe, Ariane Fátima Murawski de Mello, and Carlos Ricardo Soccol. 2022. "A Concise Update on Major Poly-Lactic Acid Bioprocessing Barriers." *Bioresource Technology Reports* 18 (May). doi:10.1016/j.biteb.2022.101094.

Onsrud, A. 2021. "Analysis of Pretreatment Methods for Bio-Based Plastic Production Based on Enzymatic Hydrolysis and Microbial Fermentation." https://ntnuopen.ntnu.no/ntnu-xmlui/handle/11250/2785454.

Otoni, Caio G., Henriette M. C. Azeredo, Bruno D. Mattos, Marco Beaumont, Daniel S. Correa, and Orlando J. Rojas. 2021. "The Food-Materials Nexus: Next Generation Bioplastics and Advanced Materials from Agri-Food Residues." *Advanced Materials* 33 (43). doi:10.1002/adma.202102520.

Pandit, Soumya, Nishit Savla, Jayesh M. Sonawane, Abubakar Muh'D Sani, Piyush Kumar Gupta, Abhilasha Singh Mathuriya, Ashutosh Kumar Rai, Dipak A. Jadhav, Sokhee P. Jung, and Ram Prasad. 2021. "Agricultural Waste and Wastewater as Feedstock for Bioelectricity Generation Using Microbial Fuel Cells: Recent Advances." *Fermentation* 7 (3). doi:10.3390/fermentation7030169.

Park, Sung ho, Gi Bae Kim, Hyun Uk Kim, Si Jae Park, and Jong il Choi. 2019. "Enhanced Production of Poly-3-hydroxybutyrate (PHB) by Expression of Response Regulator DR1558 in Recombinant Escherichia Coli." *International Journal of Biological Macromolecules* 131: 29–35. doi: 10.1016/j.ijbiomac.2019.03.044.Patermann, Christian, and Alfredo Aguilar. 2021. "A Bioeconomy for the Next Decade." *EFB Bioeconomy Journal* 1 (March): 100005. doi:10.1016/j.bioeco.2021.100005.

PHAXTEC. 2022. "PHAXTEC ®" *PHAXTEC*. www.phaxtec.com/.

Rahman, Asif, Elisabeth Linton, Alex D. Hatch, Ronald C. Sims, and Charles D. Miller. 2013. "Secretion of Polyhydroxybutyrate in *Escherichia coli* Using a Synthetic Biological Engineering Approach." *Journal of Biological Engineering* 7 (1): 1–9. doi:10.1186/1754-1611-7-24.

Rahman, Asif, Ryan J. Putman, Kadriye Inan, Fulya Ay Sal, Ashik Sathish, Terence Smith, Chad Nielsen, Ronald C. Sims, and Charles D. Miller. 2015. "Polyhydroxybutyrate Production Using a Wastewater Microalgae Based Media." *Algal Research* 8: 95–98. doi:10.1016/j.algal.2015.01.009.

Rakkan, Thanaphorn, Netnapa Chana, and Kanokphorn Sangkharak. 2022. "The Integration of Textile Wastewater Treatment with Polyhydroxyalkanoate Production Using Newly Isolated *Enterobacter* Strain TS3." *Waste and Biomass Valorization* 13 (1): 571–82. doi:10.1007/s12649-021-01504-z.

Riedel, Sebastian L., Christopher J. Brigham, Charles F. Budde, Johannes Bader, Chokyun Rha, Ulf Stahl, and Anthony J. Sinskey. 2013. "Recovery of Poly(3-Hydroxybutyrate-Co-3-Hydroxyhexanoate) from *Ralstonia eutropha* Cultures with Non-Halogenated Solvents." *Biotechnology and Bioengineering* 110 (2): 461–70. doi:10.1002/bit.24713.

Salameh, Yusra Fuad Abed-al-hafiz. 2009. "Methods of Extracting Cellulosic Material from Olive Pulp." *An-Najah National University Faculty*, 1–72.

Salim, Sina, Rouke Bosma, Marian H. Vermuë, and R. H. Wijffels. 2011. "Harvesting of Microalgae by Bio-Flocculation." *Journal of Applied Phycology* 23 (5): 849–55. doi:10.1007/s10811-010-9591-x.

Schütte, Georg. 2018. "What Kind of Innovation Policy Does the Bioeconomy Need?" *New Biotechnology* 40: 82–86. doi: 10.1016/j.nbt.2017.04.003.Silvello, Maria Augusta de Carvalho, Aline Frumi Camargo, Thamarys Scapini, Shukra Raj Paudel, Helen Treichel, and Rosana Goldbeck. 2022. "Enzymatic Hydrolysis Intensification of Lignocellulolytic Enzymes through Ultrasonic Treatment." *Bioenergy Research* 15 (2): 875–88. doi:10.1007/s12155-021-10334-9.

Soccol, Carlos Ricardo, Vincenza Faraco, Susan Grace Karp, Luciana Porto de Souza Vandenberghe, Vanete Thomaz-Soccol, Adenise Woiciechowski, and Ashok Pandey. 2011. Lignocellulosic Bioethanol: Current Status and Future Perspectives. In *Biofuels: Alternative Feedstocks and Conversion Processes*. Elsevier Inc. doi:10.1016/B978-0-12-385099-7.00005-X.

Tabasum, Shazia, Muhammad Younas, Muhammad Ansab Zaeem, Irfan Majeed, Muzamil Majeed, Aqdas Noreen, Muhammad Naeem Iqbal, and Khalid Mahmood Zia. 2019. "A Review on Blending of Corn Starch with Natural and Synthetic Polymers, and Inorganic Nanoparticles with Mathematical Modeling." *International Journal of Biological Macromolecules* 122: 969–96. doi:10.1016/j.ijbiomac.2018.10.092.

Tapia-Blácido, Delia Rita, Guilherme José Aguilar, Mateus Teixeira de Andrade, Márcio F. Rodrigues-Júnior, and Fernanda C. Guareschi-Martins. 2022. "Trends and Challenges of Starch-Based Foams for Use as Food Packaging and Food Container." *Trends in Food Science and Technology* 119 (December 2021): 257–71. doi:10.1016/j.tifs.2021.12.005.

Tomé, Luciana I. N., Vanessa Baião, Wanderson da Silva, and Christopher M. A. Brett. 2018. "Deep Eutectic Solvents for the Production and Application of New Materials." *Applied Materials Today* 10: 30–50. doi:10.1016/j.apmt.2017.11.005.

Tran, Tam T., and Trevor C. Charles. 2020. "Lactic Acid Containing Polymers Produced in Engineered *Sinorhizobium meliloti* and *Pseudomonas putida*." *PLoS ONE* 15 (3): 1–15. doi:10.1371/journal.pone.0218302.

Tu, Weiming, Dandan Zhang, and Hui Wang. 2019. "Polyhydroxyalkanoates (PHA) Production from Fermented Thermal-Hydrolyzed Sludge by Mixed Microbial Cultures: The Link between Phosphorus and PHA Yields." *Waste Management* 96: 149–57. doi:10.1016/j.wasman.2019.07.021.

Vandenberghe, Luciana Porto de Souza, Priscilla Zwiercheczewski de Oliveira, Gustavo Amaro Bittencourt, Ariane Fátima Murawski de Mello, Zulma Sarmiento Vásquez, Susan Grace Karp, and Carlos Ricardo Soccol. 2021. "The 2G and 3G Bioplastics: An Overview." *Biotechnology Research and Innovation* 5 (1): e2021004. doi:10.4322/biori.202104.

Vandi, Luigi Jules, Clement Matthew Chan, Alan Werker, Des Richardson, Bronwyn Laycock, and Steven Pratt. 2018. "Wood-PHA Composites: Mapping Opportunities." *Polymers* 10 (7): 1–15. doi:10.3390/polym10070751.

WIPO. 2020. *World Intellectual Property Indicators 2020*. Vol. 1. World Intellectual Property Organization. www.wipo.int/export/sites/www/freepublications/en/intproperty/941/wipo_pub_941_2013.pdf.

WIPO. 2022. "Statistical Country Profiles Patents, Utility Models, Trademarks and Industrial." *WIPO*. www.wipo.int/ipstats/en/statistics/country_profile/.

Yasotha, K., M. K. Aroua, K. B. Ramachandran, and I. K.P. Tan. 2006. "Recovery of Medium-Chain-Length Polyhydroxyalkanoates (PHAs) through Enzymatic Digestion Treatments and Ultrafiltration." *Biochemical Engineering Journal* 30 (3): 260–68. doi:10.1016/j.bej.2006.05.008.

Yu, Luna, Hasitha De Alwis Weerasekera, Marcos Forattini Lemos Igreja, Vani Sankar, Michael James Williamson, And Kaitlyn Soman, Sudhanshu Sanjay Chow. 2020. *Method for Producing Polyhydroxyalkanoates (PHA) from Organic Waste*. WO2020252582. World Intellectual Property Organization.

Abbreviations

1G	first generation
2G	second generation
3G	third generation
3HB	3-hydroxybutyrate
3HB-CoA	3-hydroxybutyrate–coenzyme A
3HD	3-hydroxydecanoic acid
3HHo	3-hydroxyheptanoate
3HHx	3-hydroxyhexanoate
3HHx-CoA	3-hydroxyhexanoate–coenzyme A
3HO	3-hydroxyocatanoate
4HB	4-hydroxybutyrate
5-HMF	5-hydroxy methyl furfural
6HHx	6-hydroxyhexanoate
ABS	acrylonitrile butadiene styrene
AD	anaerobic digestion
Ag+	silver ion
AGU	anhydro-d-glucopyranose units
ASTM	American Society for Testing and Materials
ATCC	American Type Culture Collection
ATP	adenosine triphosphate
BC	bacterial cellulose
bioH_2	biohydrogen
Bio-PA	bio-aliphatic polyamides
Bio-PE	bio-polyethylene
Bio-PET	bio-polyethylene terephthalate
Bio-PBS	bio-polybutylene succinate
Bio-PP	bio-polypropylene
Bio-PTT	bio-polytrimethylene-terephthalate
Bio-PU	bio-polyurethane
Bio-PVC	bio-polyvinyl chloride
BNF	bacterial nanofibrils
BOD	biological oxygen demand
BOPP	biaxially oriented polypropylene
BUE	biomass utilization efficiency
CA	cellulose acetate
$CaCl_2.2H_2O$	calcium chloride dihydrate
CAGR	compound annual growth rate
CAPEX	capital expenditures
CBP	consolidated bioprocessing
Cd2+	cadmium (II) ion
CH_4	methane
CMC	carboxymethyl cellulose
CMF	cellulose microfibrils
CNC	cellulose nanocrystals
CNF	nanofibrillar cellulose
CO_2	carbon dioxide
CO_2-eq	carbon dioxide equivalent

CoA	coenzyme A
$CoCl_2$	cobalt chloride
COD	chemical oxygen demand
COVID-19	coronavirus disease
CS_2	carbon disulfide
CSCT	Centre of Sustainable Chemical Technologies
Cu2+	copper ion
$CuSO_4.5H_2O$	copper sulfate pentahydrate
CW	cheese whey
DCB	defatted *Chlorella* biomass
DCW	dry cell weight
DES	deep eutectic solvents
DII	Derwent Innovations Index
DMA	dynamic mechanical analysis
DP	degree of polymerization
DSC	differential scanning calorimeter
EDTA	ethylenediaminetetraacetic acid
EG	ethylene glycol
ELCA	environmental life cycle assessment
EU	European Union
EVA	ethylene vinyl acetate
ExA	externality assessment
FCW	fermented cheese whey
FDA	US Food and Drug Administration
FDCA	furandicarboxylic acid
Fe2+	iron(II) ion
Fe3+	iron(III) ion
$FeCl_3.6H_2O$	ferric chloride hexahydrate
GHG	greenhouse gas
H_2S	hydrogen sulfide
H_2SO_4	sulfuric acid
H_3BO_3	boric acid
HA	hydroxyalkanoic acids
HAI	health care–acquired infection
HCl	hydrochloric acid
HDPE	high-density polyethylene
HEC	hydroxyethyl cellulose
HIV	human immunodeficiency virus
HMF	hydroxymethyl furfural
IMG/M	data management and analysis system for microbial community genomes
IPC	International Patent Classification
IRR	internal rate of return
ISO	International Organization for Standardization
IUPAC	International Union of Pure and Applied Chemistry
KCl	potassium chloride
KH_2PO_4	potassium dihydrogen phosphate
KOH	potassium hydroxide
LAB	lactic acid bacteria
LCA	life cycle assessment
LCC	life cycle cost

LDPE	low-density polyethylene
LPE	LA-polymerizing-enzyme
MA	maleic anhydride
MC	methyl cellulose
$MgCl_2.6H_2O$	magnesium chloride hexahydrate
$MgSO_4.7H_2O$	magnesium sulfate heptahydrate
MHEC	methyl hydroxyethyl cellulose
MHPC	methyl hydroxypropyl cellulose
MMC	mixed microbial culture
$MnCl_2.4H_2O$	manganese(II) chloride
MSP	minimum selling price
MT	million tonnes
Na_2HPO_4	disodium hydrogen phosphate
$Na_2MoO_4.2H_2O$	sodium molybdate dihydrate
NaCl	sodium chloride
NADPH	nicotinamide adenine dinucleotide phosphate
$NaHCO_3$	sodium bicarbonate
NaOCl	sodium hypochlorite
NaOH	sodium hydroxide
NC	nanocellulose
NH_4Cl	ammonium chloride
$(NH_4)_2SO_4-$	ammonium sulfate
$NiCl_2.6H_2O$	nickel(II) chloride
NO_x	nitrogen oxides
NPCM	non-PHA cellular materials
NPV	net present value
OF-MSW	organic fraction of municipal solid wastes
OLR	organic loading rate
OPEX	operating expenses or expenditure
P3HB	poly-3-hydroxybutyrate
P3HB-co-4HB	poly-3-hydroxybutyrate-co-4-hydroxybutyrate
P3HB-co-HH	poly-3-hydroxybutyrate-co-hexanoate
P3HB-co-HV	poly-3-hydroxybutyrate-co-valerate
P3HB4HB	poly[(R)-3-hydroxybutyrate-co-4-hydroxybutyrate]
PA	aliphatic polyamides
PA-	polyamide-based
PAM 11	polyamide 11/nylon 11
PBA	polybutylene adipate
PBAT	polybutylene-adipate-co-terephthalate
PBR	photobioreactor
PBS	polybutylene succinate
PBT	polybutylene terephthalate
PCL	polycaprolactone
PCL-	polycaprolactone-based
PDO	1,3-propanediol
PE	polyethylene
PE-g-MA	polyethylene-graft-maleic anhydride
PES	polyethylene succinate
PET	polyethylene terephthalate
PGA	poly(glycolic acid)

PHA	polyhydroxyalkanoates
PHB	poly-3-hydroxybutyrate
PHBV	poly(3-hydroxybutyrate-co-3-hydroxy-valerate)
PHHX	polyhydroxyhexanoate
PHO	polyhydroxyoctanoate
PHU	polyhydroxyurethane
PHV	polyhydroxyvaleric acid
PI	polydispersity index
PKO	palm kernel oil
pKw	log of the water ion product
PLA	polylactic acid
POME	palm oil mill effluent
PP	polypropylene
PS	polystyrene
PTA	poly(thioacrylate)
PTT	poly-1,3-propylene terephthalate
PU	polyurethane
PVA	polyvinyl alcohol
PVAc	polyvinylacetate
PVC	polyvinyl chloride
RDI	research, development and innovation
ROI	return over investment
ROP	ring-opening polymerization
s-CNC	surfactant-modified cellulose nanocrystals
$SCCO_2$	supercritical carbon dioxide
SDGs	Sustainable Development Goals
SDS	sodium dodecyl sulfate
SEM	scanning electron microscopy
SHF	separate hydrolysis and fermentation
SOC	seed oil cake
SO_x	sulfur oxides
SSF	simultaneous saccharification and fermentation
TCI	total capital investment
TGA	thermogravimetric analysis
TPS	thermoplastic starch
UHMW-PE	ultra-high-molecular-weight polyethylene
UN	United Nations
US	United States
USD	US dollars
UV	ultraviolet
VFA	volatile fatty acids
VOW	vegetable oil waste
WCO	waste cooking oil
WWTP	wastewater treatment plant
$ZnSO_4.7H_2O$	zinc sulfate heptahydrate

Index

R

S

T

V

W